Lectures on
Dynamical Systems,
Structural Stability
and their Applications

Lectures on
Dynamical Systems,
Structural Stability
and their Applications

KOTIK K. LEE

Dept. of Electrical and Computer Engineering
University of Colorado

World Scientific
Singapore • New Jersey • London • Hong Kong

Published by

World Scientific Publishing Co. Pte. Ltd.

P O Box 128, Farrer Road, Singapore 9128

USA office: Suite 1B, 1060 Main Street, River Edge, NJ 07661

UK office: 73 Lynton Mead, Totteridge, London N20 8DH

**LECTURES ON DYNAMICAL SYSTEMS, STRUCTURAL
STABILITY AND THEIR APPLICATIONS**

ISBN 9971-50-965-2

Printed in Singapore by JBW Printers and Binders Pte. Ltd.

Dedicated to
Peter G. Bergmann and Heinz Helfenstein,
who taught me physics and mathematics,
and to Lydia,
together they taught me humanity.

PREFACE

In this past decade, we have witnessed the enormous growth of an interdisciplinary field of study, namely, dynamical systems. Yet, dynamical systems, as a subfield of mathematics, has been established since late last century by Poincaré [1881], and reinforced by Liapunov [1892]. Parts of such a surge are due to many major advances in differential topology, the geometric theory of differential equations, algebraic geometry, nonlinear functional analysis, and nonlinear global analysis, just to name a few. This is partly due to computers being readily available. Nowadays, any second-year college student can use a personal computer to program the evolution of a nonlinear difference equation and find all kinds of chaotic behavior. With the new level of sophistication of graphics, it can evolve into "mathematical entertainment" or "arts". Each popular science or engineering magazine has at least one article on nonlinear dynamical systems a year. It not only caught the fascination of the scientists, but also attracted the attention of the enlightened public. Ten to fifteen years ago, there were but a dozen papers on dynamical systems in each major mathematics or physics journal per year. At present, the majority of these journals contain sections on nonlinear sciences, and indeed there are several new journals solely devoted to this subject. Nonetheless, the communication and infusion of knowledge between the mathematicians working on the analytic approach and the scientists and engineers working mostly on the applications and numerical simulations have been less than ideal.

Part of the reason is cultural. Mathematicians tend to approach the problem in a more generic sense, that is, they tend to ask questions and look for answers for more general properties and the underlying structures of the systems. Books written by mathematicians usually treat the subject

with mathematical rigor, but lack of some motivation for
wanting to study the underlying mathematical structures of
the system. The treatment by scientists and engineers
usually encompasses too many details and misses the
underlying structures which may transcend the usual
boundaries of various disciplines. Thus, scientists and
engineers may not be aware of the advances of the same type
of problems in other disciplines. We shall give examples of
such underlying mathematical structures for diverse
disciplines. This volume intends to bridge the gap between
these two categories of books treating nonlinear dynamical
systems. Thus, we would like to bridge the gap and foster
communication between scientists and mathematicians. In the
following, proofs of theorems are usually not given.
Instead, examples are provided so that the readers can get
the meaning of the theorems and definitions. In other words,
we would like the readers to get some sense of the concepts
and techniques of the mathematics as well as its "culture".
This volume is based on the lecture notes of a graduate
course I gave in 1983-4 while I was with TRW.

Chapter 1 introduces the concept of dynamical systems
and stability with examples from physics, biology, and
economics. We want to point out that even though the
differential or difference equations governing those
phenomena are different, nonetheless, the mathematical
procedures in analyzing them are the same. We also try to
motivate the reader about the concept and the need to study
the structural stability.

In Chapter 2, we assemble most of the definitions and
theorems about basic properties of algebra, points set (also
called general) topology, algebraic and differential
topology, differentiable manifolds and differential
geometry, which are needed in the course of the lectures.
Our main purpose is to establish notations, terminology,
concepts and structures. We provide definitions, examples,
and essential results without giving the arguments of proof.
The material presented is slightly more than absolutely

necessary for the course of this lecture series. For instance, we certainly can get by without explicitly talking about algebraic topology, such as homotopy and homology. Nonetheless, we do utilize the concepts of orientable manifolds or spaces so that an everywhere non-zero volume element can be found. We also discuss the connected components and connected sums of a state space. Nor do we have to discuss tubular neighborhoods, even though when dealing with return maps, the stability of orbits and periodic orbits, etc. we implicitly use the concept of tubular neighborhoods. Nonetheless, these and many other concepts and terminology are frequently used in research literature. It is also our intent to introduce the reader to a more sophisticated mathematical framework so that when the reader ventures to research literature or further reading, he will not feel totally lost due to different "culture" or terminology. Furthermore, in global theory, the state spaces may be differentiable manifolds with nontrivial topology. In such cases, concepts from algebraic topology are at times essential to the understanding.

In light of the above remarks, for the first reading the reader may want to skip Section 2.3, part of Section 2.4, de Rham cohomology part of Section 2.8 and Section 2.9.

The subjects discussed in Chapter 3 are not extensively utilized in the subsequent chapters, at least not explicitly. Nonetheless, some of the concepts and even terminology do find their way to our later discussions. This chapter is included, and indeed is lectured, to prepare the readers with some concepts and understanding about global analysis in general, and some techniques important to the global theory of dynamical systems. In particular, the reader may find it useful when they venture to theoretically oriented research literature.

In the next chapter, we shall discuss the general theory of dynamical systems. Most of the machinary developed in the last chapter is not used immediately. Only in the last section of Chapter 4, the idea of linearization of nonlinear differential operators will be utilized for the discussion of linearization of dynamical systems. Nor does Chapter 5 depend on the material of Chapter 3. As a consequence, one may want to proceed from Chapter 2 to Chapters 4 and 5, except Section 6 of Chapter 4. Then come back to Chapter 3 and continue to Chapter 6.

In Section 4.7 we briefly discuss the linearization process based on some results from Chapter 3. For most scientists and engineers, the linearization process is "trivial". Everyone has done this since their freshman year many times. Unfortunately, the linearization process is the most misunderstood and frequently mistaken procedure for scientists and engineers in dealing with nonlinear phenomena and nonlinear dynamical systems at large. This is particularly true for highly nonlinear systems. Two illustrations from our earlier training in mathematics can make the point clear. First, in our freshman calculus course, we learned about the condition of continuity of a function at a point, and its derivative at that point. For the derivative to have a meaning, not only the change of the independent variable, say, Δx, has to be very small (i.e., local), but also the limits $\lim_{x \to 0+} \Delta y / \Delta x$ and $\lim_{x \to 0-} \Delta y / \Delta x$ agree at $x = x_o$. In other words, the usual linearization makes sense, only when it is done locally, and any "displacement" from the point has to be consistent. The consistency can be illustrated by the following example. As we shall discuss in Chapter 2, a differentiable manifold M is a topological manifold (a topological space with certain nice properties) endowed with a differentiable structure. For any two points p and q in M, A, B are sufficiently small neighborhoods of p and q in M respectively. At p and q in A and B, a rectangular coordinate system can be attached to each of p and q, and every variable looks linear in A and B. What makes the coordinate system or the differentiable structure go beyond the confine of A or B is that if the intersection of A and B is non-empty, then the coordinate values in A and B have to agree at all points in the

intersection of A and B. That is, the differentiable
structure has to be consistently and continuously agreed
between A to B. Consequently, for a highly nonlinear system
when one linearizes such a system, one has to be certain
that the linearization procedure is "consistent" in the
above sense. Otherwise, except at or very near the
equilibrium point(s) in the phase space, one has no
assurance as to the correctness of the results of
linearization.

Related to the above discussion of the linearization
procedure, we would also like to caution the reader about
numerical simulation of nonlinear dynamical systems, even
when only dealing with a local situation. First and foremost,
do not just code the differential equations and let the
computer do the rest. One has to analyze the characteristics
of the system, namely, how many and what kind of fixed
points, periodic orbits, attractors, etc. Then, let the
computer do the dirty work, and compare with the analysis to
see whether or not the numerical simulation agrees with the
analysis on the number and type of characteristics.
Otherwise, there is no way one can be sure the simulation is
correct and meaningful.

Another important point sometimes scientists and
engineers have overlooked is whether or not the system has a
Cauchy data set. If the system does not admit a Cauchy data
set, it is tantamount to say that the system does not admit
a unique time. Thus, the time evolution of the system loses
its meaning. One may argue that who cares about $t \to \infty$. But
even for a finite t, the system may become unpredictable,
with or without chaos, etc.

It is also appropriate to point out a practical point
which relates to the numerical schemes for studying the
nonlinear dynamical system. The implicit method is a very
popular and efficient scheme to study the linear or weakly
nonlinear differential equations. But for highly nonlinear
systems, this scheme may leads to erroneous results. This is
because the interpolation definitely introduces errors,

which initially are small, but their growth rate usually is also the largest. After propagating for a short time, the amplitudes of the errors can be as large as the main variables. Small scale self-focusing in nonlinear optics is a very good example [Bespalov and Talanov 1966; Fleck et al 1976; Lee 1977; Brown 1981]. The two volume set by R. Bellman (**Methods of Nonlinear Analysis**, Academic Press, 1973) provides eloquent motivation and many techniques particularly useful for numerical simulation of nonlinear systems. This set is highly recommended for anyone seriously interested in the numerical simulation of nonlinear dynamical systems.

In Chapter 5 we introduce the Liapunov's direct (or second) method for analyzing stability properties of nonlinear dynamnical systems. In this chapter, we are still dealing with the local theory of stability for nonlinear dynamical systems.

The first half of Chapter 6 discusses the global theory of stability and the very important concept of structural stability. Sections 6.6 on bifurcations and 6.7 on chaos could be grouped together with the local theory of stability. Nonetheless, because some of the concepts and techniques are also useful for the global theory, we put these two sections in this chapter and deal with global theory concerned with bifurcations and chaos. A recent new definition of stability proposed by Zeeman, which is intended to replace the original structural stability, is very interesting and has a great potential in analyzing the global stability of many practical physics problems. We did not include any discussion on quantum chaos, because this author does not understand it. We only mention fractals in passing, not that it is unimportant, (on the contrary it is a very important topic), but mainly due to lack of space.

Chapter 7 discusses applications of stability analyses to various disciplines, and we also indicate the commonality and the similarities of the mathematical structures for various problems in diverse disciplines. For instance, in

Section 7.5, we not only discuss the dynamical processes of competitive interacting population processes in biology and population ecology and biochemical autocatalysis processes, but we have also pointed out that these equations also describe some processes in laser physics and semiconductor physics, to name a few. As another example, the discussion of permanence in Section 7.5 with minor modifications, can also be applied to mode structures and phased arrays in lasers. The point is that scientists and engineers can learn, or even directly apply results, from different disciplines to solve their problems. Putting it differently, physical scientists can learn from, and should communicate with, biological scientists, and vice versa. One way for the physical scientists to establish such communication is to read some biological journals. The conventional wisdom has it that papers in biological journals are not very sophisticated mathematically. The fact is, in the past decade or so, there has been an influx of mathematicians working on mathematical biology. Consequently, the whole landscape of the theoretical or mathematical biology journals has changed dramatically. Nowadays, there are many deep and far reaching mathematical papers in a much broader context being published in those journals.

The second goal of this volume is to draw the attention for such "lateral" interactions between physical and biological scientists.

The third goal is to provide the reader a very personal guide to study the global nonlinear dynamical systems. We provide the concepts and methods of analyzing problems, but we do not provide "recipes". After all, it is not intended to be a "cookbook". Should a cookbook be contemplated, the "menu" would be very limited.

We have tried to include some of the references the author has benefited from. Surely they are far from comprehensive, and the list is very subjective to say the least. The omission of some of the important works indicates the ignorance of the author.

The following books are highly recommended for
undergraduate or beginners on dynamical systems: Hirsch and
Smale [1974], Iooss and Joseph [1980], Irwin [1980], Ruelle
[1989], Thompson and Stewart [1986].

Without inspiring teachers like Peter Bergmann, Heinz
Helfenstein, John Klauder, and Douglas Anderson, I would not
have the opportunity to learn as much. My colleagues and
friends, David Brown, Ying-Chih Chen, Da-Wen Chen, Gerrit
Smith, and many others, are also my teachers. They not only
have taught me various subjects, but they also provided me
with enjoyable learning experiences throughout my
professional career. I would also like to acknowledge the
understanding and support of my two childern, Jennifer and
Peter, without their encouragement I would not be able to
complete this project. It is my pleasure to acknowledge the
constant encouragement, support, suggestions for
improvements, and patience, of the editorial staff of the
publisher, in particular, Mr. K. L. Choy.

Kotik K. Lee
Colorado Springs

CONTENTS

xvi

Chapter 1   Introduction

1.1   What is a dynamical system?

A dynamical system can be thought of as any set of
equations giving the time evolution of the state of the
system from the knowledge of its previous history. Nearly
all observed phenomena in scientific investigation or in our
daily lives have important dynamical aspects. Examples are:
(a) in physical sciences: Newton's equations of motion for a
particle with suitably specified forces, Maxwell's equations
for electrodynamics, Navier-Stokes equations for fluid
motions, time-dependent Schrodinger's equation in quantum
mechanics, and chemical kinetics; (b) in life systems:
genetic transference, embryology, ecological decay, and
population growth; (c) and in social systems: economical
structure, the arms race, or promotion within an
organizational hierarchy. Although these examples illustrate
the pervasiveness of dynamic situations and the potential
value of developing the facility for modeling (representing)
and analyzing the dynamic behavior, it should be emphasized
that the general concept of dynamics and the treatment of
dynamical systems transcends the particular origin or the
setting of the processes.

In our daily lives we often quite effectively deal with
many simple dynamic situations which can be understood and
analyzed intuitively (i.e., by experience) without resorting
to mathematics and the general theory of dynamical systems.
Nonetheless, in order to approach complex and unfamiliar
situations efficiently, it is necessary to proceed
systematically. Mathematics can provide the required
conceptual framework and proper language to analyze such
complex and unfamiliar dynamic situations.

In view of its mathematical structure, the term dynamics
takes on a dual meaning. First, as stated earlier, it is a
term for the time-evolutionary phenomena around us and about
us; and second, it is a term for the pact of mathematics

1

which is used to represent and analyze such phenomena, and the interplay between both aspects.

Although there are numerous examples of interesting dynamic situations arising in various areas, the number of corresponding general forms for mathematical representation is limited. Most commonly, dynamical systems are represented mathematically in terms of either differential or difference equations. In fact, in terms of the mathematical content, the elementary study of dynamics is almost synonymous with the theory of differential and difference equations.

Before proceeding to the quantitative description of dynamical systems, one should note that there are qualitative structures of dynamical systems which are of fundamental importance, as will be discussed later. At the moment, it is suffice to note that even though there are many different disciplines in the natural sciences, let alone many more subfields in each discipline, Nature seems to follow the economical principle that a tremendous number of results can be condensed into a few simple laws which summarize our knowledge. These laws are qualitative in nature. It should be emphasized that here qualitative does not mean poorly quantitative, rather topologically invariant, i.e., independent of local and detail descriptions. Furthermore, common to many natural phenomena, besides their qualitative similarity, is their universality where the details of the interactions of systems undergoing spontaneous transitions are often irrelevent. This calls for topological descriptions of the phenomena under consideration. Hence, the concept of structural stability and the theories of singularity and bifurcation undoubtedly lead the way.

Simply stated, the use of either differential or difference equations to represent dynamic behavior corresponds to whether the behavior is viewed as occurring in continuous or discrete time respectively. Continuous time corresponds to our usual perception that time is often

2

viewed as flowing smoothly past us. In mathematical terms, continuous time is quantified by the continuum of real numbers and usually denoted by the parameter t. Dynamic behavior viewed in continuous time is usually described by differential equations.

Discrete time consists of an ordered set rather than a continuous parameter represented by real numbers. Usually it is convenient to introduce discrete time when events occur or are accounted for only at discrete time periods. For instance, when developing a population model, it may be convenient to work with annual population changes, and the data is normally available annually, rather than continually. Discrete time is usually labeled by simple indexing of variables in order and starting at a convenient reference point. Thus, dynamic behavior in discrete time is usually described by equations relating the value of a variable at one time to the values of variables at adjacent times. Such equations are called difference equations. Furthermore, in order to calculate the dynamics of a system, which are normally represented by differential equations, with an infinite degree of freedom (we shall come to this shortly) such as fluids, it is more convenient to numerically break down the system into small but finite cells in space and discrete periods of time. Thus one also uses the difference method to solve differential equations.

In what follows, we shall concentrate on the aspect of continuous time, i.e., the differential equations aspect of dynamical systems.

In the late 1800's, Henri Poincaré initiated the qualitative theory of ordinary differential equations in his famous menoir [1881, 1882]. Ever since then, differential topology, a modern development of calculus, has provided the proper setting for this qualitative theory. As we know, ordinary differential equations appear in many different disciplines, and the qualitative theory often gives some important insight into the physical, biological, or social realities of the situations studied. And the qualitative

3

theory also has a strong appeal, for it is one of the main areas of inter-disciplinary studies between pure mathematics and applied science.

If we are studying processes that evolve with time and we wish to model them mathematically, then the possible states of the systems in which the processes are taking place may often be represented by points of differentiable manifolds known as state spaces of the models. For instance, if the system is a single particle constrained to move on a plane, then the state space is the Euclidean R^4 and the point (x_1, x_2, v_1, v_2) represents the position of the particle at $\mathbf{x} = (x_1, x_2)$ with the velocity $\mathbf{v} = (v_1, v_2)$. Note that the state space of a model can be finite dimensional, as in the above example, or it may be infinite dimensional, such as in fluid dynamics. Furthermore, it may occur that all past and future states of the system are completely determined by the equations governing the system and its state at any one particular instance. In such a case, the system is said to be deterministic, such as in Newtonian mechanics. The systems modeled in quantum mechanics are not.

In the context of deterministic processes, usually they are governed by a smooth vector field on the state space. In classical mechanics, the vector field is just another way of describing the system. When we say a vector field governing a process we mean that as the process develops with time the point representing the state of the system moves along a curve (integral curve) in the state space. The velocity at any position $\mathbf{x}$ on the curve is a tangent vector to the state space based at $\mathbf{x}$. We say the process is governed by the vector field, if this tangent vector is the value of the vector field at $\mathbf{x}$, for all $\mathbf{x}$ on the curve.

In the qualitative theory (or geometric theory) of differential equations, we study the smooth vector fields on differentiable manifolds, focusing on the collection of parameterized curves on the manifold which have the vector fields as the tangents of the curves (integral curves). Hopefully, a geometric feature of the curves and vector

4

fields will correspond to a significant physical phenomenon and also is part of a good mathematical model for such a physical situation.

In the following we shall provide some simple examples of qualitative theory of differential equations and illustrate the approaches we will be taking later. Let us first examine some familiar examples in classical mechanics from a geometrical viewpoint.

First, let us consider a pendulum which is the simplest and the most well-known dynamical system. For simplicity, let us scale the mass m and the length l of the pendulum to be unity, i.e., m = l = 1. We shall not dwell here on the physical basis of the equation of motion for the pendulum, which is $d^2\theta/dt^2 = -g\sin\theta$ except that there is no air resistance and no friction at the pivot. By using the definition of the angular velocity $\omega = d\theta/dt$, we can replace the equation of motion by a pair of first order differential equations:

$$d\theta/dt = \omega,$$
$$d\omega/dt = -g\sin\theta \qquad\qquad (1.1-1)$$

The solution of Eq.(1.1-1) is an integral curve in the (θ,ω) plane parameterized by t, and the parametrized coordinates of the curve are $(\theta(t),\omega(t))$. The tangent vector to the curve at t is $(\omega(t), -g\sin\theta(t))$. From various initial values of θ and ω at t = 0, one obtains corresponding integral curves and these curves form the phase-portrait of the system. It can be shown that the phase-portrait of the pendulum looks like Fig.1-1a,b. One can easily distinguish five distinct types of integral curves. They can be interpreted as follows:

(a) the pendulum hangs vertically downward and is at rest,
(b) the pendulum swings between two positions of instantaneous rest which are equally inclined to the vertical,
(c) the pendulum continuously rotates in the same direction and never at rest,
(d) the pendulum stands vertically upward and is at rest,

5

(e) the limiting case between (b) and (c) when the pendulum
takes an infinitely long time to swing from one upright
position to another.

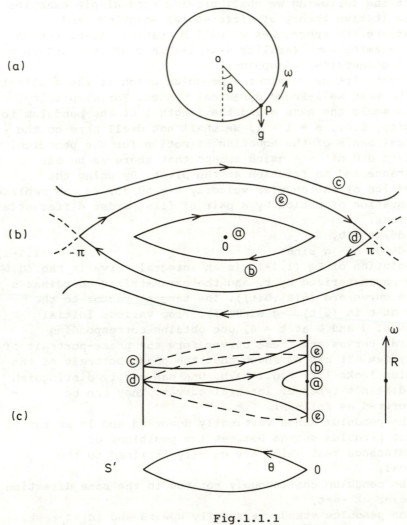

Fig.1.1.1

6

There are certain features in the phase-portrait which are unsatisfactory. First of all, the pendulum has only two equilibrium positions, the one which hangs downward is stable and the other which stands upward is unstable. Secondly, solutions of type (c) are periodic motions of the pendulum but it appears as nonperiodic curves in this particular form of phase-portrait. In fact, we ought to regard $\theta = \theta_o$ and $\theta = 2n\pi + \theta_o$ for any given integer n as giving the same position of the pendulum. In other words, the configuration space, which is the differentiable manifold representing the spatial positions, of a pendulum is really a circle rather than a straight line. Thus, we should replace the first factor R of R^2 by the circle S^1, which is the real module 2π. By keeping θ and as two parameters, we obtain the phase-portrait on the cylinder S^1xR as shown in Fig.1.1c.

Consider the kinetic energy T and the potential energy V of the pendulum, and $T = 1/2\dot{\theta}^2$ and $V = g(1- \cos\theta)$. Let the total energy of the pendulum be $E = T + V$. Then it is clear from Eq.(1.1-1) that $dE/dt = 0$, i.e., E is a constant on any integral curve. Any system with constant total energy is called a conservative or Hamiltonian system.

In fact, for a pendulum, one can easily construct the phase- portrait by determining the energy levels (i.e., energy contours). One can represent the state space cylinder S^1 x R as a bent tube and interpret the height as energy. This is illustrated in Fig.1.1.2a. The two arms of the tube represent solutions of the same energy $E > 2g$, where 2g is the potential energy of the unstable equilibrium, with the pendulum rotating in the opposite direction.

The stability properties of individual solutions are particularly apparent from the above picture. Any integral curve through a point close to the stable equilibrium position A remains close to A at all times, since the energy function E attains its absolute minimum at A and is stationary at B. In fact, B is a saddle point. Thus, there are points arbitrarily close to the unstable equilibrium point B such that integral curves through them depart from a given small neighborhood of B.

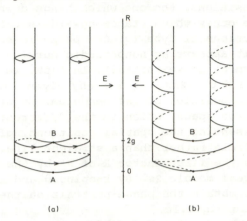

(a)   (b)

Fig.1.1.2

The above example does not include the effects of air resistance and friction at the pivot of the pendulum. Let us now take these dissipative forces into consideration, and for simplicity let us assume they are directly proportional to the angular velocity. Thus, Eqs.(1.1-1) become

$$d\theta/dt = \omega \quad \text{and} \quad d\omega/dt = -g\sin\theta - a \qquad (1.1-2)$$

where a is a positive constant. Now we find that the energy no longer remains constant along any integral curve and the system is called <u>dissipative</u>. This is because for $\omega \neq 0$, $dE/dt = -a\omega^2$ which is negative and the energy is dissipated away along integral curves. If we represent E as a height function as before, the inequality $E < 0$ implies that the integral curves cross the horizontal contours of E "downward" as in Fig.1.1.2b.

Now the stable equilibrium becomes asymptotically stable in the sense that nearby solutions tend toward equilibrium solution A as time goes by. Yet, we still have the unstable equilibrium solution B and other solutions that tend either toward or away from B. Nonetheless, we would not expect to realize any such solutions, since we could not hope to

8

satisfy the precise initial conditions needed.

By comparing the systems of Eqs.(1.1-1) and (1.1-2), one obtains some hint of what is involved in the important notion of structural stability. Roughly speaking, a system is structurally stable if the phase-portrait remains qualitatively (or topologically) the same when the system is modified by any sufficiently small perturbation. By qualitatively (or topologically) the same, we mean that some homeomorphism of the state space map integral curves of the one onto integral curves of the other. The existence of systems (1.1-2) shows that the original system (1.1-1) is not structurally stable since the constant a can be as small as we want. Yet, the systems (1.1-2) are themselves structurally stable. To distinguish between the systems (1.1-1) and (1.1-2), we observe that most solutions of the former are periodic whereas the only periodic solutions of the latter are the equilibria [We shall discuss this in Chapter 6]. In fact, this last property holds true for any dissipative system, since E is decreasing along integral curves.

In the above example, it is more convenient and desirable to use a state space other than Euclidean space, but it is not essential. In studying more complicated systems, the need for non-Euclidean state spaces becomes more apparent. Indeed, it is often impossible to study complicated systems globally using only Euclidean state spaces. We need non-Euclidean spaces on which systems of differential equations are defined globally, and this is one of the reasons for studying differentiable manifolds. We shall give a brief outline of it in the next chapter.

To illustrate the necessity of a non-Euclidean state space globally, let us consider the spherical pendulum. We get the spherical pendulum from the pendulum by removing the restriction that the rod moves in a plane through the pivot. Thus the pendulum is constrained to move on a unit 2-sphere of radius one in Euclidean 3-space. Here once again we assume that the length of the rod is unity. The 2-sphere can be represented parametrically by Euler angles θ and ϕ. The equations of motion for the spherical pendulum are:

$$d^2\theta/dt^2 = \sin\theta \cos\theta (d\phi/dt)^2 + g \sin\theta \qquad (1.1-3)$$
$$d^2\phi/dt^2 = -2(\cot\theta) \, d\theta/dt \, d\phi/dt.$$

We can replace this system of second order equations by the equivalent system of four first order equations.

$$d\theta/dt = \omega,$$
$$d\phi/dt = \mu, \qquad (1.1-4)$$
$$d\omega/dt = \mu^2 \sin\theta \cos\theta + g \sin\theta$$
$$d\mu/dt = -2\omega\mu \cot\theta.$$

The state of the system is determined by the position of the pendulum on the sphere, together with the velocity which is specified by a point in the 2-dim tangent plane of S^2 at the position of the pendulum. In fact, the state space is not homeomorphic to R^4, nor to $S^2 \times R^2$, but is the tangent bundle of S^2, TS^2. This is the set of all planes tangent to S^2 and it is an example of a non-trivial vector bundle. We shall discuss these concepts in the next chapter. Locally, TS^2 is topologically indistinguishable from R^4 and one can use the four variables θ, ϕ, ω, μ as local coordinates in TS^2 except at the north and south poles of S^2.

The total energy of the system in terms of local coordinates is

$$E = (\omega^2 + \mu^2 \sin\theta)/2 + g(1 + \cos\theta)$$

and it is straightforward to show that $E = 0$ along integral curves, i.e., the system is conservative. Thus, every solution is contained in a contour of E = constant. $E = 0$ is again a single point at which E is the absolute minimum, corresponding to the pendulum hanging vertically downwards in stable equilibrium. Again $E = 2g$ contains the other equilibrium point, where the pendulum stands vertically upward in unstable equilibrium, and at this point E is stationary but not minimal. From Morse theory (see Chapter 2 and Hirsch [1976], Milnor [1963]) one knows that for $0 < c < 2g$ the contour $E^{-1}(c)$ is homeomorphic to S^3.

The spherical pendulum is symmetrical about the vertical axis through the pivot point. This symmetry manifests itself in Eq.(1.1-4), for they are unaltered if we replace ϕ by ϕ + k (k = constant) or if we replace ϕ and μ by $-\phi$ and $-\mu$. Thus the orthogonal group O(2) acts on the system as a group of symmetries about the vertical axis. Such symmetries can reveal important features of the phase-portrait. Here for any c with $0 < c < 2g$, the 3-sphere $E^{-1}(c)$ is partitioned into a family of tori, together with two exceptional circles as in Fig.1.1.3. This decomposes R^3 into a family of tori, with a circle through p and q and the line l. Combining with a "point at ∞ " turns R^3 into a topological 3-sphere and the line l into a topological circle. The submanifolds of this partition are each generated by a single integral curve under the action of SO(2) and they are the intersections of $E^{-1}(c)$ with the contours of the angular momentum function on TS^2. The two exceptional circles correspond to the pendulum revolving in a horizontal circle in two opposite directions. Between them comes a form corresponding to that of a pendulum in the various planes through l.

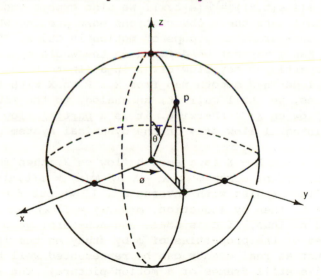

Fig. 1.1.3

11

In the examples given above, the dynamical state of the system is represented by a point of the state space which is the tangent bundle (such as S^1 x R^1 or $T(S^2)$) of the configuration space (S^1 or S^2 respectively). And the equation of motion represented by a vector field on the state space and its integral curves give the possible motions of the system.

A useful way of visualizing a vector field **v** on an arbitrary manifold X is to imagine a fluid flowing on X. Let us assume that the velocity of the fluid at each point x ϵ X is independent of time and equals to the value v(x). Then the integral curves of **v** are precisely the paths followed by particles of the fluid. Let $\phi(t,x)$ be the point of X reached at time t by a particle of the fluid that leaves X at time 0. Obviously, $\phi(0,x)$ is always x. Since the velocity is assumed to be independent of time, $\phi(s,y)$ is the point reached at time s+t by a particle starting at y at time t. if we set y = $\phi(t,x)$, as the particle started from x at time 0, then $\phi(s,\phi(t,x)) = \phi(s+t,x)$. We also expect smoothness of ϕ. We shall make these observations more precise when we discuss one-parameter groups of motion in Ch. 2.

The map ϕ may not be defined on the whole space X x R, because particles may flow off X in a finite time. But, if ϕ is a well-defined smooth map from X x R to X with the above properties, we shall call it, an analogy to the fluid, a smooth flow on X; othherwise, it is a partial flow on X. ϕ is the integral flow of **v** or the dynamical system given by **v**.

If ϕ : X x R → X is a smooth flow on X, then for any t ϵ R, we may define a map ϕ^t : X → X by $\phi^t(x) = \phi(t,x)$ and it is a diffeomorphism with inverse ϕ^{-t}. If we put f = ϕ^a for some a ϵ R, then by induction, $\phi(na,x) = f^n(x)$ for all integers n. Thus, if a is small and non-zero, we often get a good idea of the properties of ϕ by studying the iterates f^n of f (just as real events can be represented well by the successive still frames of a motion picture). The theory of discrete dynamical systems or discrete flow resembles the

12

theory of flow in many ways; and we shall cover both of them.

Now let us turn our attention to an example of a dynamic system in the context of population growth. Let us look at the simplest example of the rich theory of interacting populations, the predator-prey model. Let us imagine an island populated by goats and wolves only. The goats survive by eating the island vegetation and the wolves survive by eating the goats. The modeling of this kind of population system goes back to Volterra in response to the observation that populations of species often oscillate. Let $N_1(t)$ and $N_2(t)$ represent the populations of the prey (goats) and predators (wolves), respectively. Volterra described the situation in the following way:

$$dN_1(t)/dt = aN_1(t) - bN_1(t)N_2(t)$$
$$dN_2(t)/dt = -cN_2(t) + dN_1(t)N_2(t) \qquad (1.1-5)$$

where the constants a, b, c and d are positive. The model is based on the assumption that in the absence of predators (wolves), the prey (goats) population will increase exponentially with a growth rate a. Likewise, in the absence of prey, the predator population will diminish at a rate c. When both populations are present, the frequency of "encounters" is assumed to be proportional to the product of the two populations. Each encounter decreases the prey (goats) population and increases the predator (wolves) population. The effects of these encounters are accounted for by the second terms in the differential equations.

Of course, these equations are highly simplified and do not take into account a number of external factors such as general environment conditions, supply of other food for both predator and prey, migration of the populations, disease, and crowding. An important application of a model of this type is the study and control of pests and feed on agricultural crops. The pest population is often controlled by introducing predators, and such a predator-prey model often forms the foundation of ecological intervention.

The nonlinear dynamic Eq.(1.1-5) cannot be solved

analytically in terms of elementary functions. Nonetheless, it is easy to see that there are equilibrium points. For the steady state situation, by setting $dN_1/dt = dN_2/dt = 0$, we have one equilibrium point at $N_1 = N_2 = 0$ and another at $N_1 = c/d$, $N_2 = a/b$. It is convenient to normalize variables by letting

$$x_1 = dN_1/c, \qquad x_2 = bN_2/a.$$

Then the dynamic Equations (1.1-5) become

$$dx_1/dt = ax_1(1 - x_2)$$
$$dx_2/dt = -cx_2(1 - x_1). \tag{1.1-6}$$

Clearly the nontrivial equilibrium point is at $x_1 = 1$, $x_2 = 1$.

Let us study the stability of the two equilibrium points $(0,0)$ and $(1,1)$. It is clear that $(0,0)$ is unstable, for if x_1 is increased slightly it will grow exponentially. The point $(1,1)$ requires more elaborate analysis. A linearization of the system in terms of displacements x_1, x_2 from the equilibrium point $(1,1)$ can be obtained by evaluating the first partial derivatives of Eq.(1.1-6) at $(1,1)$, and we have

$$(\Delta dx_1/dt) = -a(\Delta x_2)$$
$$(\Delta dx_2/dt) = c(\Delta x_1).$$

The linearized system has eigenvalues $\pm iac$ representing a marginally stable system. From linear analysis (such as the first method of Liapunov) it is not possible to infer whether the equilibrium point is stable or unstable. Therefore we have to study the nonlinearity more explicitly.

We can find a constant of motion by writing Eq.(1.1-6)

$$dx_2/dt/dx_1/dt = [-cx_2(1-x_1)]/[ax_1(1-x_2)]$$

and rearranging terms leads to

$$cdx_1/dt - cdx_1/dt/x_1 + adx_2/dt - adx_2/dt/x_2 = 0.$$

Integrating, we have

$$cx_1 - c \log x_1 + ax_2 - a \log x_2 = \log k$$

where k is a constant. For $x_1 > 0$, $x_2 > 0$, we can define the function

$$V(x_1,x_2) = cx_1 - c \log x_1 + ax_2 - a \log x_2.$$

Clearly V is a constant of motion. Thus, the trajectory of

14

population distribution lies on a curve defined by V = k.
The predator-prey cycles look like Fig.1.1.4. Since the
trajectories circle around the equilibrium point, it is
stable but not asymptotically so. The function V, similar to
the energy function E in the case of a pendulum, attains a
minimum at the equilibrium point (1,1). V also serves as a
Liapunov function for the predator-prey system and
establishes stability. We shall discuss Liapunov functions
in Ch. 5.

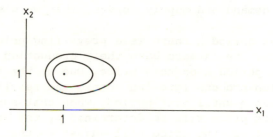

Fig.1.1.4

We shall discuss crowding, multispecies cases, and their
interrelationships with biochemical reactions, semiconductor
physics and laser physics in Chapter 7. For an introduction,
an interested reader should consult another text [e.g.,
Luenberger].

Next we shall discuss a classical dynamic model of
supply and demand interaction, which also serves as an
example of a difference equation. The model is concerned
with a single commodity, say corn. The demand d for the
commodity depends on the price p through a function d(p) .
If the price increases, consumers will buy less, thus d(p)
decreases as p increases. For simplicity, in this example we
assume that the demand function is linear, i.e., $d(p) = d_o -
ap$, where d_o and a are positive constants. Likewise, the
supply of the commodity, s, also depends on the price p
through a function S(p). Usually, the supply increases when
the price increases. For instance, a higher price will
induce farmers to plant more corn. Note, there is a time lag

15

involved (we shall come to this point shortly). Let us assume that the supply function is also linear, i.e., $s(p) = s_o + bp$ where b is positive and s_o can have any value, but usually negative.

In equilibrium, the demand must equal the supply, this corresponds to the point where these two lines intersect. But the equilibrium price is attained only after a series of adjustments made by both consumers and producers. It is the dynamics of this adjustment process, movement along the appropriate demand and supply curves, that we wish to describe.

Assume at period k there is a prevailing price $p(k)$ for the commodity. The farmers base their production (or planting) in period k on this price. Due to the time lag in the production process (growing corn), the resulting supply is not available until next period, when that supply is available, its price will be determined by the demand function. That is, the price will adjust so that all of the available supply will be sold. This new price at period $k+1$ will determine the production for the next period. Thus a new cycle begins.

Let us set up the supply and demand equations according to the cycles described above. The supply equation can be written as

$$s(k+1) = s_o + bp(k)$$

and the demand equation

$$d(k+1) = d_o - ap(k+1).$$

The condition of equilibrium leads to the dynamic equation

$$s_o + bp(k) = d_o - ap(k+1) .$$

This equation can be restated in the standard form for difference equation $p(k+1) = - bp(k)/a + (d_o - s_o)/a$.
By setting $p(k) = p(k+1)$, one obtains the equilibrium price,

$$p = (d_o - s_o)/(a + b),$$

which would persist indefinitely. It is natural to ask whether this price will ever be established or even if over successive periods the price will tend toward this

16

equilibrium price and not diverge away from it. From the general solution of the first- order equation we have

$p(k) = (-b/a)^k p(0) + [1-(-b/a)^k](d_o - s_o)/(a + b)$.

If $b < a$, it follows that as $k \to \infty$ the solution will tend toward the equilibrium value since all $(-b/a)^k$ terms go to zero and the equilibrium value is independent of the initial price. Clearly, $b < a$ is both necessary and sufficient for this convergence property to hold.

Let us trace the path of supply and demand over successive periods on graphs and interpret the results. The graphs are shown in Fig.1.1.5b and 1.1.5c, which represent a converging and a diverging situation, respectively. The initial price $p(0)$ determines the supply s that will be available in the next period. This supply determines the demand d and thus the price $p(1)$, and so on. Thus we are led to trace out a rectangular spiral. If $b < a$, the spiral will converge inward, but if $b > a$, it will diverge outward.

From this stability analysis, we can deduce an important conclusion for the economic model we have been considering. In order for the equilibrium to be attained, the slope b of the supply curve must be less than the slope a of the demand curve. In other words, the producers must be less sensitive to price changes than the consumers.

Strangely enough, some dynamical behavior of Boolean networks have the same structure of the supply and demand problem [Martland 1989].

17

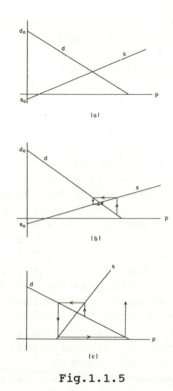

Fig.1.1.5

1.2   What is stability, and why should we care about it?

Very early in scientific history, the stability concept
was specialized in mathematics to describe some types of
equilibrium of a material particle or system. For instance,
a particle subject to some forces and possessing an
equilibrium point p_o. The equilibrium is called <u>stable</u> if,
after any sufficiently small perturbations of its position
and velocity, the particle remains forever arbitrarily near
p_o, with arbitrarily small velocity. In Sec. 1 we have
discussed the dynamics of a pendulum and its stability in
this light.

When formulated in precise mathematical terms, this

18

mechanical definition of stability was found useful in many situations, but inadequate in many others. This is why, over the years, a host of other concepts have been introduced, each of them related to the first mechanical definition and to the common sense of stability.

In contrast to the mechanical definition of stability, the concept known as Liapunov's stability has the following characteristics: (i) it pertains not to a material particle or the particular equation, but to a general differential equation; (ii) it applies to a solution, thus not only to an equilibrium or critical point. More precisely, let

$$d\mathbf{x}/dt = \mathbf{f}(t,\mathbf{x})$$

(1.2-1) where $\mathbf{x}$ and $\mathbf{f}$ are real n-vectors, $t \in R$ is the time, $\mathbf{f}$ is defined on $R \times R^n$. We also assume $\mathbf{f}$ is smooth enough to ensure the exsistence, uniqueness and continuous dependence of the solutions of the initial value problem associated with Eq.(1.2-1) over $R \times R^n$. Let $\|\cdot\|$ denote any norm on R^n. A solution $\mathbf{x}(t)$ of Eq.(1.2-1) is stable at t_o, or at $t = t_o$ in the sense of Liapunov if, for every $\epsilon > 0$, there is a $\delta > 0$ such that if $\mathbf{x}'(t)$ is any other solution with $\|\mathbf{x}'(t_o) - \mathbf{x}(t_o)\| < \delta$, then $\|\mathbf{x}'(t) - \mathbf{x}(t)\| < \epsilon$ for all $t > t_o$. Otherwise, $\mathbf{x}(t)$ is unstable at t_o. Thus, stability at t_o is nothing but continuous dependence of the solutions on $\mathbf{x}'_o = \mathbf{x}'(t_o)$, uniform with respect to $t \in [t,\infty)$.

Notice that in the case of the pendulum, the equilibrium point $(0,0)$ in the phase space is such that no neighboring solution approaches it when $t \to \infty$, except if some friction were present. In many practical situations, it is useful to require, in addition to Liapunov stability of a solution $\mathbf{x}(t)$, that all neighboring solutions $\mathbf{x}'(t)$ tend to $\mathbf{x}(t)$ when $t \to \infty$. This leads to the notion of asymptotic stability.

Many other examples can illustrate the necessity of creating new specific concepts. Indeed, the stability of relative equilibria in celestial mechanics is subtle, to say the least, depending on deep properties of Hamiltonian systems (as has been shown by Kolmogorov [1954], Moser [1962], Arnold [1963a,b] and Rüssman [1970]), it is known as

the Kolmogorov-Arnold-Moser (KAM) theorem. We shall briefly
discuss this theorem in Section 2.8. For further reading of
KAM theorem, consult, for instance, Abraham and Marsden
[1978], or Arnold [1978]. From common sense, the solar
system is considered stable because it is durable, i.e.,
none of its planets escapes to infinity, nor do any two such
planets collide. But the velocities are unbounded iff two
bodies approach each other. Therefore, the Lagrange
stability simply means that the coordinates and velocities
of the bodies are bounded. Thus boundedness of the solution
appears as a legitimate and natural type of stability. For
many other definitions of stability and attractivity, please
see Rouche, Habets & Laloy [1977].

The most comprehensive of many different notions of
stability is the problem of structural stability. This
problem asks: If a dynamical system X has a known phase
portrait P, and X is then perturbed to a slightly different
system X' (such as, changing the coefficients in the
differential equations slightly), then is the new phase
portrait P' close to P in some topological sense? This
problem has obvious importance, because in practice the
qualitative information obtained for P is not applied to X
but to some nearby system X'. This is because the
coefficients of the equation are determined experimentally ,
thus approximately.

An important role physics plays in various disciplines
of science is that most systems and structures in nature
enjoy an inherent "physical stability", i.e., they preserve
their quality under slight perturbations - i.e., they are
structurally stable. Otherwise we could hardly think about
or describe them, and reproducibility and confirmation of
experiments would not be possible. Thus we have to accept
structural stability as a fundamental principle, which not
only complements the known physical laws, but also serves as
a foundation upon which these physical laws are built. Thus
universal phenomena have a common topological origin, and
they are describable and classifiable by unfoldings of

singularities, which organize the bifurcation processes
exhibited by dynamical systems. Here bifurcation refers to
the changes in the qualitative structure of solutions of
differential equations describing the governing dynamical
systems. A phenomenon is said to be structurally stable if
it persists under all allowed perturbations in the system.

Another important approach or "utility" of structural
stability analysis is the following. Since most of the
nonlinear equations of nature are not amenable to a
quantitative analysis, only a few are known. Consequently it
is often unclear which particular quantitative effects one
ought to study. Nonetheless, since the nonlinear equations
are derived from geometrical or topological invariance
principles, they must process structurally stable solutions.
In determining these stable solutions qualitatively, it will
provide us with conceptual guidance to single out the most
significant phenomena in complex systems to answer the
questions of structure formation and recognition.

Traditionally, the usefulness of a theory is judged by
the criterion of adequacy, i.e., the verifiability of the
predictions, or the quality of the agreement between the
interpreted conclusions of the model and the data of the
experiments. Duhem adds the criterion of stability. This
criterion refers to the stability or continuity of the
predictions, or their adequacy when the model is slightly
perturbed. And the general applicability of this type of
criterion has been suggested by Thom [1973]. This stability
concerns variation of the model only, the interpretation and
experimental domain being fixed. Therefore it mainly
concerns the model, and is primarily a mathematical or
logical question. It is safe to say that a clear enunciation
of this criterion in the correct generality has not yet been
made, although some progress has been made recently.

A tacit assumption (or criterion) which has been
implicitly adapted by physicists may be called the doctrine
of stability. For instance, in a model of a system with
differential equations where the model depends on some

21

parameters or some coefficients of the differential equations, each set of values corresponds to a different model. As these parameters can be determined approximately, the theory is useful only if the equations are structurally stable, which cannot be proved at present in many important cases. Thus physicists must rely on faith at this moment. Thom [1973] offered an alternative to the doctrine of stability. He suggested that stability, when precisely formulated in a specific theory, could be added to the model as an additional hypothesis. This formalization reduces the criterion of stability to an aspect of the criterion of adequacy, and may admit additional theorems or predictions in the model. Although no implications of this axiom is known for celestial mechanics as yet, Thom has described some conclusions in his model for biological systems. A careful statement of this notion of stability in the general context of physical sciences and epistemology, just to name a few disciplines, could be quite useful in technical applications of mechanics as well as in the formation of new qualitative theories in physical, biological, and social sciences.

In all fields of physics, waves are used to investigate (probe) some unknown structures. Such a structure or object impresses geometrical singularity upon smooth incident wavefields. The question is then, what information about the structure under study can be inferred from these geometrical singularities; this is the so-called inverse scattering problem. In order that the reconstruction of structures from the backscattered or transmitted curves will be physically repeatable, the scattering process has to be structurally stable, that is, qualitatively insensitive, to slight perturbations of the wavefields.

Imposing this structural stability principle on the inverse scattering process allows us to classify the geometrical singularities, impresses on the sensing wavefields by the unknown structure, into a few universal topological normal forms described by catastrophe

polynomials. Moreover, the topological singularities provide
an explanation for the similarity and universality of the
patterns encountered in geophysics and seismology [Dangelmay
and Guttinger 1982, Hilterman 1975], ocean acoustics [Keller
and Papadakis 1977], optics [Baltes 1980, Berry 1977, Nye
1978, Berry 1980], and various topography, etc.

On a more fundamental level, nonlinear dynamical systems
and their stability analyses have clearly demonstrated the
unexpected fact that systems governed by the Newtonian
dynamics do not necessarily exhibit the usual
"predictability" property as expected. Indeed, a wide
classes of even very simple systems, which satisfy those
Newtonian equations, predictability is impossible beyond a
certain definite time horizon. This failure of
predictability in Newtonian dynamics has been very well
elucidated in a review paper by Lighthill [1986]. Moreover,
recently, there have been several papers attempting to
relate such unpredictability to statistical mechanics.
Nonetheless, a much more fundamental framework together with
appropriate mathematical structure have to be established
before any such correlation can be made.

In the next chapter, we shall assemble some of the
definitions and theorems, without proof, about basic
properties of algebra, topology, and differential geometry,
which are essential to our discussion of the geometric
theory of nonlinear dynamical systems.

Chapter 2   Topics in Topology and Differential Geometry

In this chapter, we assemble most of the definitions and theorems about basic properties of algebra, points set (also called general) topology, algebraic and differential topology, differentiable manifolds, and differential geometry, which are needed in the course of the lectures. Our main purpose is to establish notation, terminology, concepts, and structures. We provide definitions, examples, and essential results without giving the arguments of proof. The material presented is slightly more than the essential knowledge for this lecture series. For instance, we certainly can get by without explicitly talking about algebraic topology, such as homotopy and homology. However, we do utilize the concepts of orientable manifolds or spaces, so that everywhere the non-zero volume element can be found. We also discuss the connected components and connected sums of a state space. Although we do discuss tubular neighborhoods at the end of this chapter, we do not explicitly use some of the results in discussing dynamical systems; nonetheless when dealing with return maps, the stability of orbits and periodic orbits, etc., we implicitly use the concept of tubular neighborhoods. Nonetheless, these and many other concepts and terminology are frequently used in research literature. It is also our intent to introduce the reader to a more sophisticated mathematical framework so that when the reader ventures to research literature or further reading, he will not feel totally lost due to different "culture" or terminology. Furthermore, in global theory, the state spaces may be differentiable manifolds with nontrivial topology. In such cases, concepts from algebraic topology at times are essential to the understanding.

In light of the above remarks, for the first reading, the reader may want to skip Section 2.3, part of Section 2.4, the de Rham cohomology part of Section 2.8, and Section 2.9.

There are several well written introductory books on general topology, algebraic topology, differential topology, and differential geometry, which can provide more examples, further details and more general results, e.g., Hicks [1971], Lefschetz [1949], Munkres [1966], Nach and Sen [1983], Singer and Thorpe [1967], Wallace [1957]. For more advanced readers, the following books are highly recommended: Bishop and Crittenden [1964], Eilenberg and Steenrod [1952], Greenberg [1967], Hirsch [1976], Kelley [1955], and Steenrod [1951].

2.1 Getting to the basics – algebra

Some results and notations from group theory will be needed later, here we sketch some of them.

Recall that a <u>group</u> is a set of elements, denoted by G, closed under an operation · (or +) usually called multiplication (or addition), satisfying the following axioms:

(1) $x \cdot (y \cdot z) = (x \cdot y) \cdot z$ for all $x, y, z \in G$;

(2)   There exists a unique element $e \in G$ called the identity such that

$x \cdot e = e \cdot x = x$ for all $x \in G$;

(3) for a given $x \in G$, there is a unique element $x^{-1} \in G$ called the inverse of x such that $x \cdot x^{-1} = x^{-1} \cdot x = e$.

A subset H of G, $H \subset G$, is called a <u>subgroup</u> if H is a group (with the same operation as in G). A subset $H \subset G$ is a subgroup of G iff $ab^{-1} \in H$ for all $a, b \in H$. Let H, G be two groups (with the same operation ·), a mapping $f : G \xrightarrow{into} H$ is a <u>homomorphism</u> if $f(x \cdot y) = f(x) \cdot f(y)$ for all $x, y \in G$. If e is the identity of H, then $f^{-1}(e) \subset G$ and $f^{-1}(e)$ is called the <u>kernel</u> of f, and $f(G)$ is the <u>image</u> of f. It can be shown that $f(G) \subset H$. Graphically, it looks like this:

f (into)

G   H

Two groups G and H are called <u>isomorphic</u> if there is a one- to-one, onto map f:G $\overset{\text{onto}}{\rightarrow}$ H such that f(x·y) = f(x)·f(y) for all x, y ∈ G, and f is called an <u>isomorphism</u> and denoted by G ≈ H. A homomorphism f: G → H is an isomorphism iff f is onto and its kernel contains only the identity element of G. Graphically,

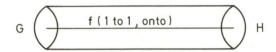

$$G \quad \text{f (1 to 1 , onto)} \quad H$$

If G is a group and S a set (finite or infinite) of elements of G such that every element of G can be expressed as a product (here we use the multiplication operation) of elements of S and their inverses, then G is said to be <u>generated</u> by S and elements of S are called the <u>generators</u> of G.

If a group G satisfies another axiom, in addition to (1), (2) and (3) stated above, specifically, the commutative law, i.e., x·y = y·x for all x, y ∈ G, then G is <u>abelian</u>. If the group operation is + instead of ·, such an abelian group is called an additive abelian group and its identity element is denoted by 0, and the inverse of x is -x, for all x ∈ G. If G is any additive abelian group and H ⊂ G, G can be split up into a family of subsets called <u>cosets of H</u>, where any two elements x,y ∈ G belong to the same coset if x-y ∈ H. The coset of H with the additive operation form a group called the <u>quotient</u> group of G with respect to H, denoted by G/H.

By considering a sequence of homomorphisms, one can calculate a group from other related groups. This is the notion of exact sequence.

Let f: A → B be a homomorphism, then from the definition of the images and kernel of f, we have Im f ≈ A/Ker f. A sequence A $\overset{f}{\rightarrow}$ B $\overset{g}{\rightarrow}$ C is <u>exact at B</u> iff Im f = Ker g. An exact sequence has the following properties:
(i) a sequence id $\overset{h}{\rightarrow}$ A $\overset{f}{\rightarrow}$ B is exact at A iff f is 1-to-1;

(ii) a sequence A $\xrightarrow{f}$ B $\xrightarrow{g}$ id is exact at B iff f is onto;
(iii) a sequence id $\to$ A $\xrightarrow{f}$ B $\to$ id is exact (everywhere) iff
    f is an isomorphism;
(iv) a short exact sequence id $\to$ A $\xrightarrow{f}$ B $\xrightarrow{g}$ C $\to$ id has the
    property that C $\approx$ B/Im f.
    We shall utilize these properties of a sequence in
Chapter 3 and thereafter. In the next section, we shall
discuss some fundamentals of general topology.

2.2   Bird's eye view of general topology

    A topological space consists of a set X with an
assignment of a non-empty family of subsets of X to each
element of X. The subsets assigned to each point p ϵ X will
be called neighborhoods of p. The assignment of
neighborhoods to each point p ϵ X must satisfy:
(1) If U is a neighborhood of p, then p ϵ U;
(2) Any subset of X containing a neighborhood of p is itself
a neighborhood of p;
(3) If U and V are neighborhoods of p, so is U $\cap$ V;
(4) If U is a neighborhood of p, there is a neighborhood V
of p such that U is a neighborhood of every point of V.
    Example: X = R^n , U = {a sphere of center p} then the
above conditions are satisfied, so X is a topological space.
    Exercise: Let X be a set and suppose that each pair of
elements x and y of X is assigned a real number d(x,y)
called the distance function between x and y and satisfying
the following conditions:
    (1)   d(x,y) $\geq$ 0   and   d(x,y) = 0   iff   x = y;
    (2)   d(x,y) = d(y,x) for all   x,y ϵ X;
    (3)   d(x,z) $\leq$ d(x,y) + d(y,z)   for any x, y, z ϵ X.
Now define an ϵ-neighborhood U of x ϵ X as U = {y ϵ X| d(x,y)
< ϵ}. Then define a neighborhood of x ϵ X to be any set in X
containing an ϵ-neighborhood of x. Prove that with these
definitions, X becomes a topological space.
    Remark: A topological space defined in such a way is
called a metric space. Also, prove that Euclidean space is a
metric space.

If A and B are any sets, then the set of all pairs (a,b) with a ∈ A and b ∈ B is denoted by A x B, and is called the product set of A and B. Let X and Y be two topological spaces. Let a set W ⊂ X x Y a neighborhood of (x,y) ∈ X x Y where x ∈ X, y ∈ Y if there is a neighborhood U of x ∈ X and a neighborhood V of y ∈ Y such that U x V ⊂ W. The product space X x Y becomes a topological space and X x Y is called the topological product of X and Y. Clearly, X = Euclidean 2-plane is the product of **R** x **R** where **R** is the space of real numbers.

Let A be a set of points in a topological space X. A point p ∈ A is an interior point of A if there is a neighborhood U of p such that U ⊂ A. The interior of A ≡ Å = {all interior points of A}.

Example: $X = R^2$. $A = \{(x \cdot y) \mid x^2 + y^2 \leq 1\}$.
$Å = \{(x \cdot y) \mid x^2 + y^2 < 1\}$.

A set A in a topological space is an open set if, for each p ∈ A, there is a neighborhood U of p such that U ⊂ A (i.e., there is no "boundary").

Example: $A = \{x \mid a < x < b$,a,b ∈ R\}. Also, any solid open n-sphere (i.e., $\Sigma x_i^2 < a^2$) in R^n is an open set.

The following theorems can also be taken as a set of axioms to give an alternative definition of a topological space.

Theorem 2.2.1 Let X be a topological space. Then (1) the union of an arbitrary collection of open sets in X is an open set; (2) the intersection of a finite collection of open sets in X is an open set; (3) the whole space X is an open set; (4) the empty set is an open set.

As a consequence, we have the following theorem:

Theorem 2.2.2 If A is any set in a topological space, then Å is an open set and it is the largest open set contained in A.

We have mentioned prior to Theorem 2.2.1 that the properties of open sets could be used to provide an alternative definition of a topological space. Such a procedure is based on:

28

Theorem 2.2.3   Let X be an abstract set and let O be a
family of subsets of X such that
(1) The union of any collection of sets belonging to O is a
    set belonging to O;
(2) The intersection of a finite collection of sets
    belonging to O is a set belonging to O;
(3) The whole set X belongs to O;
(4) The empty set O belongs to O.
Then there is one and only one way of making X into a
topological space (i.e., by assigning neighborhoods to the
elements of X) so that O is the family of open sets of X: by
requiring that a set U is a neighborhood of p ϵ X iff there
is a set W beloning to O such that p ϵ W $\subset$ U.

Intuitively, it is easy to see with the help of Theorem
2.2.1, nonetheless the proof is somewhat tedious. The
essence of this theorem is that a topological space can be
defined by giving the open sets instead of assigning
neighborhoods to each point. The original definition of a
topological space in terms of neighborhoods is probably the
most convenient and the most intuitive one, which depends on
a common sense notion of nearness. But there are advantages
from the viewpoint of open sets. First, it is more simple
logically to name one single family of open sets in X than
to name a family of sets attached to every point of X.
Moreover, many definitions can be made and theorems proven
more easily and elegantly in terms of open sets than in
terms of neighborhoods.

Let X be a topological space and S a subset. If p ϵ S, a
subset U of S is a neighborhood of p in S iff U = S $\cap$ V for
some neighborhood V of p in X. Then the neighborhoods in S
of points of S define a topology on S. This is the topology
induced on S by X, and S with this topology is a subspace of
X.

Remark: One must be careful, in speaking of a space and
a subspace, to distinguish between neighborhoods in the
space and neighborhoods in the subspace; likewise, to
distinguish between open sets in the space and open sets in

29

the subspace.

Example: Let $X = R^2$, i.e., the (x,y)-plane, and let $S = R^1$ be the x-axis. As usual, a neighborhood of a point $p \in X$ is any set containing a circular disk with center p, while a neighborhood of a point $p \in S$ is any set in S containing an interval with midpoint p. If U is an open interval in S (i.e., interval without endpoints), U is open in S but not in X because U contains no circular disk. The usual topologies on X and S make S a subspace of X. This can be seen by noting that if $p \in S$, the intersection of S with a disk of center p is an interval with midpoint at p.

Theorem 2.2.4 Let X be a topological space and S a subspace. Then a subset U of S is open in S iff there is an open set V in
X such that $U = V \cap S$.

Exercise: Let X be a topological space and S a subset of X. Define $U \subset S$ to be open in S iff $U = V \cap S$ for some set V open in X. Show that the sets so defined as being open in S satisfy the conditions of Theorem 2.2.3 thus defining a topology on S.

A sequence of points p_1, p_2 .... in a topological space X is said to have a <u>limit</u> p, or to <u>converge to</u> p, if for any preassigned neighborhood U of p, there is an integer N such that: $p_n \in U$ for all $n \geq N$.

Remark: There are two warnings to be made concerning the limits of an arbitrary topological space. First, the Cauchy criterion for the convergence of a sequence of real numbers has no analogue in a topological space in general. This is because in general topological space there is no uniform standard of nearness which can be applied to a variable pair of points. (But Cauchy criteria do have a uniform standard, i.e., the metric). Second, there is no guarantee that the limit of a sequence of points, if it exists, is unique.

It is desirable to restrict our attention to topological spaces satisfying a condition wich will enable the uniqueness of the limit of a sequence of points to be proven. The following definition will provide the required condition.

A topological space X is a <u>Hausdorff space</u>, sometimes denoted by T_2 space, if for every pair p, q ϵ X with p $\neq$ q, there is a neighborhood U of p and a neighborhood V of q such that U $\cap$ V = 0.

<u>Theorem 2.2.5</u>   Let X be a Hausdorff space, and suppose that a sequence $\{p_n\}$ has a limit p. Then this limit is unique.

Example: R^1 is T_2. In general, R^n in its usual topology is a T_2 space.

Let A be a set of points in a topological space X. Then a point p ϵ X is a <u>limit point</u> (or <u>accumulation point</u>) of A if every neighborhood of p contains a point of A different from p.

Example: Take X = R^1. Then the above definition becomes the one usually given in analysis. One can take a decreasing sequence of intervals with p as midpoint of length, say 1, 1/2, 1/3, 1/4,.... If p is a limit point of the set of numbers A, each of these intervals will contain a number x_1, x_2, ... respectively, different from p and belonging to A. If U is any neighborhood of the number p, U will contain every interval of length 1/n and midpoint p for sufficiently large n. Thus the sequence x_1, x_2, ... has p as its limit. Incidentally, every neighborhood of p contains infinitely many points of A.

There are pathological cases one should be aware of (in particular for non-T_2 spaces).

<u>Corollary 2.2.6</u> In a T_2 space, only infinite sets of points are capable of having limit points at all.

Note also that a limit point of a set A in a topological space need not be the limit of any sequence of points of A. The construction of an example is a bit complicated. Let X be the set of all real valued functions of the real variable x, defined for all values of x. If f is any point of X (i.e., a function of x) the (ϵ, x_1, x_2,....)-neighborhood of f will be defined to be the set of all points ϕ of X such that $|\phi(x_i) - f(x_i)| < \epsilon$, i = 1,2,... n. Such a neighborhood of f can be defined for every positive number ϵ, and every finite collection of number x_1, x_2,...x_n. The topology of X can be defined by saying that U is a neighborhood of f iff U contains a $(\epsilon, x_1, x_2,...,x_n)$- neighborhood of f for some ϵ, x_1, x_2,...,x_n. It can be verified that this assignment of neighborhood satisfies the conditions of a topological space. Furthermore, one can show that this space is a T_2 space.

Now let f_0 be the function of x which is identically zero, and define a subset A of X as follows. Let x_1, x_2,...,x_n be any finite collection of real numbers and let $f_{x_1 x_2...x_n}$ be the function of x defined by setting $f_{x_1 x_2...x_n}(x_i) = 0$ (i = 1,2,..,n) and $f_{x_1 x_2...x_n}(x) = 1$ for all other values of X. Let A be the set of all $f_{x_1 x_2...x_n}$ for all possible dinite sets x_1, x_2..,x_n of real numbers. f_0 certainly does not belong to A. But f_0 is a limit point of A in the topological space X. Let U be a neighborhood of f_0. Then U contains the $(\epsilon, x_1, x_2,..,x_n)$-neighborhood of f_0 for some ϵ and x_1, $_2$,..,x_n and the function $f_{x_1...x_n}$ belongs to this $(\epsilon, x_1,..,x_n)$-neighborhood, since $|f_{x_1..x_n}(x_i) - f_0(x_i)| < \epsilon$ because both terms of the difference are zero for i =1,2,..,n. Clearly $f_{x_1...x_n}$ is a member of A and belongs to the given neighborhood U of f_0. Thus f_0 is a limit point of A.

Now we want to show that f_0 is not the limit of any sequence selected from A. Suppose f_1, f_2,.. is a sequence belonging to A and having f_0 as the limit. By the definition of A, each f_n is a function of x equal to zero at a finite set S_n of values of x and equal to 1 for all other values. The union of all the S_n for n = 1,2,... is, at most, a denumerable set of points, and sothere is a real number x_0 not belonging to any of the S_n. Let V be the (ϵ, x)-neighborhood of f_0 for some $\epsilon < 1$. By the choice of x_0, $f_i(x_0) = 1$ for all i, and so none of the members of the sequence f_1, f_2,... belongs to V, it follows that f_0 cannot be the limit of this sequence.

It might be added that the topological space constructed

in the above example appears quite naturally and is used
frequently in analysis. To say that a sequence f_1, f_2,... of
points of this space converges to the limit f means exactly
that the functions f_1, f_2,.. converge pointwise to f, i.e.,
for every fixed value of x, the sequence $f_1(x)$, $f_2(x)$,...of
real number converges to $f(x)$.

Just in passing, suppose a topological space X satisfies
the following condition: For each point $p \in X$ there is a
sequence $N_1(p)$, $N_2(p)$,... of neighborhoods of p such that
$N_{n+1}(p) \subset N_n(p)$ for each n, and given any neighborhood U of p
there is an n such that $N_n(p) \subset U$. One can prove that in
this case if p is a limit point of a set A in X, p is the
limit of some sequence of points of A. The above condition
imposed on X is called the first axiom of countability. (The
second one is a condition on the family of open sets. We
shall come to that later).

Exercise: Prove that a Euclidean space satisfies the
first countability axiom. In fact, one can show that the
axiom holds for any metric space.

Let A be a set in a topological space X. Then the
closure of A, $\bar{A}$, is defined to be the set in X consisting of
all the points of A along with all the limit points of A.

Examples: (1) $X = R$ in the usual topology, $A = \{x | a < x$
$< b\}$, then $\bar{A} = \{x | a \leq x \leq b\}$. (2) $X = R^n$ and A is the set of
points at distance less than r from some fixed point p, then
$\bar{A}$ is the set of points whose distance from p are less than
or equal to r.

A set A in a topological space X with the property that
$\bar{A} = A$ is called a closed set of X.

Theorem 2.2.7   A set A in a topological space X is
closed iff the complement of A in X is an open set.

This theorem is sometimes used to define a closed set.
One can also translate Theorem 2.2.1 into a theorem about
closed sets by taking the complements of all the open sets
mentioned there.

Once again, it should be pointed out that if one is
considering a subspace S of a space X as well as S itself,

33

the notion of closure and closed sets, like those of
neighborhoods and open sets, are relative.

Up to now, we have been considering the topological
space, its subsets and their properties. In the following,
we shall briefly discuss the relationships between spaces
and the topological properties of spaces.

Let X and Y be two topological spaces. A mapping f of X
into Y is a rule which assigns to each point p ϵ X a well
defined point f(p) in Y. The mappping f is <u>continuous at p</u>
if, for each neighborhood U of f(p) in Y, there is a
neighborhood V of p in X such that f(V) $\subset$ U. The mapping f
of X into Y is <u>continuous</u> if it is continuous at all points
of X.

Let X and Y be two topological spaces, and let f: X → Y
(not necessarily continuous), and let X' be a subspace of X.
Then f induces a map f' of X' into Y, defined by setting
f'(p)= f(p) for all p ϵ X'. f' is called <u>the restriction of
f to X'</u>.

Note that f and f' are different mappings, because if
only f' is known, f is by no means uniquely determined. For
example, f:R → R defined by f(x)= x^2 for all x ϵ R, and X'=
[0,1] and f':X' → R by f'(x)= x^2 for x ϵ [0,1]. Clearly f'
is a restriction of f to X'. Let g:R → R be defined by
setting g(x) = 0 for x < 0, g(x) = x^2 for 0 $\leq$ x $\leq$ 1, and
g(x) = 1 for x > 1. Clearly f' is also a restriction of g to
X'.

<u>Theorem 2.2.8</u>   Let X and Y be two topological spaces. A
map f:X → Y is continuous iff the inverse image of every
open set in Y is open in X.

Note that the inverse image is not in general a mapping.
If it is to be a mapping, f must be one-to-one. Furthermore,
f is onto. But in topology, special interest is attached to
those mappings which are not only one-to-one and onto, but
also have the property that both the mapping and its inverse
are continuous.

Let X and Y be two topological spaces and let f be a
one-to- one mapping of X onto Y. Then if both f and f^{-1} are

34

continuous, f is a <u>homeomorphism</u> of X onto Y, and X and Y
are <u>homeomorphic</u> under f.

   Clearly, if f is a homeomorphism of X onto Y, then f^{-1}
is a homeomorphism of Y onto X since if $g = f^{-1}$, then $g^{-1}= f$.

   From Theorem 2.2.8 and the definition just given, it
follows that a homeomorphism of X onto Y sets up a
one-to-one correspondence between the open sets of X and
those of Y. The idea that these two spaces are homeomorphic
to each other under f means that not only that the points
are in one-to-one correspondence, but also the neighborhoods
of corresponding points are very similar to one another.
These are properties of a topological space X which depend
only on the definition of X as a topological space (i.e.,
depend only on the knowledge of which sets in X are open),
but depend in no way on any other properties the elements of
X may have. Such a property is characterized by the
following definition:

   A property of a topological space X is a <u>topological</u>
<u>property of X</u> if it also belongs to <u>every</u> topological space
homeomorphic to X.

   Since two homeomorphic spaces are to be regarded as
having the same topological structure, homeomorphism (of
spaces) plays the part in topology analogous to that played
by an isomorphism (of groups) in algebra.

   We shall discuss some elementary topological properties
which are of considerable importance in topology. First, we
would like to generalize the notion of a closed bounded set
in a Euclidean space to other topological spaces. An
important property of Euclidean spaces (normally mentioned
in analysis) is given by the Heine-Borel theorem: A set A in
a Euclidean space R^n is closed and bounded iff whenever A is
contained in the union of an arbitrary collection of open
sets in R^n, then it is also contained in the union of a
finite number of open sets chosen from the given collection.
One can avoid the explicit mentioning of the open sets of
the Euclidean space and the wording of the theorem can also
be tidied up with the following definition. (This definition

35

is not just for tidying up the theorem, its importance will be manifested later).

Let X be any topological space, and let F be a family of sets in X such that X is their union. Then F is a <u>covering of X</u>. If all sets in F are open, then F is an <u>open covering of X</u>. If F and F' are two coverings of X such that every set belonging to F' also belongs to F, then F' is a <u>subcovering of X</u>. Then the Heine- Borel theorem can be restated as follows: A set A in a Euclidean space is closed and bounded iff every open covering of A contains a finite subcovering.

We note that the property of being closed and bounded in a Euclidean space is equivalent to a topological property. This topological property can be formulated for any topological space.

A topological space X is a <u>compact space</u> if it is a Hausdorff space, and if every open covering of X contains a finite subcovering. If A is a set in X, then A is a <u>compact set</u> if A, as a subspace of X, is a compact space.

So the Heine-Borel theorem says that a set A in R^n is compact iff it is closed and bounded. We have just said that because compactness (or a set is closed and bounded) is defined entirely in terms of open sets, thus we refer it to be a topological property. But we want to check whether if a space is homeomorphic to a given compact space is also compact. Indeed, we have the following stronger theorem:

<u>Theorem 2.2.9</u> Let f be a continuous mapping of a topological space X onto a topological space Y. Then if X is compact, and Y is Hausdorff, Y is compact.

Proof: Let F be a given open covering of Y. Then f^{-1} of each set of F is open in X (Theorem 2.2.8), and $f^{-1}(F)$ form a covering F'of X. F' is thus an open covering of the compact space X, so it contains a finite subcovering of X. But the images under f of the sets in this subcovering of X are known to be sets of the covering of F (from the definition of F') and form a covering of Y. Thus the given covering of Y contains a finite subcovering, and Y is Hausdorff, so Y is compact.

36

There are some "trivial" corollaries:

Corollary 2.2.10   If X and Y are homeomorphic spaces and X is compact, so is Y.

Corollary 2.2.11   If f: X → Y is a continuous mapping of a compact X into a Hausdorff space Y, then f(X) is a compact set.

One of the most important properties of product spaces is given by the Tychonoff Theorem: The product of compact spaces is compact.

Next we shall dicuss and make precise the idea based on our observation that certain sets of points, say in a plane, have the property that any two of their points can be joined by a curve lying entirely in the set, while certain other sets fail to do so. E.g., if A is a disk, either open or closed, it is clear that every pair of points of A can be joined by a curve (in fact, by a line segment) lying entirely in A. But if A' be a set consisting of two disjoint circular disks, then any path joining a point of one of these disks to a point of the other disk must cross the gap between the disks, thus not lying entirely in A'. First, we have to define what is a path or a curve.

Let X be a given topological space, and I be the unit interval $0 \leq t \leq 1$, regarded as a subspace of the space of real numbers in the usual topology. Then a path in X joining two points p and q of X is defined to be a continuous mapping f of I into X such that f(0) = p and f(1) = q. The path is said to lie in a subset A of X if f(I) ⊂ A. It is important to note that the path is the mapping.

A topological space X is arcwise connected if, for every pair of points p, q ∈ X there is a path in X joining p and q. If A is a set in X, then A is arcwise connected if every pair of points of A can be joined by a path in A.

Now we shall show that arcwise connectedness is a topological property.

Theorem 2.2.12   Let X and Y be two spaces and f is a continuous mapping of X onto Y. Then if X is arcwise connected, so is Y.

To give the proof is as easy as giving an example, so we opt for the proof. Let p and q be points in Y. Since f is onto, there are points p' and q' in X such that f(p')= p and f(q')= q. Since X is arcwise connected, there is a path g (a continuous map) in X joining p' and q'. Then the composite map f·g is a continuous map of I into Y such that (f·g)(0)= p and (f·g)(1)= q. [As an exercise, one can prove that if X, Y, Z are three spaces and f:X → Y, g:Y → Z are two continuous maps, then the composite map g·f is a continuous map of X into Z.] That is, f·g is a path in Y joining p and q. Since p and q are arbitrary points of Y, the theorem is proved.

Corollary 2.2.13 If X and Y are homeomorphic spaces, then X is arcwise connected iff Y is. That is, arcwise connectedness is a topological property.

As we shall discuss shortly, arcwise connectedness is a useful condition for a physical space. But it is too restrictive for an abstract space. A much weaker and simpler condition in terms of open sets is about to be discussed.

A topological space X is <u>connected</u> if it cannot be expressed as the union of two disjoint non-empty open sets. A set A in X is connected if, when regarded as a subspace of X in the induced topology, A is a connected space.

Example: Let A consist of two circular disks in the plane such that the distance between their centers is strictly greater than the sum of their radii. Then A is not a connected set.

Theorem 2.2.14 Let X and Y be two topological spaces and f is a continuous map of X onto Y. Then if X is connected so is Y.

Corollary 2.2.15 If X and Y are homeomorphic, then X is connected iff Y is.

Thus, connectedness is also a topological property. As we have pointed out earlier, connectedness is weaker than arcwise connectedness, and it is intuitively easy to realize the following theorem.

Theorem 2.2.16 An arcwise connected space is connected.

38

Note that the converse is not true. For example: Let A be the set in the (x,y)-plane such that y = sin(1/x) for 0 ≤ x ≤ 1 along with the line segment (0,y) for -1 ≤ y ≤ 1. Clearly A is connected, nonetheless A is not arcwise connected. This is because for any map f of the interval 0 ≤ t ≤ 1 into A such that f(0) is a point of A with x ≠ 0 and f(1) is a point with x = 0 is necessarily discontinuous at t = 1. But arcwise connectedness requires the mapping f to be continuous for 0 ≤ t ≤ 1.

A simpler example is the following:
A = {(x,y)|(0,y): -1 ≤ y ≤ 1; (x,0): 0 ≤ x ≤ 1}.

We have introduced some basic concepts in point set (or general) topology. Before we go on introducing some basic notions of differential geometry, let us reflect on those concepts we have just discussed and try to relate them to the world models (be they physical, biological, or social) we want to construct and understand.

As we have discussed in the introduction, any world model must be a topological space. Before put more structure on it, intuition as well as some reality dictates the elimination of a few classes of spaces.

(1) One would not wish to construct a world model based on a non- Hausdorff space. This is because non-Hausdorff space will not allow us to describe "distinct events", which are of fundamental importance in physical sciences. Nor will non-T_2 spaces allow any statistical inference because there is no distinct sampling, (let alone discrete or continuous sampling spaces), nor can one construct such concepts as distribution. So the Hausdorff property of the space is equally important for biological and social sciences.

(2) Nor can one build a model of nature based on a non-connected space. This is because no "communication" (or influence) can be carried out between separated components or distinct events. It is clear that such a model is unacceptable.

As we go along, we shall discuss a few more restrictions as to the acceptable spaces upon which world models can be built. Next, we shall introduce some basic concepts in albegraic topology and familiarize ourselves with some of the terminology and fundamental results.

2.3   Algebraic topology

Roughly speaking, algebraic topology is a branch of mathematics which deals with some "equivalence" in algebraic fashion. A typical process in algebraic topology is to associate certain groups with a given space. It studies homotopy, homology, cohomology, exact sequences, spectral sequences, excision, obstruction, characteristic classses, duality etc. It has a close relationship with differential topology, which differs from differential geometry by studying the differentiable structure of the space, instead of being particularly interested in the geometric construct of the space.

In the following pages we shall briefly describe some of the concepts in algebraic topology and useful results for future use.

2.3.1. Homology theory

If a simple closed curve, such as an ellipse or a polygon, is drawn on the plane, then it has an "inside" and "outside". That is to say the closed curve forms the common boundary of these two portions of the plane. Similarly, if a closed curve is drawn on the surface of a sphere, the curve is the boundary of two portions of that surface. In contrast to this situation, by drawing the closed curve α on the surface torus, α does not necessarily divide the surface into two portions, or α is not the boundary of any portion of the surface of the torus. The possibility of drawing a closed curve on a surface or the maximum number of closed curves along which the surface may be cut without dividing the surface into disjoint portions is clearly a topological property.

A 1-chain may be a line segment, a curve, or a closed

loop. The boundary of a line segment is its endpoints. A 1-chain without endpoints, thus has no boundary, is called a 1-cycle. A 1-chain which is a boundary of some 2-dim surface lying in the domain D is called a 1-boundary. Two cycles are said to be homologous (equivalent) if their difference is a boundary. Thus c_2 is homologous to c_3.

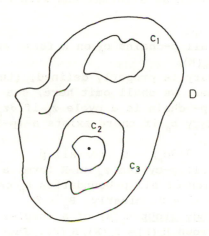

The constituent elements in the simplicial homology theory are n-simplexes in R^n. A p-dimensional simplex, p-simplex, is denoted by σ^p. σ^0 is a point (vertex), σ^1 is a line interval whose two end points are excluded, σ^2 is the interior of a triangle, σ^3 is the interior of a solid tetrahedron, and so on. Since a p-simplex can be uniquely determined by p+1 distinct vertices, one may write a p-simplex σ^p as $\sigma^p = <v_0\ v_1\ \ldots\ v_p>$ where v_i is the i-th vertex of the simplex. Clearly, any subset of k+1 vertices of σ^p forms a k-simplex. Each such sub-simplex is called a face and will be denoted by $<v_0\ \ldots\ v'_i\ \ldots\ v_j\ \ldots\ v_p>$, where v'_i means that the v_i vertex is absent in the subset. It is easy to see that the faces of σ^2 are its three sides (1-faces) and three vertices (0-faces).

    A simplicial p-complex K^p is a collection of simplexes

$\{\sigma^0{}_i,\ \sigma^1{}_j,\ \sigma^2{}_k,\ \dots,\ \sigma^p{}_m\}$ satisfying the following conditions: (a) The simplexes of K^p are disjoint and no two have all the same vertices. (b) If a simplex is in K^p, all its faces are also in K^p. From this definition, it is clear that a simplex cannot cross or end in the interior of another simplex.

A <u>chain c_p</u> on complex K is a finite collection of p-simplexes $\sigma_i{}^p$ written as a formal sum with constant coefficients g_i:

$$c_p = \Sigma\ g_i \sigma_i{}^p.$$

The collection of all p-chains c_p on K forms an Abelian group denoted by $C_p(K)$.

Assume a boundary is properly defined, (this involves the incidence number, we shall omit here, see Wallace[1957]), a p- chain is a cycle z_p if $\partial z_p = 0$. A p-chain is a <u>boundary</u> b_p if there exists a (p+1)-chain c_{p+1} such that $\partial c_{p+1} = b_p$.

<u>Theorem 2.3.1</u>: $\partial(\partial c_p) = 0$ for all p.

The collection of all p-cycles z_p on K forms a group $Z_p(K) \subset C_p(K)$. The collection of all p-boundaries b_p on K forms the group $B_p(K)$. Since $\partial\partial = 0$, clearly $B_p \subset Z_p$

A p-dim. <u>homology group</u> $H_p(K)$ of a complex K is defined to be the factor group $H_p(K) = Z_p(K)/B_p(K)$. Thus each element of $H_p(K)$ is an equivalence class of p-cycles; two cycles are said to be <u>homologous</u> if they differ by a boundary. The general form of $H_p(K)$ is $Z \oplus \dots \oplus Z \oplus G_T{}^p$. The number of generators of $H_p(K)$ is called the <u>p-th Betti number</u> $= \beta_p$. $G_T{}^p$ is the torsion subgroup of $H_p(K)$ - an Abelian group with only finite elements.

2.3.2. Homotopy:

What topological property of a space can be used to distinguish between, say a closed disc and an annulus (a disc with a hole in it)? The natural answer to this question is by considering the possibility of shrinking closed loops drawn on the two spaces to a point. (For an arcwise connected space, it is independent of the base point. We shall come to this point later.)

42

Let σ, τ be paths in arcwise connected topological space X (i.e., maps of I into X) with the same endpoints (i.e., $\sigma(0) = \tau(0) = x_0$, $\sigma(1) = \tau(1) = x_1$). We say σ and τ are homotopic with endpoints held fixed $\sigma \approx \tau$ rel $(0,1)$, if there is a map $F: I \times I \to X$ such that

$$F(s,0) = \sigma(s) \qquad \text{all } s$$
$$F(s,1) = \tau(s) \qquad \text{all } s$$
$$F(0,t) = x_0 \qquad \text{all } t$$
$$F(1,t) = x_t \qquad \text{all } t$$

F is called a homotopy from σ to τ. For each t, $s \to F(s,t)$ is a path F_t from x_0 to x_1, and $F_0 = \sigma$, $F_1 = \tau$. We write $F_t: \sigma \approx \tau$ rel $(0,1)$. Pictorially,

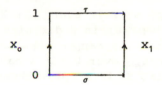

In particular if σ is a loop at x_0 (i.e., $x_1 = x_0$) and τ is the constant loop $\tau(s) = x_0$ for all s, and if $\sigma \approx \tau$ rel$(0,1)$ we say that "σ can be shrunk to a point", or is homotopically trivial.

For instance, the correct statement of Cauchy's theorem in complex analysis is that: $\int_c f(z)dz = 0$ for all loops C in the domain D of analyticity of f which are homotopically trivial.

Let a path σ start from x_0 and end at x_1, and another path τ starts from x_1 and ends at x_2, then the composite path $\sigma\tau$ (multiplication operation) starts from x_0 and ends at x_2.

Theorem 2.3.2 Let $\pi_1(X,x_0)$ be the set of homotopy classes of loops in X at x_0. If multiplication in $\pi_1(X,x_0)$ is defined as above, $\pi_1(X,x_0)$ becomes a group, the neutral element is the class of the constant loop at x_0 and the inverse of a class $[\sigma]$ is the class of the loop σ^{-1} defined by $\sigma^{-1}(t) = \sigma(1 - t)$, $0 \le t \le 1$ (i.e., travel backward along

43

σ).

Theorem 2.3.3 If X is pathwise connected, the group $\pi_1(X,x_0)$ is independent of x_0, up to isomorphism. It is denoted by $\pi_1(X)$ - the fundamental group of X.

A space X is simply-connected if it is pathwise connected and $\pi_1(X) = 0$ (i.e., any closed loop in X can be shrunk to a point).

Theorem 2.3.4 $\pi_1(S^1) = Z$, $\pi_1(S^n) = 0$, n ≥ 2.

Proposition 2.3.5 $\pi_1(X \times Y) = \pi_1(X) \times \pi_1(Y)$.

In Section 2.2 we have defined covering and subcovering to define the concept of compact space. Here we shall briefly discuss the concept of covering space and its relation to homotopy.

E $\xrightarrow{P}$ X is a covering space of X if every x ϵ X has an open neighborhood U such that $p^{-1}(U)$ is a disjoint union of open sets S_i in E, each of which is mapped homeomorphically onto U by p. S_i are called sheets over U. If X has a covering space $\tilde{X} \to X$ such that $\tilde{X}$ is simply-connected, then $\tilde{X}$ is unique up to equivalence, and it is called the universal covering space of X.

Theorem 2.3.6 Every connected space (manifold) has a universal covering space (manifold).

Remark: We shall discuss the covering manifold later.

Examples: (i) SO(3), group of rotations of R^3, $\pi_1(SO(3))$ = Z/2. The universal covering space of SO(3) is S^3. (ii) Proper Lorantz group $L_t \approx P^3 \times R^3$ its universal covering group is S L(2,C) $\approx S^3 \times R^3$.

Higher homotopy groups are commutative ($\pi_1(X)$ may not). In fact:

Theorem 2.3.7 (Hurewicz) If n ≥ 2 and $\pi_q(X) = 0$ for all q < n, then $\pi_q(X) \approx H_q(X)$ for all q ≤ n.

Theorem 2.3.8 If X is pathwise connected, then θ : $\pi_1(X) \to H_1(X)$ is an isomorphism iff $\pi_1(X)$ is commutative.

In algebraic topology, in order to calculate a particular group G, one often proceeds by first finding an exact sequence with G in it, then evaluating all the easier groups in the sequence neighboring G, then determining G by

44

using the properties of an exact sequence. In the next section, we shall briefly discuss some of the properties of a differentiable manifold and some geometric structures.

2.4   Elementary differential topology and differential geometry

A <u>topological manifold</u> of dimension m is a topological space M, (we are going to define manifolds, so instead of denoting the topological space by X, we shall use M, N,... to denote manifolds), which satisfies: (i) M is a Hausdorff space; (ii) if $p \in M$, then there is an open set $U \subset M$, $p \in U$ such that U is homeomorphic to R^m; (iii) M has a countable basis for its topology, i.e., there is a countable family of open sets $\{U_\alpha\}$ such that every open set is a union of some of the U_α's.

A topological manifold M of dimension m has some set-theoretical and topological properties, such as:

(i)   M is locally compact. That is, if $p \in M$, any open neighborhood U of p which is homeomorphic to R^m, then centered at p in U we may choose an open ball of finite radius. The closure of such an open ball is of course compact.

(ii)   M is separable.

(iii) M is regular.

(iv)   M is a normal space [Kelley 1955].

(v)   With the help of the Urysohn's metrization theorem, one can show that M is a metric space [Kelley 1955].

A T_2 space X is called <u>paracompact</u> if for each covering $\{U_\alpha\}_{\alpha \in A}$ of X, there exists a locally finite covering $\{V_\beta\}_{\beta \in B}$, which is a refinement of $\{U_\alpha\}_{\alpha \in A}$ (i.e., each V_β is contained in some U_α). The following theorems are the results from general topology:

Theorem 2.4.1   A paracompact T_2 space is normal [Kelley 1955].

Theorem 2.4.2   A topological space is paracompact iff

45

every open cover has a subordinate partition of unity
[Kelley 1955].

Theorem 2.4.3 Every metric space is paracompact [Kelley 1955].

Since "metric" notion is fundamental to any world model, one does not want to construct a world model which is non-paracompact.

Recall Theorem 2.2.16 and the subsequent example, an arcwise connected, T_2 space is necessarily to be connected. But the strcuture of a topological manifold is rich enough such that:

Theorem 2.4.4 A topological manifold is connected iff it is arcwise connected.

Nonetheless, a topological manifold need not be connected. For example, the group space of O(n) is a topological manifold, which has two connected components corresponding to the positive and negative determinants.

Let U and U' be open subsets of R^m and R^n respectively, and let f be a mapping of U into U'. The map f is <u>differentiable</u> if the coordinates $y_i(f(p))$ of f(p) are differentiable (sometimes call smooth, i.e., infinitely differentiable, denoted by C^∞) functions of the coordinates $x_i(p)$, p ∈ U. A differentiable map f: U → U' is a <u>diffeomorphism</u> of U onto U' if f(U) = U', f is one-to-one, and the inverse map f^{-1} is differentiable.

Let M be a Hausdorff space. An <u>open chart</u> on M is a pair (U,ϕ) where U ⊂ M and open, and ϕ is a homeomorphism of U onto an open subset of R^m. So, the dimension of M is m. A <u>differentiable structure</u> on M of dimension m is a collection of open charts $(U_\alpha,\ \phi_\alpha)_{\alpha \in A}$ on M where $\phi_\alpha(U_\alpha)$ is an open subset of R^m such that (i) M = U U_α; (ii) for each pair α, β ∈ A the composite map $\phi_\beta \cdot \phi_\alpha^{-1}$ is a differentiable map of $\phi_\alpha(U_\alpha \cap U_\beta)$ onto $\phi_\beta(U_\alpha \cap U_\beta)$ (see the following figure); (iii) the collection $(U_\alpha,\ \phi_\alpha)$ is a maximal family of open charts for which (i) and (ii) hold. Note: (iii) means that the set of open charts {U_α} covers M.

46

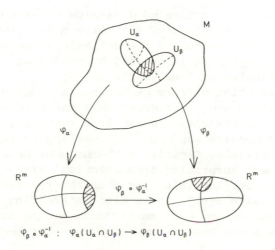

$$\varphi_\beta \circ \varphi_\alpha^{-1} : \quad \varphi_\alpha(U_\alpha \cap U_\beta) \longrightarrow \varphi_\beta(U_\alpha \cap U_\beta)$$

Fig.2.4.1

Theorem 2.4.5 [Whitney 1936] Each C^p-structure is C^p equivalent to a C^∞-structure ($p \geq 1$).

In other words, if a manifold has a C^p-structure, then a compatible C^∞-structure can be found. Consequently, one usually considers the C^∞-structure except where otherwise stated. When one considers a coarser or finer topology of a set of additional structures, one may want to specify the definite differentiability as we shall see later. For convenience, we assume the differentiable structure to be C^∞ unless stated otherwise.

A related notion to paracompactness is: a family of differentiable functions $\{f_\alpha\}_{\alpha \in A}$ on a C^∞-manifold M is called a **partition of unity** subordinate to the covering $\{U_\alpha\}_{\alpha \in A}$ if:

(i) $0 \leq f_\alpha \leq 1$ on M for every $\alpha \in A$;

(ii)   the support of each f_α, i.e., the closure of the set $\{p \in M \,|\, f_\alpha(p) \neq 0\}$ is contained in the corresponding U_α;

(iii) $\Sigma_\alpha f_\alpha(p) = 1$.

A **differentiable manifold** (or C^∞ manifold, or smooth manifold) of dimension m is a Hausdorff space with a differentiable structure of dimension m. In other words, a differentiable manifold of dimension m is a topological manifold endowed with a differentiable structure. Similarly, an <u>analytic manifold</u> is defined by replacing "differentiable" by "analytic".

The essential difference between a topological manifold of dimension m and a differentiable manifold of dimension m is that the differentiable manifold requires, in addition, that the composite map $\phi_\beta \cdot \phi_\alpha^{-1}$ of overlapping region $U_\alpha \cap U_\beta$ is differentiable. Clearly, a C^∞-manifold is a topological manifold. Not surprising from a logical viewpoint, but remarkable from intuition, is that the reverse assertion is false [Kervaire 1961]. It is also interesting to ask whether the C^∞-structure on a C^∞- manifold is unique or not. Milnor [1956] has shown that S^7 has more than one distinct C^∞-structure.

Recall from earlier, we have defined a differentiable mapping in relation to open sets in R^m and R^n. Here we use the same concept to define a differentiable map between smooth manifolds.

Let M and N are smooth manifolds of dimensions m and n respectively and f : M → N be continuous. Then f is a <u>differentiable</u> (or smooth, or C^∞) map if for every p ϵ M and every coordinate charts ϕ_α: U_α → R^m with p ϵ U_α and η_β: U_β → R^n with f(p) ϵ U_β, the composite map $\eta_\beta \cdot f \cdot \phi_\alpha^{-1}$ is differentiable.

The intuitive outcome of this definition is that we can use the coordinate charts to transfer various notions from manifolds to the easily understood framework in Euclidean spaces.

Let M_1 and M_2 be two C^∞-manifolds. They are <u>diffeomorphic</u> if there are C^∞-maps f: M_1 → M_2 and g: M_2 → M_1 such that g·f = id_1 and f·g = id_2 where id_1 and id_2 are identity maps on M_1 and M_2 respectively. The maps f and g are <u>diffeomorphisms</u>. If the manifolds and mappings are C°, then the diffeomorphism is just a homeomorphism. Furthermore, we do not need to explicitly assume that the manifolds have the same dimension. This follows from the fact that if two manifolds are homeomorphic, then they have the same dimension.

The trivial example of diffeomorphism is a coordinate transformation. Moreover, the composition of two diffeomorphisms is a diffeomorphism, thus diffeomorphisms form a group called the group of diffeomorphisms. We shall meet it again later.

It is easy to convince ourselves that equivalently one defines a diffeomorphism of M_1 onto M_2 as a one-to-one map $f: M_1 \rightarrow M_2$ such that f and f^{-1} are C^∞. Without requiring f^{-1} be also C^∞, f is just a C^∞ homeomorphism. But C^∞-homeomorphism needs not be a diffeomorphism. This can be easily seen by the following example. For $n > 1$, let $u^{2n-1}: R \rightarrow R$. The inverse is not differentiable at the origin. This also shows that if f is C^∞, f^{-1} needs not be C^∞.

Since our goal is to discuss the dynamical systems, differential equations, their structural stability and their various applications, we shall deal with either the configuration spaces or their phase spaces, which are at least $C^k(k \geq 1)$, thus without any confusion, a manifold shall mean a differentiable manifold hereafter except otherwise stated.

Let M be a manifold, $p \in M$ and denote $F(M,p)$ be the set of all C^∞ real functions with domain in a neighborhood of p. A C^k curve in M is a map of a closed interval $[a,b]$ into M. A tangent vector at p is a real function X on $F(M,p)$ (i.e., $X : F(M,p) \rightarrow R$) having the following properties:

(i) $X(f + g) = Xf + Xg$,
(ii) $X(af) = a(Xf)$,
(iii) $X(fg) = (Xf) g(p) + f(p)(Xg)$,

where $f, g \in F(M,p)$ and $a \in R$. The above rules are the definition of a derivation, thus a tangent vector is often called a derivation of $F(M,p)$. All the tangents at p form a linear space denoted by M_p.

If $\phi = (x_1,..,x_m)$ is a coordinate syatem, the partial

49

derivative at p with respect to x_i, $D_x (p)$, is the tangent defined by $(D_x (p))f = (\partial(f \cdot \phi^{-1})/\partial u_i)(\phi(p))$, which is also denoted by $D_x f(p)$, where $x_i = u_i \cdot \phi$. In the more conventional notation, it will be denoted by

$$(\partial/\partial x_i)_p f = [\partial(f \cdot \phi^{-1})/\partial u_i]\phi(p).$$

It is easy to see that $D_x X_j(p) = \delta_{ij}$, the Kronecker delta, and hence $\{D_x (p)\}$ is linearly independent.

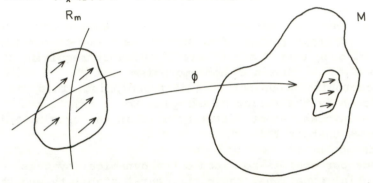

Theorem 2.4.6 Let M be a C^∞ n-manifold and let $(x_1, .., x_n)$ be a coordinate system about p ϵ M. Then if X in M_p, where $X_p = (Xx_i)D_x (p)$ (or $X_p = (Xx_i)(\partial/\partial x_i)_p$) where repeated indices are summed, and the coordinate vectors $D_x (p)$ (or $(\partial/\partial x_i)_p$) form a basis for M_p which has dimension n.

A <u>vector field</u> X on a subset A of M is a mapping that assigns to each point p ϵ A a vector X_p in M_p. If X is a vector field on an open set A, and f ϵ F(M,p) on B then the real function $(Xf)(p) = X_p f$ is C^∞ on A ∩ B. If X is a vector field with its domain contained in the coordinate system, we may write in terms of its basis $X = f_i D_x$ (or $X = f_i \partial/\partial x_i$), where the f_i are real functions. It is easy to show that X is C^∞ iff f_i is C^∞. If f is a C^∞ map of M into R^k, so f = $(f_1, .., f_k)$ with f_i real functions, and X a vector field on M, then we write Xf for $(Xf_1, .., Xf_k)$. It is clear that if X is C^∞, then Xf is C^∞.

Let V(M) denotes the set of all C^∞ vector fields on M. If X, Y, Z ϵ V(M), we define a C^∞ vector field [X,Y], the

50

bracket of X and Y, on the intersection of their domains by
$[X,Y] = XY - YX$. Let us also denote $F(M)$ be the set of C^∞
real valued functions on M. If $f,g \in F(M)$, it is trivial
that $[X,Y](f+g) = [X,Y]f + [X,Y]g$ and $[X,Y](af) = a[X,Y]f$
for $a \in R$. One only has to check that the product property
$[X,Y](fg) = f[X,Y]g + g[X,Y]f$ to establish $[X,Y] \in V(M)$. It
is clear that $[X,Y]= -[Y,X]$ and $[X,X]= 0$. Moreover, $[fX,gY]$
$= fg[X,Y] + f(Xg)Y - g(Yf)X$. The bracket operation also
satisfies the Jacobi identity,
$$[X,[Y,Z]] + [Z,[X,Y]] + [Y,[Z,X]]= 0.$$
   Let M and N be C^∞ manifolds of dimension m and n
respectively. We have defined the concept of a C^∞ map f
from M into N. Such a map induces a linear map from each
tangent space M_p into the tangent space $N_{f(p)}$. This linear map
is called the differential of f and it is defined by: if $t \in$
M_p, $\phi \in F(N,f(p))$, then $df(t)(\phi)= t(\phi \cdot f)$. By selecting a
coordinate system $(x_1,..,x_m)$ about p and $(y_1,..,y_n)$ about
$f(p)$, we can determine a matrix representation for df, which
is called the Jacobian matrix of df, with respect to the
chosen coordinate systems and the bases $\{D_x(p)\}$ and $\{D_y$
$(f(p))\}$, by the Jacobian $(D_x(y_i \cdot f)(p))$ for $1 \le j \le m$ and $1 \le$
$i \le n$. We call a C^∞ map $f:M \to N$ non- singular at p if df at
p is non-singular i.e., the Jacobian at p is non-singular.
   Let $f:M \to N$ be C^∞ into map.
   (a) f is an immersion if df_p is non-singular for each p
$\in$ M.
   (b) The pair (M,f) is a submanifold of N if f is an
one-to-one immersion.
   (c) f is an imbedding if f is an one-to-one immersion,
which is also a homeomorphism into, i.e., f is open as a map
into $f(M)$ with the relative topology.
   For example, one can immerse the real line R into the
Euclidean plane as illustrated in the following figure. Note
that the first case is an immersion but not a submanifold,
the second is a submanifold but not an imbedding, and the
third is an imbedding.

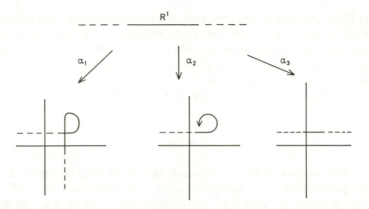

immersion but   submanifold but   imbedding
not submanifold   not imbedding

The ideas in imbedding or immersion arose from the
desire to view manifolds as submanifolds of a simpler one,
namely the Euclidean space. The notion of an imbedding is
like this: a manifold M^m may be imbedded in a "larger"
manifold N^n if M^m is identical to a submanifold of N^n. On
the other hand, an immersion is a map that locally appears
to be an imbedding failing to be one- to-one. That is, after
one moves a distance away from a given point, the mapping
begins to fold back on itself. The importance of putting a
manifold into a simpler one (either by an imbedding or
immersion) cannot be overemphasized. Once a manifold is a
submanifold of a Euclidean space, one can introduce various
analytic and differential geometric concepts which may not
be clear or accessible otherwise. Furthermore, sometimes an
imbedding or immersion gives the best intuitive hold on a
construction. Here we should mention the remarkable result:

<u>Theorem 2.4.7</u> A C^∞ manifold of dimension m can always
be C^∞-immersed in R^{2m}, and C^∞-imbedded in R^{2m+1}.

Whitney [1944] showed that this theorem can be
sharpened: for m > 0, every paracompact T_2 m-manifold imbeds
in R^{2m} and immerses in R^{2m-1} if m > 1. Of course, if the
manifold has further properties, such as orientable,

52

parallelizable, (these will be discussed later), then the imbedding (or immersion) theory can be improved even further. E.g., a theorem due to Hirsch [1959] states that any parallelizable manifold can be immersed in an Euclidean space of one higher dimension. For a nice survey of work on imbeddings, see Lashof [1965].

   Inverse Function Theorem 2.4.8   Let $(x_1,..,x_m)$ be a coordinate system at $p \in M$, $f_1,..,f_m \in F(M,p)$. Then $\phi = (f_1,..,f_m)$ restricts to a coordinate system at p iff $\det(D_{x_j}f_i(p)) \neq 0$, i.e., $d\phi$ is non-singular on M_p.

   We can restate the Inverse Function Theorem in a more familier form:

   Theorem 2.4.9   Let $f: R^m \to R^m$ be a C^∞ map. For $p \in R^m$ if $D_{x_j}f_i(p)$ is non-singular, i.e., as a linear transformation $D_{x_j}f_i(p)$ is onto. Then there are open subsets U and V of R^m, and $p \in U$ and $q = f(p) \in V$, such that the restriction $f|U$ is a diffeomorphism from U onto V; i.e., there is a C^∞ map $g:V \to U$ which is an inverse to $f|U$, and $D_{x_j}g_i(q)$ is the inverse to the matrix $D_{x_j}f_i(p)$.

   For example, in functions of several variables, say when we deal with curvilinear coordinates, a pair of functions $x = f(u,v)$, $y = g(u,v)$ can be regarded as mappings from the uv-plane to the xy-plane, and the inverse function $u = \phi(x,y)$, $v = \psi(x,y)$ with proper domains.

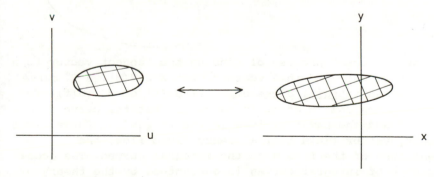

The well-known result $[\partial(x,y)/\partial(u,v)] \cdot [\partial(u,v)/\partial(x,y)] = 1$, i.e., the Jacobian of the inverse mapping is the reciprocal of the Jacobian of the mapping.

Corollary 2.4.10 (Inverse Function Theorem for Manifolds). If $p \in M$, ϕ is a C^∞ map from $M \rightarrow N$, then ϕ is a diffeomorphism of an open neighborhood of p onto an open neighborhood of $\phi(p)$ iff $d\phi$ is an isomorphism onto at p.

In defining arcwise connectedness, we have defined a path as a mapping. Likewise in these notes, curves will also be viewed as a special case of mappings. In particular, we will deal almost exclusively with parameterized curves, in particular, we shall discuss the various trajectories and orbits in the phase space of a dynamical system.

A C^∞ mapping $\sigma:[a,b] \rightarrow M$ is a C^∞ curve in M. Let $t \in [a,b]$. Then the tangent vector to the curve σ at t is the vector $d\sigma(t)/dt$ where $d\sigma(t)/dt = d\sigma(d/dt)_t \in M_{\sigma(t)}$. Note that $d\sigma(t)/dt$ is the usual "velocity" vector associated with a parameterized curve in R^3.

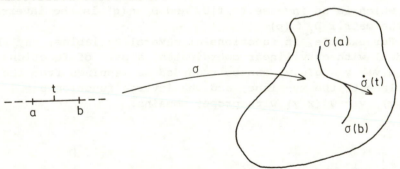

The "reverse" process of finding the tangent vector to a curve is to "filling in a vector". Let $X \in V(M)$. A C^∞ curve σ in M is an integral curve of X iff $d\sigma(t)/dt = X(\sigma(t))$ for each t in the domain of σ. It is clear that the curve σ "fits" X and the physical idea is the following. For a given velocity vector field X of a steady fluid flow, the streamlines of the flow give the integral curves. The local existence of integral curves is guaranteed by the theory of ordinary differential equations via the following theorem.

Theorem 2.4.11 Let X ∈ V(M) and let p be a point in the domain of X. Then for any real number a there exists a real number r > 0 and a unique curve σ:(a-r, a+r) → M such that σ(a) = p, and σ is an integral curve of X.

As we shall see, dynamical systems are often governed by the type of equations for integral curves, i.e., dσ(t)/dt = X(σ(T)). An integral curve is called a trajectory or orbit of the system. We shall come to these again later.

It is also convenient to define a broken C^∞ curve σ on an interval [a,b] to be a continuous map σ from [a,b] into M which is C^∞ on each of a finite number of subintervals [a,b_1], [b_1,b_2],..,[b_{k-1},b].

Let X ∈ V(M), we associate with X a local one-parameter group of transformations T, which for every p ∈ M and t ∈ R sufficiently close to φ assigns the points T(p,t)= σ(t) where σ is the integral curve of X starting at p. Theorem 2.4.11 tells us that for every p there is a positive number r and a neighborhood U of p such that T is defined and C^∞ on Ux(-r,r). From our notation, since the real numbers used as the second variable of T, are parameter value along a curve, they must satisfy additive property, that is: if q ∈ U, t, s, s+t ∈ (-r,r) then T(T(q,t),s) = T(q,s+t).

The set of pairs (p,t), p ∈ M, t ∈ I, is an open subset of MxR containing p, hence a smooth manifold Σ_x of dimension m+1. The mapping σ: Σ_x → M by (p,t) → σ(t) is the flow of X. Since M and X are C^∞, so the flow is also C^∞.

Let us look at this description in terms of fluid flow again. As before, let us suppose that the fluid is steady state, i.e., the velocity of the fluid at each point p ∈ M is independent of time and equal to the value X(p) of the vector field. In this case, the integral curves of X(p) are the paths followed by the particles of the fluid. Now let φ(p,t) be the point of M reached at time t by a particle of the fluid which leaves p at time 0. We notice that φ(p,0) is always p. Since velocity is independent of time, φ(q,s) is the point reached at time s+t by particle starting at q at time t. If we put q = φ(p,t), so the particle started from

55

the point p at time 0, we can conclude that
$\phi(\phi(p,t),s) = \phi(p,s+t)$. Also, the smoothness of ϕ, as
functions of p and t, will be influenced by the smoothness
of X.

A p-dimensional distribution on a manifold M (p ≤ dim M)
is a function D defined on M which assigns to each m ∈ M a
p- dimensional linear subspace D(m) of M_m. A p-dimensional
distribution D on M is of class C^∞ at m ∈ M if there are C^∞
vector fields $X_1, \ldots, X_p$ defined in a neighborhood U of m
and such that for every n ∈ U, $X_1(n), \ldots, X_p(n)$ span D(n).
An integral manifold N of D is a submanifold of M such that
$di(N_n) = D(i(n))$ for every n ∈ N. We say that a vector field
X belongs to the distribution D and write X ∈ D, if for
every m in the domain of X, X(m) ∈ D(m). A distribution D is
involutive if for all C^∞ vector fields X, Y which belong to
D, we have [X,Y] ∈ D. A distribution D is integrable if for
every m ∈ M there is an integral manifold of D containing m.
It is easy to see that an integrable C^∞ distribution is
involutive. Clearly, every one- dimensional C^∞ distribution
is both involutive and integrable, by the existence of
integral curves. We would like to mention the classical
theorem of Frobenius:

Theorem 2.4.11 A C^∞ involutive distribution D on M is
integrable. Furthermore, through every m ∈ M there passes a
unique maximal connected integral manifold of D and every
other connected integral manifold containing m is an open
submanifold of this maximal one.

The following local theorem gives more information as to
how the integral manifolds are situated with respect to each
other:

Theorem 2.4.12 If D is a C^∞ involutive distribution on
M, and m ∈ M, then there is a coordinate system $(x_1, \ldots,$
$x_d)$ on a neighborhood of m, such that $x_i(m) = 0$ and for
every m' in the coordinate neighborhood the slice {p ∈ M|
$x_i(p) = x_i(m')$ for every i > dim D} is an integral manifold
of D, when given the obvious manifold structure induced by
the coordinate map.

Before we take off from the concept of flow to the basic idea of dynamical systems and structural stability, we should prepare ourselves with more conceptual notions and tools in differential geometry so that when we are facing the geometric theory of differential equations, which is an integral part of dynamical systems as we have pointed out earlier, we will be ready for it. There are several topics we would like to briefly discuss, namely critical values, Morse Lemma, groups and group action on spaces, fiber bundles and jets, and differential operators on manifolds. These last two subjects will be discussed in Chapter 3.

2.5   Critical points, Morse theory, and transversality

The idea of critical points to be introduced here is an extension of the concept of maxima and minima of a function. As we know in calculus, if a differentiable function f of one variable x has a maximum of minimum for $x = x_0$, then $df/dx = 0$ at x_0. Similarly, if a function of two variables x, y has a maximum or minimum at (x_0, y_0), then $\partial f/\partial x = \partial f/\partial y = 0$ at this point. Geometrically, what we are saying is that the tangent plane to the surface $z = f(x,y)$ is horizontal at (x_0, y_0). Of course the same condition is also satisfied at a saddle point, a point that behaves like a maximum when approached in one way and like a minimum when approached in another. Moreover, this situation can be thought of as corresponding to an embedding of M in 3-space such that the function f is identified with one of the coordinates z, and the horizontal plane $z = f(x_0, y_0)$ is a tangent plane to M at $(x_0, y_0, f(x_0,y_0))$. More precisely, we have the following definitions.

Let M^m, N^n are C^∞-manifolds, and f is a C^∞ map. A point $a \in M^m$ is a _critical point_ of f if $df_a = 0$, (i.e., df is not onto at a, or the Jacobian matrix representing df has rank less than the maximum (n)). $b \in N^n$ is a _critical value_, if $b = f(a)$ for $a \in M^m$. A value b is a _regular value_ if $f^{-1}(b)$ contains no critical points. Thus f maps the set of critical points onto the set of critical values.

For example, if $M = N = R^1$, then a critical point is a point where the derivative vanishes. In calculus or advanced calculus, the main interest in critical point and critical value is centered on the search for extrema. Although they are important in their applications, they are of equal importance in answering geometric questions such as the immersions, submanifolds, and hypersurfaces as the following theorem illustrates.

Theorem 2.5.1   Let f: $M^m \rightarrow N^n$ be a C^∞ map and b ϵ N^n be a regular value. Then $f^{-1}(b)$ is a submanifold of M^m whose dimension is $(m-n)$.

Next we use a special but remarkably simple notion of Lebesque measure in real analysis. This particular notion of measure zero gives us a very simple yet intuitive definition for our purpose without resorting to a host of machinery.

Let W_i be a cube in R^n and denote its volume by $\mu(W_i)$. A set $S \subseteq R^n$ is said to have measure zero, $\mu(S) = 0$, if for any given $\epsilon > 0$, there is a countable family of W_i such that (i) $S \subseteq U_i W_i$; (ii) $\Sigma_i \mu(W_i) < \epsilon$.

It should be noted that it is possible for the continuous (C°) image of a set of measure zero to have positive measure [Royden 1963]. Nonetheless, such a possibility is excluded when the maps are C^∞ as the following theeorem shows.

Theorem 2.5.2   Let $S \subseteq U \subseteq R^n$, where $\mu(S) = 0$ and U is open, and let f: $U \rightarrow R^m$ be C^r $(r \geq 1)$. Then $\mu(f(S)) = 0$.

Theorem 2.5.3 (Sard)   Let f: $M \rightarrow N$ be C^∞. Then the set of critical values of f has measure zero in N.

Let C be the set of critical points of f, then f(C) is the set of critical values of f, and the complement $N - f(C)$ is the set of regular values of f. Since M can be covered by countable neighborhoods each diffeomorphic to an open subset of R^m, we have

Corollary 2.5.4 (Brown)   The set of regular values of a C^∞ map f: $M \rightarrow N$ is everywhere dense in N.

Corollary 2.5.5   Let f: $M^m \rightarrow N^n$ $(n \geq 1)$ be onto and C^∞. Then except for a subset of N^n of measure zero, for all y ϵ

58

N^n, $f^{-1}(y)$ is a submanifold of M. Moreover, there is always some $y \in N^n$ such that $f^{-1}(y)$ is a proper submanifold of M.

   Corollary 2.5.6   Let the n-disk be $D^n = \{x \in R^n| \ \|x\| \leq 1\}$ and its boundary $\partial D^n = S^{n-1}$, an (n-1)-sphere. Let i: $S^{n-1} \to D^n$ be the inclusion map. Then there is <u>no</u> continuous map r: $D^n \to S^{n-1}$ such that $r \cdot i = id$ on S^{n-1}, i.e., no continuous r such that for each $x \in S^{n-1}$, $r(i(x)) = x$.

   If such an r exists, it is called a <u>retraction</u> of D^n onto S^{n-1}. For n = 2, this corollary can be worded as follows: The circle is not a retraction of the closed unit disk (normally a theorem in elementary homotopic theory). As a corollary: Any continuous map f of the closed disk into itself has a fixed point, i.e., $f(x_0) = x_0$ for some $x_0 \in D^2$. This is the n = 2 case of the <u>Brouwer Fixed Point Theorem</u>.

   Corollary 2.5.7 (Brouwer Fixed Point Theorem)   Let $D^n = \{x \in R^n| \ \|x\| \leq 1\}$ be the n-disk, let f: $D^n \to D^n$ be continuous. Then f has a fixed point, i.e., there is some $x_0 \in D^n$ such that $f(x_0) = x_0$.

   It has been realized for some time that a topological space can often be characterized by the properties of continuous functions on it. But it was Morse [1934] who first called attention to the importance of nondegenerate critical points and invariant index, which completely characterizes local behavior near that point. Moreover, the number of critical points of different indices relates to the topology of the manifold by means of the Morse inequalities. In addition, a sufficiently isolated critical point indicates the addition of a cell to the cell decomposition of the manifold. Consequently, this shows how a manifold is put together, as a cell complex, in terms of the critical points of a sufficiently well behaved function. On the other hand, Morse theory also treats geodesics on a Riemannian manifold. Although Morse did a great deal more, here we shall only touch on a few items directly concerning our main emphasis. There is some material from algebraic topology, such as homology, Betti numbers, Euler characteristics, which will be needed when we get to the

Morse inequalities. At the appropriate places, we shall
state all the basic facts without proof.

Let us recall the concept of a critical point. Let M be
a m-dimension C^∞ manifold and f: M → R be a C^∞ function.
Then a ∈ M is a <u>critical point</u> of f if f is not onto at a.
Since the range of df is a 1-dimension vector space at a, a
is a critical point when df is the zero map at a. From a
more conventional viewpoint, a ∈ M is a critical point if
there is a coordinate chart ϕ_α : U_α → R^m, x ∈ U_α such that
all first partial derivatives of $f \cdot \phi_\alpha^{-1}$ vanish at $\phi_\alpha(a)$. And
a real number b = f(a), where a is a critical point, is
called a <u>critical value</u>.

Clearly, the first partial derivatives at a critical
point have degenerate behavior. Nonetheless, when the second
partial derivatives are better behaved, it is called a
nondegenerate critical point. More precisely, if a ∈ M is a
critical point for f: M → R, f ∈ F(M), and $\partial^2 (f \cdot \phi_\alpha^{-1})/\partial x_i \partial x_j$;
the <u>Hessian</u> at a, is non-singular, then a is a <u>nondegenerate</u>
<u>critical point of f</u>. It can be shown that this definition is
independent of the choice of the coordinate chart.

For example, let S^2 be the unit sphere centered at the
origin in R^3, and let f assign to any point its z = constant
planes, i.e., f(x,y,z) = z. It is easy to see that there are
only two critical points $a_1(0,0,1)$ and $a_2(0,0,-1)$ and their
critical values z = ±1. Moreover, both of the critical
points are nondegenerate.

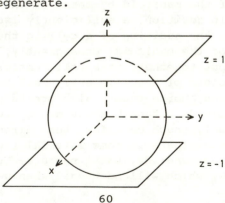

As another example, let T^2 be a 2-dimension torus imbedded as a submanifold of R^3. This T^2 can be thought of as the surface traced by the circle of center $(2,0)$ and radius 1 in the (x,y)- plane as this plane is rotated about the y-axis.

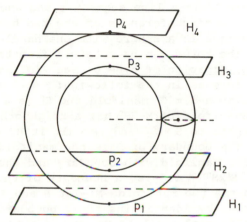

The surface has the equation
$$(x^2 + y^2 + z^2 + 3)^2 = 16(x^2 + z^2).$$
It is easy to show (and easy to see from the figure) that there are just four z = constant horizontal planes H_1, H_2, H_3, and H_4 that are tangent planes of T^2 at p_1, p_2, p_3 and p_4 respectively, coresponding to four critical points for the function z on T^2 and H_i $(i=1,2,3,4)$ are critical levels. Furthermore, one can show that these four critical points p_i $(i=1,2,3,4)$ are nondegenerate. Notice that, in this example, if N_c is a non-critical level of z, it is surrounded by neighboring noncritical levels, all of which are homeomorphic to each other. For example, see the above figure, between H_1 and H_2 all the noncritical levels are circles. But as soon as we cross a critical level, a change takes place. The noncritical levels immediately below H_2 are quite different from those immediately above. In fact, this observation is valid in general.

61

Let the differentiable manifold M be a 2-sphere and take a zero-dimension sphere S^0 in M. S^0 has a neighborhood consisting of two disjoint disks. This is of the form $S^0 \times E^2$. (E^2 is a 2- dim disk). Let us call this neighborhood B. Then M -Int B is a sphere with two holes in it. $E^1 \times S^1$ is a cylinder (here E^1 is a line segment), and when its ends are attached to the circumferences of the two holes, the resulting surface is a sphere with one handle, i.e., a torus. Thus the torus can be obtained from the 2-sphere by a <u>spherical modification of type 0</u>. See Fig.2.5.1. Let us define this term as in the following:

Let N be an n-dim C^∞ manifold and S^r is a directly embedded submanifold of M. S^r has a neighborhood in M which is diffeomorphic to $S^r \times E^{n-r}$ and we call it B, where E^{n-r} is a (n-r)-cell. The boundary of B is the manifold $S^r \times S^{n-r-1}$. Thus M - Int B is a manifold with boundary and the boundary is $S^r \times S^{n-r-1}$. But $S^r \times S^{n-r-1}$ is also a boundary of the C^∞ manifold, $E^{r+1} \times S^{n-r-1}$. So the two manifolds M -Int B and $E^{r+1} \times S^{n-r-1}$ can be joined together by identifying their boundaries. Such a joined space is a C^∞ manifold M'. M' is said to be obtained from M by a <u>spherical modification of type r</u>.

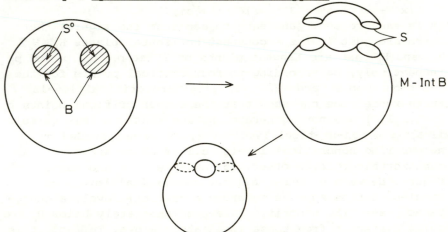

T^2 obtained by spherical modification of type 0 from S^2.

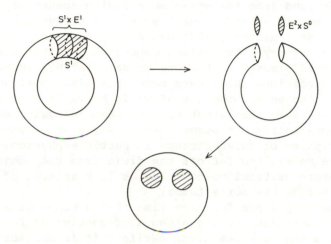

S^2 obtained by spherical modification of type 1 from T^2.

Fig.2.5.1

From the examples above, it is clear that if M' is
obtained from M by a spherical modification, then M can be
obtained from M' by another spherical modification.
Furthermore, we have noticed from the examples that if M is
a C^∞ manifold and f a C^∞ function on it and if M_a and M_b
are noncritical level manifolds of f separated by one
critical level, then M_b can be obtained from M_a by a
spherical modification. This fascinating subfield of
algebraic and differential topology is called <u>surgery</u> and it
has been an active field since 1960. There are some quite
far reaching results. Related to this subject is the concept
of cobounding of a manifold. If two compact differentiable
manifolds M_0 and M_1 are said to <u>cobound</u> if there is a
compact differentiable manifold M such that the boundary of
M is the disjoint union $M_0 \cup M_1$. In general, to testing
whether a pair of manifolds are cobounding is very
complicated, nonetheless, the following very interesting
result can be stated: If M_0 and M_1 are compact
differentiable manifolds, then they are cobounding iff each

63

can be obtained from the other by a finite number of spherical modifications. A well written introductory book is Wallace [1968]. Hirsch [1976] has a couple of chapters discussing surgery (also called cobordism in literature). A very advanced classic is the one by Wall [1970]. There may be revised edition. There were some potential applications in physics the author thought of while he was a graduate student. However, the author has not been following the developments lately. It seems that it may also be useful for the description of super-strings in particle physics.

Before we are too far off the field from the subject of nondegenerate critical points, let us look at some of its properties from the Morse Lemma.

It has been known for some time that a topological space may be characterized by the algebra of continuous functions on it. From the examples given earlier, it is not surprising that one can learn a great deal about a smooth manifold from the smooth real functions defined on it. Morse [1934] first realized the importance of nondegenerate critical points and the numerical invariant called the index, which completely characterizes the local behavior near that point. Moreover, the number of critical points of various indexes relates to the topology of a manifold by means of the Morse inequalities. A sufficiently isolated critical point also signals the addition of a cell to the decomposition of the manifold. Thus, this shows how a manifold is put together as a cell complex, in terms of the critical points of a sufficiently well-behaved function. Furthermore, from Sard's theorem (Theorem 2.5.3), these well-behaved functions are actually very common. In the rest of this section, we will discuss the Morse lemma and inequalities, and transversality properties.

A symmetric bilinear form represented by a matrix B has index i if B has i negative eigenvalues. We asy that B has nullity k if k of the eigenvalues are zero. Thus, p is a nondegenerate critical point for f when p is a critical point and the nullity of the Hessian of f at p is zero.

Lemma 2.5.8 (Morse)   Let M be n-dim C^∞ manifold, f: M →
R be smooth and $x_o \in$ M a nondegenerate critical point. Then
there is a coordinate chart U_α containing x_o with $\phi_\alpha(x_o) = 0$
and such that $f(\phi_\alpha^{-1}(u)) = f(\phi_\alpha^{-1}(0)) - u_1^2 - \ldots - u_i^2 + u_{i+1}^2$
$+ \ldots + u_n^2$.
where $u = (u_1, \ldots, u_n) \in R^n$ and i is the Morse index of f at
x_o.

Corollary 2.5.9   The non-degenerate critical points of a
smooth function are isolated.

Corollary 2.5.10   If f is smooth on a compact smooth
manifold with all critical points non-degenerate, then f has
only finitely many critical points.

As an example, consider the family of surfaces in R^3,
$$x^2 - y^2 - z^2 = c$$
with $-1 \leq c \leq 1$. Note that the two surfaces obtained by
putting c = -1 and 1 are hyperboloids of one and two sheets
respectively. Then the critical points are easily obtained.

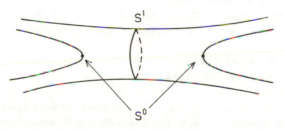

These two S° are nondegenerate critical points which are
isolated. These infinitely many critical points of S^1 are
degenerate.

If all critical points of f are non-degenerate and all
critical values are distinct, then f is called a Morse
function. A fundamentally important result is the following
set of Morse inequalities:

Theorem 2.5.11 (Morse inequalities)   Let M be a C^∞
compact m- manifold (without boundary) and f: M → R be a
Morse function. Let c_k, k = 0,1,..,m, denote the number of
critical points of f of index k and β_k be the kth Betti

65

number of M (i.e., the number of independent generators of the kth homology group of M). Then

$$c_0 \geq \beta_0$$
$$c_1 - c_0 \geq \beta_1 - \beta_0$$
$$c_2 - c_1 + c_0 \geq \beta_2 - \beta_1 + \beta_0$$
$$\dots\dots\dots\dots\dots\dots\dots\dots$$
$$c_m - c_{m-1} + \dots + (-1)^m c_0 \geq \beta_m - \beta_{m-1} + \dots + (-1)^m \beta_0.$$

Before we continue, we would like to briefly discuss orientation, duality, and Euler characteristics, and some of their geometric interpretations. Orientation of a manifold is of fundamental importance. Duality relates (or more appropriately pairing) the Betti numbers of an orientable, compact manifold. Euler characteristic is a topological invariant quantity built upon from Betti numbers, and its existence on a manifold has profound geometric implications. For lower dimensional manifolds, in particular, two-dimensional manifolds, it is directly related to the curvature of the 2-manifold.

<u>Proposition 2.5.12</u> Every manifold has a unique Z/2 − orientation (the number of orientation = no. of elements in $H^0(X, Z/2)$).

There are several ways to define orientation. The above is a topological one. We shall discuss a geometric one later.

<u>Theorem 2.5.13</u> Let X be a connected non-orientable manifold. Then there is a 2-fold connected covering space E $\xrightarrow{p}$ X such that E is orientable.

This theorem tells us that if we are interested in the detailed local geometry of the manifold (assumed to be arcwise- connected) as a model of physical state space, then we may as well assume the manifold to be orientable. This is because one can always find an orientable covering space which has same local geometry. Indeed, as the next corollary shows, one can always goes to its universal covering space, which is simply-connected.

<u>Corollary 2.5.14</u> Every simply-connected X is orientable.

Proposition 2.5.15 $H_n(X) = 0$ if X is connected and non-compact, (dim X = n).

Poincaré Duality Theorem 2.5.16 If X is an oriented, n-dim manifold, then the homomorphism D : $H_c^q(X) \to H_{n-q}(X)$ is an isomorphism for all q, where $H_c^q(X)$ is the cohomology group with compact support.

We shall discuss differential forms, and the de Rham cohomology, and the cohomology groups of a manifold.

Corollary 2.5.17 If X is compact, orientable, then the Betti numbers of X satisfy $b_q = b_{n-q}$ for all q.

The <u>Euler characteristic of X</u> $\chi(X)$ is defined as an alternating sum of all the Betti numbers of X, i.e.,

$\chi(X) = \Sigma \ (-1)^q b_q$, n = dim X.

Examples: For $q \geq 1$, $n \geq 1$

$$H_q(S^n) = \begin{cases} R, & q = n \\ 0, & q \neq n. \end{cases}$$

Thus, $\chi(S^n) = \begin{cases} 0, & n = \text{odd} \\ 2, & n = \text{even.} \end{cases}$

Remark: $\chi(X)$ is a very useful topological invariant. E.g.:

Theorem 2.5.18 A differentiable manifold (any dimension) admits a non-zero continuous vector field iff its Euler characteristics are zero [Steenrod 1951].

As a consequence, we have: For compact manifold M, there exists a non-vanishing vector field iff $\chi(M) = 0$. Thus only odd- dim spheres admit non-vanishing vector fields. Indeed, one can prove that this is equivalent to that the tangent bundle of M splits, which is also equivalent to the manifold admits a metric of Lorentz signature (i.e.,pseudo-Riemannian metric). Since it is well-known that any differentiable manifold admits a Riemannian metric, the Lorentz metric can be constructed by

$\eta(Y,Z) = g(Y,Z) - 2g(X,Y)g(X,Z)/g(X,X)$

or $\eta_{ij} = g_{ij} - 2 \ X_i X_j / |X|^2$.

where X, Y, and Z are non-zero vector fields on M. We shall come back to this later, when we discuss characteristic classes.

Gauss-Bonnet theorem 2.5.19:   Let M be a compact
connected oriented Riemannian 2-manifold with Riemannian
(Gausssian) curvature function K. Then $\int K = 2\pi\chi(M)$ [Hicks
1971].

Generalized Gauss-Bonnet Theorem 2.5.20:   If M is an
even dimension (n = 2k) compact connected oriented
Riemannian manifold, then $\int_M Q = 2^n \pi^k (k!)\chi(M)$ where

$$Q = \Sigma(-1)^{\pi} R_{\pi(1)\pi(2)} \Lambda R_{\pi(3)\pi(4)} \Lambda \cdots \Lambda R_{\pi(n-1)\pi(n)} \quad \epsilon \; \Lambda^n(M)$$

where $\pi(i)$ are permutations [Chern 1951].

Let us get back to the Morse lemma and Morse
inequalities. If furthermore, M is orientable, then,
applying the Poincaré duality theorem, we can further
simplify the inequalities. As an example, if $M = T^2$, the
2-dim torus, then any Morse function on T^2 has at least four
distinct critical points since $\beta_0 = \beta_2 = 1$ and $\beta_1 = 2$. We
have already demonstrated and discussed this earlier.

Let us get back to Sard's theorem. Although many
arguments in imbedding and immersion can be reformulated and
occasionally made more precise by using Sard's theorem, we
shall turn to a very important concept of transversality.
This is a theory which investigates the way submanifolds of
a manifold cross each other.

Let f be C^∞ map of C^∞ manifolds, $f: M^m \rightarrow N^n$, and W^p be a
submanifold of N. Roughly speaking, f is transverse to W at
$x \in M$, $f \pitchfork_x W$, means that the intersection in N of f(M) and
W has the lowest possible dimension in a neighborhood of
$f(x) \in W$, but the sum of the dimensions of f(M) and W is at
least n. More precisely: Given a C^∞ map $f: M^m \rightarrow N^n$ between
two C^∞ manifolds and a submanifold W^p of N, we say f is
transverse to W if for each $x \in M$, $y = f(x) \in W$ such that
$df(M_x) + W_y = N_y$. Here + means that we take the set of all
vectors in N_y that are sums of a vector in the image of df
and a tangent vector to the submanifold W. In other words,
the tangent space N_y is spanned by W_y and the image $df(W_x)$.
If f(M) does not intersect W, i.e., $f(M) \cap W = 0$, then f is
automatically transverse to W.

As a simple example, let $M = S^1$, $N = R^2$, W = x-axis in

68

R. Then this position is transverse

but this position is not.

It seems that the concept of transversality requires the
intersection be in the most general position.

Some more examples of transversality:

(a) Let M = R = W, N = R^2, f(x)=(x,x^2). Then f ⋔ W at
all nonzero x.

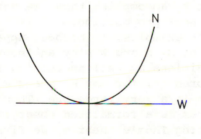

Note that f can be slightly perturbed so that it is
transversal to W; e.g.,

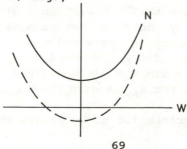

(b)   M, N, W as in (a) and f is defined by the graph

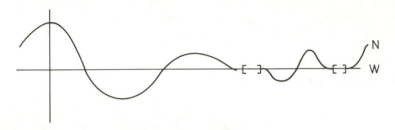

Then f ⋔ W everywhere except on the segments within the brackets.

(c)   If M = W = R, N = R³, then if f is any mapping of M → N, it is transversal to W only if f(M) ∩ W = 0. Note that here a nontransversal mapping can also be approximated closely by a transversal mapping because in 3-space f can avoid W even easier by just going around it and f only has to move a little bit to accomplish this. We shall make this more precise in the next proposition.

From these simple examples, it becomes apparent that the relative dimensions of M, N and W play an important role in determining the conditions as well as meaning for f to be transversal to W. Moreover, for any M, N and W, the set of transversal mappings is a very large one. Thom's transversality theorem is a formalized observation of this fact. Before discussing Thom's theorem, we first give some properties of the set of maps which is transversal to W.

Proposition 2.5.21 Let M and N be smooth manifolds, W ⊂ N a submanifold. Suppose dim W + dim M < dim N. Let f : M → N be smooth and suppose that f ⋔ W. Then f(M) ∩ W = 0.

This can be seen by the fact that suppose f(p) ∈ W, then by the definition of tangent space and the assumption,
$$\dim(W_{f(p)} + (df)(M_p)) \leq \dim W_{f(p)} + \dim M_p$$
$$= \dim W + \dim M < \dim N = \dim N_{f(x)},$$
thus it is impossible for $W_{f(x)} + df(M_p) = N_{f(x)}$. Thus if f ⋔ W at p, then f(x) ∉ W.

It is also appropriate for us to relate the notion of

70

transversality with Sard's theorem. The simplest example is letting $f : R \to R$ be C^1. If y_0 is a regular value, then the horizontal line $R \times \{y_0\} \subset R \times R$ (i.e., $y = y_0$ line) is transverse to the graph of f.

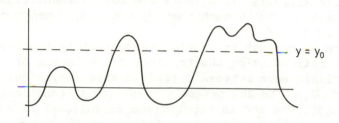

Thus, the Sard's theorem or its corollary (Brown's theorem) implies that "most" horizontal lines are transverse to the graph. For $f : R^2 \to R^1$, the Sard's theorem says that most horizontal planes $R^2 \times \{z_0\} \subset R^2 \times R^1$ are transverse to the graph of f. If we replace C^∞ in the Sard's theorem by C^r, we may want to know whether the theorem will change or not. For $f : R^2 \to R^1$, the theorem seems plausible for f to be C^2. In fact, intuitively it even seems plausible for f being only C^1. But Whitney [1935] has found an ingenious and very interesting counter-example. He constructed a C^1 map $f : R^2 \to R^1$ whose critical set contains a topological arc r, yet $f|r$ is not constant. Thus $f(C_f)$ contains an open subset of R, where C_f = {critical points of f}. This leads to an interesting paradox. The graph of f is a surface $S \subset R^3$ on which there is an arc r such that at every point of r the surface has a horizontal tangent plane, yet r is not at a constant height! We shall not go into any more detail about this example, but to say that for C^r mappings, there is a differentiability condition for Sard's theorem. Let us state Sard's theorem for C^r ($r < \infty$) maps.

Theorem 2.5.22 (Sard) Let M and N be smooth manifolds of dimensions m and n respectively, and $f : M \to N$ be a C^r map. If $r > \max \{0, m-n\}$, then $f(C_f)$ (the set of critical values of f) has measure zero in N. The set of regular values of f is dense..

Theorem 2.5.23 Let f, M, N and W as before. If f is transverse to W, then $f^{-1}(W)$ is a submanifold of M with dimension m - n + p.

Corollary 2.5.24 If M and W are both submanifolds of N and for each $x \in M \cap W$ such that $W_x + M_x = N_x$. Then $M \cap W$ is a submanifold of N.

Before we can get to Thom's Transversality theorem, we need some refinement on the topology of the space of differentiable maps between differentiable manifolds.

Let $C^\infty(R^n, R^k)$ be the set of C^∞-maps (or C^∞-functions) from R^n to R^k. The set is topologized as follows: If $\epsilon(x)$ is a positive, continuous function defined on R^n, and $p > 0$ is any integer, let
$$B(0, \epsilon(x), p) = \{f \in C^\infty(R^n, R^k) \mid |D^\alpha f_j(x)| < \epsilon(x)$$
$$\text{for all } |\alpha| \leq p \text{ and } j\}$$
where f_j is the j-th coordinate of f. This set forms a basis for the neighborhood of the constant function 0. A similar basis neighborhood for $g \in C^\infty(R^n, R^k)$ can be defined by $B(g, \epsilon(x), p) = \{f \in C^\infty(R^n, R^k) \mid (f-g) \in B(0, \epsilon(x), p)\}$. Here the integer p is allowed to vary. To generalize the topology of $C^\infty(M^m, N^n)$, where M^m, N^n are two C^∞-manifolds, one can proceed as above by choosing coordinate charts to cover N^n and demand that the above construction holds in all the coordinate charts near any point. Of course one may also reduce this problem to the above construction in much larger Euclidean spaces by using the imbedding theorem.

We want to point out that the above construction provides a rather fine topology on $C^\infty(M^m, N^n)$, this is because function such as $\epsilon(x)$ may decrease to zero rapidly, even though $\epsilon(x) > 0$ for all x.

Theorem 2.5.25 Let M^m, N^n be C^∞ manifolds, and W^p N^n a submanifold. The set $F_W(M^m, N^n)$, consisting all maps in $C^\infty(M^m, N^n)$ that are transverse to W, is an open subset of $C^\infty(M^m, N^n)$.

Theorem 2.5.26 (Thom's transversality theorem) $F_W(M^m, N^n)$ is dense in $C^\infty(M^m, N^n)$.

The transversality theorem in its many variations are not only of fundamental importance to structural stability and bifurcations, but are also of crucial importance to many areas of differential topology such as Thom's construction of cobordism theory. Two compact manifolds M_1, M_2 are called <u>cobordant</u>, if there is a compact manifold N such that $\partial N \approx$ $\{M_1 \times 0\} \cup \{M_2 \times 1\}$. Loosely speaking, this means that the disjoint union of M_1 and M_2 is the boundary of N. We call N a cobordism from M_1 to M_2. The spherical modification we mentioned earlier is a special situation of cobordism theory.

In the next section, we shall discuss more geometric details of a differentiable manifold, in particular, the group actions on manifolds. This will provide the fundation for the discussion of fiber bundles, which is the section after next.

2.6   Group and group actions on manifolds, Lie groups

Lie groups exist naturally in many areas of mathematics and physics where natural group structures may be found on certain manifolds. Lie groups are also very useful in physics, in particular, as transformation groups, internal symmetry groups and their representations for particle classifications, gauge groups, etc. Even in classical mechanics, the usual Lagrange and Poisson brackets not only provide the system's symmetry or conservation laws, but also provide a geometric manifestation of the dynamical processes by noticing that a Poisson bracket of a pair of vector fields is the dragging of a vector field along the integral curve of another vector field. There are many well written books on Lie groups. As a beginning, many differential geometry books have a chapter or two on Lie groups, their geometry, and representations, e.g., Bishop and Crittenden [1964]. There are several sections in various chapters of Choquet-Bruhat, De Witt-Morette and Dillard-Bleick [1977]. For more advanced readers, Chevalley [1946] is still the best. Pontgryagin [1966] is also a very good book. Helgason [1962] is a detailed treatment of the differential geometry of the group spaces.

A <u>topological group</u> G is a group which is also endowed with the structure of a topological space and the map ϕ : GxG into G defined by $\phi(x,y) = xy^{-1}$ is continuous. Clearly, the multiplication and inverse maps defined by $(x,y) \to xy$ and $x \to x^{-1}$ are both continuous.

For example, the additive group of real numbers is a topological group, and the group of invertible nxn real matrices is also a topological group.

Let G be a topological group and X a topological space. <u>G acts on X to the left</u> if there is a continuous map ϕ : G x X $\to$ X and we write $\phi(g,x) = gx$ such that: (i) $\phi(g, \phi(h,x)) = \phi(gh,x)$ or $(gh)x = g(hx)$ for all $g,h \in G$, $x \in X$ and (ii) if $e \in G$ is the identity element in G, and $x \in X$, $\phi(e,x) = ex = x$. Sometimes G is called a <u>left transformation group</u>. Note, a right action would be defined by a map τ : X x G $\to$ X with the appropriate properties.

Given an action ϕ : G x X $\to$ X and a set $S \subseteq X$, then $G_S =$ {$g \in G|$ $\phi(g,y) = y$ for all $y \in S$}. If $G_x = \{e\}$, i.e., only the identity element leaves X fixed, we say the action is <u>effective</u>. If x, y $\in$ X, there is some $g \in G \to \phi(g,x) = y$, then we say the action is <u>transitive</u>. A trivial example of a transitive action which is not effective, is when G is nontrivial but x is a single point. Of course, if G = {e}, then G is effective on any X.

Given an action ϕ : G x X $\to$ X and a subgroup H $\subseteq$ G, then there is an induced action ϕ_H : H x X $\to$ X defined by $\phi(h,x)$ where h $\in$ H. If ϕ is effective, so is ϕ_H, and if ϕ_H is transitive, so is ϕ.

Let X be a Hausdorff space, ϕ : G x X $\to$ X a transitive action. Fix some $x_0 \in$ X, and let H = {$g \in G|$ $\phi(g,x_0) = x_0$}, the <u>isotropy group</u> of x_0, then H is a closed subgroup of G. The <u>coset space</u>, G/H, is defined to be the quotient space of G by the equivalence relation g - h iff $g^{-1}h \in$ H. G/H is Hausdorff because the mapping π : G $\to$ G/H is continuous and open. If G/H is compact, then the map G/H → X is a homeomorphism. If H is a closed subgroup of G, then G acts transitively on G/H. A space with a transitive group of operators is called <u>homogeneous</u>, such as G/H for H being a closed subgroup of G.

For example, the action of O(n), the group of nxn orthogonal matrices, on S^{n-1}. The subgroup of O(n) which leaves a unit vector v = (1,0,..,0) fixed is isomorphic to O(n-1). The coset space is O(n)/O(n-1) = S^{n-1} which is a homogeneous space. More generally, a Stiefel manifold $V_{m,k}$ is defined to be O(m)/O(k). Thus, $V_{m+1,m}$ = S^m [Chevalley 1946, Steenrod 1951].

A <u>Lie group</u> G is a differentiable (or analytic) manifold and also endowed with a group structure such that ϕ : G X G → G defined by (g,h) → gh^{-1}, where g,h ∈ G, is C^∞ (or analytic).

For example:

(i) R^n is a Lie group under vector addition.

(ii) The manifold GL(n,R) of all nxn non-singular real matrices is a Lie group under matrix multiplication.

(iii) The product G X H of two Lie groups is itself a Lie group with the product manifold structure and the direct product group structure, i.e., $(g_1,h_1)(g_2,h_2)$ = $(g_1g_2,(h_1h_2)^{-1})$ where g_1,g_2 ∈ G, h_1,h_2 ∈ H.

(iv) The unit circle S^1 is a Lie group with the addition of angles.

(v) The n-torus T^n (n an integer > 0) is the Lie group which is the product of the Lie group S^1 with itself n times.

If f is C^∞, then we may be able to extend Taylor's formula into a convergent series. If we can, then f is said to be <u>analytic</u>. A standard example of a C^∞ function which is not analytic is the following.

$$\phi(x) = \begin{cases} \exp\left(-1/(1-|x|^2)\right), & |x| < 1, \\ 0, & |x| \geq 1. \end{cases}$$

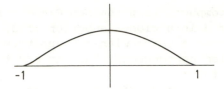

Let $g \in G$. <u>Left translation by g</u> and <u>right translation by g</u> are the diffeomorphisms l_g and r_g of G defined by $l_g(h) = gh$, $r_g(h) = hg$ for all $h \in G$. If U is a subset of G, we denote $l_g(U)$ and $r_g(U)$ by gU and Ug respectively. A vector field X on G is called <u>left invariant</u> if for each $g \in G$, X is l_g-related to itself, i.e., if $l_g: G \to G$
  $dl_g: G_h \to G_{gh}$,   the tangent space of G
  $dl_g \cdot X = X \cdot l_g$.

Clearly, the translation map gives a prescribed way of translating the tangent space at one point to the tangent space at another point on a Lie group G. We shall come back to this later after we have introduced the concept of tangent bundles and fiber bundles.

Recall that we have defined the tangent vector fields as derivatives and the bracket operation of two vector fields earlier, now we shall use these concepts to define the Lie algebra.

A <u>Lie algebra</u> **g** over R is a real vector space **g** together with a bilinear bracket operator [,] : **g** x **g** $\to$ **g** such that for all X, Y, Z $\in$ **g**,

  (i) the bracket is anti-commuting, i.e., $[X,Y] = -[Y,X]$,

  (ii) the Jacobian identity is satisfied, i.e.,
      $[[X,Y],Z] + [[Y,Z],X] + [[Z,X],Y] = 0$.

The importance of the concept of Lie algebra is its intimate association with a Lie group. For instance, the connected, simply connected Lie group are completely determined (up to isomorphism) by their Lie algebras. Thus the study of these Lie groups reduces in large part to the study of their Lie algebras.

Examples of Lie algebra:

76

(i)   The vector space of all smooth vector fields on the manifold M forms a Lie algebra under the bracket operation on vector fields.

(ii)   Any vector space becomes a Lie algebra if all brackets are set equal to 0. Such a Lie algebra is called abelian.

(iii) The vector space $\mathbf{gl}(n,R)$ of all nxn matrices form a Lie algebra if we set $[A,B] = AB - BA$ where A, $B \in \mathbf{gl}(n,R)$.

(iv) R^3 with the bilinear operatioon $X \times Y$ of the vector cross product is a Lie algebra.

<u>Theorem 2.6.1</u>   Let G be a Lie group and $\mathbf{g}$ its set of left invariant vector fields. Then,

(i) $\mathbf{g}$ is a real vector space, and the map $\alpha : \mathbf{g} \rightarrow G_e$ defined by $\alpha(X) = X(e)$ is an isomorphism of $\mathbf{g}$ with the tangent space G_e of G at the identity. Thus, dim $\mathbf{g}$ = dim G_e = dim G.

(ii) Left invariant vector fields are smooth.

(iii) The Lie bracket of two left invariant vector fields is itself a left invariant vector field.

(iv) $\mathbf{g}$ forms a Lie algebra under the Lie bracket operation on vector fields.

It should be noted that the correspondence between Lie groups and Lie algebras of the above theorem is not unique. For instance, all 1-dim Lie groups such as S^1 or R^1 have the same Lie algebra. Similarly, the plane R^2, torus T^2, and cylinder $S^1 \times R^1$ all have the same abelian Lie algebra. It is not difficult to see that two Lie groups that have isomorphic Lie algebra are themselves isomorphic at least in a sufficiently small neighborhood of their identity elements. Such a notion can be formulated more precisely in terms of a local Lie group in a neighborhood of e. Then the correspondence between such local Lie groups and Lie algebras is indeed one-to-one and onto. We have

<u>Theorem 2.6.2</u>: Two Lie groups are locally isomorphic iff their Lie algebras are isomorphic.

Let us now consider the subgroups and subalgebras of a

given Lie group and its algebra.

Let G be a Lie group. A subgroup H ⊆ G is a <u>Lie subgroup</u> of G if the set of elements of H is a submanifold of the smooth manifold G. Let g be a Lie algebra. h ⊆ g is a Lie <u>subalgebra</u> of g if (i) h is a vector subspace of g, and (ii) if X,Y ϵ h, then [X,Y] ϵ h.

A Lie subgroup is clearly a Lie group with the structure inherited from the bigger group. A Lie subgroup is also closed. Conversely, it is known that a closed subgroup of a Lie group is a Lie subgroup.

It is obvious that if g is an abelian Lie algebra, then any vector subspace of g is a Lie subalgebra. Clearly, any one- dimensional subspace of a Lie algebra is a Lie subalgebra.

<u>Theorem 2.6.3</u> (a) Let G be a Lie group and H a Lie subgroup. Then the Lie algebra of H, h, is a subalgebra of the Lie algebra of G, g. (b) Let G be a Lie group with Lie algebra g. Suppose h is a Lie subalgebra of g. Then there is a Lie group H whose Lie algebra is isomorphic to h and a 1-1 map of H to G.

Nonetheless, the image of H and G need not be a submanifold of G, thus H is not necessarily a Lie subgroup of G.

<u>Theorem 2.6.4</u> A closed subgroup of a Lie group is a Lie subgroup.

From the idea of the imbedding theorem, one may be tempted to think that any given Lie group is isomorphic to a Lie subgroup of GL(n,R) for sufficiently large n. But this is not to be the case. Certainly, there are Lie groups with the same Lie algebra which are mapped to some GL(n,R) by a one-to-one map (Theorem 2.5.3(b)), moreover, the image need not be a Lie subgroup as pointed out earlier.

Don't despair, as a corollary of a famous theorem of Peter and Weyl[1927] for representations of Lie groups, one is assured that a compact Lie group is isomorphic to a subgroup of some GL(n,R) for sufficiently large n. By using a corollary of the duality theorem of Tannaka [Chevalley

1946, Hochschild 1965] and the reduction theorem of Lie groups, one can sharpen the result by stating that a compact Lie group is isomorphic to some Lie subgroup of O(k) (see also, Chevalley 1946, 1951, Hochschild 1965].

Let G be a Lie group. A continuous homomorphism $\sigma : G \to GL(n,R)$, for some positive integer n, is called a <u>representation of G</u>. σ is <u>trivial</u> if the image of σ is constantly the identity matrix and σ is <u>faithful</u> if σ is one-to-one.

Roughly speaking, an application of the theorem of Peter and Weyl asserts that compact Lie groups always have at least one faithful representation.

The theory of representation is an interesting subfield of mathematics, for those who are interested in this subject, the theory of characters, orthogonality relations, theorem of Peter and Weyl, can be found in [Chevalley 1951, Pontryagin 1966]. There are many important applications of representation theory in physics such as in quantum mechanics, atomic spectra, solid state physics, nuclear and particle physics [Bargmann 1970, Hamermesh 1962, Hermann 1966, Am. Math. Soc. Translation Ser. 1, Vol 9, 1962]. We shall not go into this any further.

Let us take a closer look at closed subgroups of Lie groups. As a preliminary, we shall discuss the exponential map which is a generalization of the power series for e^x where x is a matrix.

Let G be a Lie group, and let $\mathbf{g}$ be its Lie algebra. A homomorphism $\phi: R \to G$ is called a <u>one-parameter subgroup of G</u>. Let $X \in \mathbf{g}$, then $\lambda d/dt \to \lambda X$ is a homomorphism of the Lie algebra of R into $\mathbf{g}$. Since the real line is simply-connected, then there exists a unique one-parameter subgroup $\exp_X : R \to G$ such that $d \exp_X(\lambda d/dt) = \lambda X$. That is, $t \to \exp_X(t)$ is the unique one-parameter subgroup of G whose tangent vector at 0 is X(e). The <u>exponential map</u> is defined as exp: $\mathbf{g} \to G$ by setting $\exp(X) = \exp_X(1)$.

We shall show that the exponential map for the GL(n,R) is actually given by exponentiation of matrices.

In a more geometric setting the exponential map can be defined by: Let G be a Lie group and g its Lie algebra. Let $X \epsilon g$. Let Γ_X be the integral curve of X starting at the identity. The <u>exponential map</u> exp: $g \to G$ is the map which assigns $\Gamma_X(1)$ to X, i.e., exp $X = \Gamma_X(1)$.

There is also an exponential map from the tangent space to the base manifold. Let M be a C^∞ manifold, $p \epsilon M$, the <u>exponential map</u> $\exp_p : M_p \to M$ is defined as: if $X \epsilon M_p$, there is a unique geodesic Γ such that the tangent of Γ at 0 is X. Then $\exp_p X = \Gamma(1)$. In a local sense, a geodesic is a curve of shortest distance between two points on a manifold. We will not have time and space to discuss this interesting and important topic in differential geometry. It is also a foundation of general relativity. Interested readers should consult, for instance [Burke 1985, Hicks 1971, Willmore 1959].

Theorem 2.6.5 Let $X \epsilon g$ the Lie algebra of the Lie group G. Then:

(i) $\exp(aX) = \exp_X(a)$,   for all $a \epsilon R$.

(ii) $\exp(a_1 + a_2)X = (\exp a_1 X)(\exp a_2 X)$, for all $a_1, a_2 \epsilon$ R.

(iii) $\exp(-aX) = (\exp aX)^{-1}$ for all $a \epsilon R$.

(iv)   exp : $g \to G$ is C^∞ and d exp: $g_0 \to G_e$ is the identity map, so exp gives a diffeomorphism of a neighborhood of 0 in g onto a neighborhood of e in G.

(v) $l_\sigma \cdot \exp_X$ is the unique integral curve of X which takes the value σ at 0. As a consequence, left invariant vector fields are always complete.

(vi) The one-parameter group of diffeomorphism X_a associated with the left invariant vector field X is given by $X_a = \Gamma_{\exp(a)}$.

<u>Theorem 2.6.6</u> Let ϕ: H → G be a homomorphism, then the following diagram commutes:

$$
\begin{array}{ccc}
H & -\ ^{\phi}\ \to & G \\
\exp \uparrow & & \uparrow \exp \\
h & -\ _{d\phi}\ \to & g
\end{array}
$$

In the case of groups of matrices, we shall see that the exponential map exp : $\mathbf{gl}$(n,R) → GL(n,R) for the general linear group is given by exponentiation of matrices. Let I (rather than e) denote the identity matrix in GL(n,R), let

$$e^A = I + A + A^2/2! + \ldots + A^k/k! + \ldots \qquad (2.5\text{-}1)$$

for A ϵ $\mathbf{gl}$(n,R). We want to show that the right hand side of the series converges. In fact, the series converges uniformly for A in a bounded region of $\mathbf{gl}$(n,R). For a given bounded region Ω of $\mathbf{gl}$(n,R), there is a $\mu > 0$ such that for any matrix A in Ω, $|A_{ij}| \leq \mu$ for each element (or component) of A. It follows by induction that $|(A^k)_{ij}| \leq n^{(k-1)}\mu^k$. Then by the Weierstrass M- test, each of the series of components

$$\Sigma_{k=0}^{\infty}\ (A^k)_{ij}/k!, \qquad (1 \leq i,\ j \leq n) \qquad (2.5\text{-}2)$$

converges uniformly for A in Ω. Thus the series in (2.5.1) converges uniformly for A in Ω. Let $S_k(A)$ be the k-th partial sum of the series (2.5.1), i.e.,

$$S_k(A) = I + A + A^2/2! + \ldots + A^k/k! \qquad (2.5\text{-}3)$$

and let B,C ϵ $\mathbf{gl}$(n,R). Since C → BC is a continuous map of $\mathbf{gl}$(n,R) into itself, it follows that

B ($\lim_{k \to \infty} S_k(A)$) = $\lim_{k \to \infty}$ (B$S_k(A)$).

In particular, if B is non-singular, then

B ($\lim_{k \to \infty} S_k(A)$) B^{-1} = $\lim_{k \to \infty}$ B$S_k(A)$B^{-1}.

Because BA^{k}B^{-1} = (BAB^{-1})^k, it then follows that Be^{A}B^{-1}= e^{BAB}.

From this, it is easy to show that if the eigenvalues of A are $\lambda_1, \ldots \lambda_n$, then the eigenvalues of e^A are e$^{\lambda_1}, \ldots, e^{\lambda_n}$: e$^{\lambda_1}$ is an eigenvalue of e^A follows easily if the upper left corner of A is λ_1 and the rest of the first column are zero. Likewise for the others. The determinant of e^A is e raised to the trace of A, by observing that the determinant is the product of the eigenvalues and the trace is the sum of the eigenvalues. e^A is always non-singular. If A and B commute, then $e^{A+B} = e^A e^B$. Consider the map $\phi_A : t \to tA$ of R into GL(n,R). It is smooth because e^{tA} can be represented by a power series in t with infinite radii of convergence. Its tangent vector at 0 is A (simply differentiate the power series term by term), and this map is a homomorphism, because $\phi_A(t) = e^{tA}$,

$$\phi_A(t_1 + t_2) = e^{(t_1+t_2)A} = e^{t_1 A + t_2 A} = \phi_A(t_1)\phi_A(t_2), \text{ etc.}$$

Thus $\phi_A : t \to e^{tA}$ is the unique one-parameter subgroup of GL(n,R) whose tangent vector at 0 is A. So the exponential map exp: $\mathbf{gl}(n,R) \to Gl(n,R)$ is given by exponentiation of matrices: $\exp(A) = \phi_A(1) = e^A$ where $A \in \mathbf{gl}(n,R)$. Thus this is the historical justification for the terminology.

It should be pointed out that exp : $\mathbf{g} \to G$ need not be onto even when G is connected. An example for $\mathbf{gl}(2,R)$ has been constructed to demonstrate this [Kahn 1980, p.272].

2.7 Fiber bundles

Fiber bundles provide a convenient framework for discussing the concepts of relativity, invariance, gauge transformations and group representations. Fiber bundles were originally introduced in order to formulate and solve global topological problems. Nonetheless, the notion of a fiber bundle is also very appropriate for local problems of differential geometry. The concept of induced representation of Lie groups, which is important for particle physics and field theory, can be very easily represented and explained by using fiber bundle language. The canonical formalism of classical mechanics assumes the cotangent bundle of the configuration space as the underlying manifold. Classical electrodynamics may be interpreted as the theory of an infinitesimal connection in a principal fiber bundle with structure group U(1). A similar interpretation can be given to the Yang-Mills fields, and in general, to all fields resulting from "gauge transformations".

Bundles are a generalization of the concept of Cartesian product. An example from the evolution of the framework of physics can clarify as well as provide the need for such a generalization.

In Aristotelian physics, space and time are considered to be absolute, i.e., every physical event being defined by an instant of time and a location in space. This is equivalent to say that space-time M is a Cartesian product of T (time) and S (space). In Galilean (or Newtonian) physics, the time remains absolute, but space is not. This can be described by saying that there is a projection map π : M $\to$ T which to any event p ϵ M associates the corresponding instant of time t = π(p). T is called the <u>base space</u>. The inverse image of t, $\pi^{-1}(t)$, is called a <u>fiber</u>. Each fiber is isomorphic to R^3, is therefore called a <u>typical fiber</u>. This is the usual spatial part of the space-time, where the Galilean transformations (translations and rotations) map a point on a typical fiber to another point on the same fiber. And the Galilean invariant quantities make sense. In special relativity, neither the space nor the time is absolute. Nonetheless, the bundle of linear frames of the space-time is a product bundle. But in general relativity, it is so complicated that only the principal bundle of the bundle of frames of the space-time is a product bundle.

The word "global" came as a specific mathematical notion when differential geometry broke away from its historical confinement as a "local" discipline. The contrast between the local and global aspects of differential geometry is perhaps best illustrated by the analytic tools that are being used for studying local and global aspects of surfaces or manifolds in general.

The analytic machinery of local differential geometry manifests itself as a collection of differential expressions

83

relating to local geometric properties in the neighborhood of points, lines, surfaces or manifolds in general. In global differential geometry, by contrast, one is interested in the integrals of those local differential relations and questions arise whether those integrals exist and whether they are unique.

For lower dimensional spaces ($n \leq 3$) there is a well balanced relationship between local and global results in differential geometry (e.g., Gauss-Bonnet theorem). However, for higher dimensional space, it is much more difficult to generalize the local results to global ones.

In Newtonian or Poincaré-invariant theories the space-time is considered to be given a priori, and the physical dynamics are defined on this background. In general relativity, the topological and geometric structure of space-time is to be established as part of the dynamics. Global structures place restrictions on the class of differentiable manifolds suitable for space-times. Global structures are presented in mathematical form but their imposition is usually based upon physical intuition and on observations. Once they are imposed, however, the resultant class of admissible spaces further clarifies the significance of any given global structure and could even lead to its rejection.

Before getting into formal definitions and major results, let us describe the concepts we shall encounter in an intuitive way. We have defined the tangent space of a manifold M at a point $p \in M$, i.e., M_p. The <u>tangent bundle of M</u>, denoted by TM, is defined as the union of all tangent spaces of M, i.e., TM = U M_p for all $p \in M$. A <u>vector bundle of a manifold M</u> is a family of vector spaces V each attached to a point of M such that locally the vector bundle is homeomorphic to U x V where U is a neighborhood of $p \in M$. The <u>principal bundle of M</u> with structural group G is locally homeomorphic to the attachment to each point in M a different copy of G, i.e., the bundle is locally homeomorphic to U x G where U is a neighborhood of $p \in M$. We

say a bundle is <u>trivial</u>, we mean that instead of the bundle
space "is locally homeomorphic to" by "is homeomorphic to".

In the following, we shall discuss fiber bundles on a
smooth manifold with smooth mappings. These notions are very
useful for our discussion as well as for differential
geometry. But it should be pointed out that fiber bundles
can be defined on topological manifolds only with all maps
continuous [Steenrod 1951].

For example, let B, X and F are topological manifolds
and let a continuous map π : B → X of B onto X called the
projection and B be the <u>bundle space</u> and X the <u>base space</u>.
The B is a <u>fiber bundle over X with fiber F</u>, and projection
π if for every p ϵ X, $\pi^{-1}(p)$ is homeomorphic to F, and there
exists a neighborhood U of p and a homeomorphism ϕ_U : B_U →
UxF where $B_U = \pi^{-1}(U)$ such that the following diagram
commutes

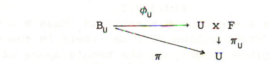

where π_U is an obvious projection.

A C^∞ <u>principal fiber bundle</u> is a set (B, M, G) where B,
M are C^∞ manifolds, G a Lie group

(i)   G acts freely (and C^∞) to the right on B, i.e., B
x G → B defined by (b,g) → bg = $R_g b \epsilon$ B where b ϵ B, g ϵ G.

(ii)   M is the quotient space of B by equivalence under
G and the projection π : B → M is C^∞, so for p ϵ M, G is
simply transitive on $\pi^{-1}(p)$ where $\pi^{-1}(p)$ is a fiber over p ϵ
M.

(iii) B is locally trivial, i.e., for any p ϵ M, there
is a neighborhood U of p and C^∞ map F_U : $\pi^{-1}(U)$ → G such that
F_U commutes with R_g for every g ϵ G and the map of $\pi^{-1}(U)$ →
UxG given by b → $(\pi(b), F_U(b))$ is a diffeomorphism. Here B is
called the <u>bundle space</u>, M the <u>base space</u>, and G the
<u>structural group</u>. Note, the fibers $\pi^{-1}(p)$ are diffeomorphic
to G in a special way, i.e., via the map b : G → $\pi^{-1}(\pi(p))$ ⊂
B defined by b(g)= $R_g b$. This definition can be graphically

85

illustrated by the following figure (Fig.2.7.1).

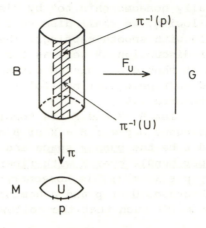

Fig.2.7.1

As before, G a Lie group, M a manifold, then B = M X G provided with the right action of G on itself in the second factor, i.e., (p,g)h = (p,gh), is the bundle space of a principal bundle which is called the <u>trivial bundle</u>. A <u>cross-section</u> of a bundle is a continuous map s : M → B such that $\pi \cdot s$ = id on M (i.e., $\pi(s(p))$ = p for all p $\in$ M).

Let us discuss some examples of bundles: (i) From the above pictorial representation of a fiber bundle, we can let the base space be S^1, the fiber being a line segment [0,1], and the bundle space being S^1 x [0,1]. Clearly, this bundle is a trivial bundle. (ii) For a flexible rectangular sheet, we hold one side of the sheet fixed and twist the other side through 180° and then identify (glue together) these two opposing sides. Then the resulting two-dimensional space is a bundle space, and is the well-known Möbius band. The base space is the circle (S^1), the fiber is a line segment [0,1], and the group action is the "twist" (Z_2). Aside from the "seam", it is clear that the inverse of the projection map of a neighborhood of a point on the circle is a small rectangle (i.e., locally a product bundle). But clearly,

86

this fiber bundle is not a trivial bundle.

Theorem 2.7.1 A bundle is isomorphic (or homeomorphic) to a trivial bundle iff there is a C^∞ (or continuous) cross-section.

The above theorem means that a trivial bundle has global cross-sections. The two simple examples illustrate exactly this point. Both bundles have local cross-sections, but only the first one has a global cross-section. By the way, the global cross- section, if it exists, is homeomorphic to the base space.

Let M^m be a C^∞ manifold, B be the set of $(m+1)$-tuples $(p,e_1,..,e_m)$, where $p \in M$, and $\{e_1,..,e_m\}$ is a basis of M_p, and let $\pi : B \to M$ be defined by $\pi(p,e_1,..,e_m) = p$. Let $g \in GL(m,R) = G$, then $GL(m,R)$ acts to the right on B by $R_g(p,e_1,..,e_m) = (p, \Sigma g_{i1}e_i,.., \Sigma g_{im}e_i)$ where g is viewed as a matrix $g = (g_{ij})$. Let $(x_1,..,x_m)$ be a coordinate system defined in a neighborhood U of p, Then F_U is defined by letting $F_U(p',f_1,...,f_m) = (dx_j(f_i)) = (g_{ij}) \in GL(m,R)$, where $p' \in U$. Using the C^∞ structure given to B by the local product representation (π,F_U), we see that B is the bundle space of a principal bundle, called the <u>bundle of bases</u> of M, B(M).

For a classical notion of fiber bundles or tangent bundles one can find in Anslander and MacKenzie [1963]. It is also convenient at times to view B as the set of nonsingular linear transformations of R^m into the tangent spaces of M, i.e., we identify $b = (p,e_1,..,e_m)$ with the map $b : (r_1,..,r_m) \to \Sigma r_i e_i$. If this is done, it is natural to consider $GL(m,R)$ as the nonsingular linear transformations of R^m, where $bg(r_1,..,r_m) = \Sigma_{i,j} r_i g_{ji} e_j = \Sigma_j (\Sigma_i r_i g_{ji}) e_j = b(g(r_1,..,r_m))$. In other words, bg (as a map) = b (as a map)$\cdot$g.

If G is a Lie group, H a closed subgroup, then there is a principal bundle with base space G/H (left cosets), bundle space G, and the structure group H such that $\pi: G \to G/H$ is the projection and right action is given by $(g,h) \to gh$. Thus a homogeneous space is an example of a principal fiber bundle. [Helgason 1962, Steenrod 1951].

Let (B,M,G) be a principal bundle and let F be a manifold on which G acts to the left. The <u>fiber bundle associated to (B,M,G) with fiber F</u> is defined as follows: Let P'= B x F, andconsider the right action of G on P' by (b,f)g = (bg,g^{-1}f) where b ϵ B, f ϵ F, g ϵ G. Let P = P'/G, the quotient space under equivalence by G, then P is the bundle space of the associated bundle. The projection π': P → M is defined by π'((b,f)G) = π(b). For p ϵ M, we take a neighborhood U of p as in (iii) of the definition of a principal bundle, with F$_U$: π^{-1}(U) → G. Likewise we have F$_U$': π'(U) → F by F$_U$'((b,f)G) = F$_U$(b)f so that (π')$^{-1}$(U) is homeomorphic to U x F, and define P as a manifold by requiring these homeomorphisms to be diffeomorphisms. Thus π' and the projection ϕ: P'→ P are C$^\infty$.

First of all, (B,M,G) as above, let G act on itself by left translation, then (B,M,G) is the bundle associated to itself with fiber G.

Let us look at a tangent bundle as a bundle associated to the bundle of basis B(M) with fiber R^m. Since GL(m,R) is the group of nonsingular linear transformations of R^m, and hence act on R^m to the left. The bundle space of the associated bundle with fiber R^m is denoted by TM and it is called the <u>tangent bundle of M</u>. TM can be identified with the space of all pairs (p,t) where p ϵ M, t ϵ M$_p$ as follows:

$$((p,e_1,..,e_m),(r_1,..,r_m))GL(m,R) \rightarrow (p, \Sigma\ r_i e_i).$$

Hence the fiber of TM above p ϵ M may be viewed as the linear space of tangents at p, i.e., M$_p$, and TM as the union of all the tangent spaces together with a manifold structure. Moreover, the coordinates of TM can easily be adopted by letting U be a coordinate neighborhood in M with coordinates x$_1$,..,x$_m$. Define coordinates y$_1$,..,y$_{2m}$ on (π')$^{-1}$(U) in such a way that if (p,t) ϵ (π')$^{-1}$(U), then y$_i$(p,t) = x$_i$(p), y$_{m+i}$(p,t) = dx$_i$(t) where i = 1,..,m. Clearly, a C$^\infty$ vector field may be regarded as a cross- section of π'. For more on tangent bundles see [Bishop and Crittenden 1964, Yano and Ishihara 1973, and other modern differential geometry books]. It is interesting to note the following theorem.

Theorem 2.7.2   TM is orientable even if M is not.

When R^m, in the tangent bundle of M, is replaced by a vector space constructed from R^m via multilinear algebra, i.e., the tensor product of R^m and its dual with various multiplicities, we get a tensor bundle. A cross-section which is C^∞ on an open set is called a C^∞ tensor field, and the type is given according to the number of times R^m and its dual occur. The structural group of a tensor bundle is, of course, GL(m,R) and it acts on each factor of the tensor product independently. GL(m,R) acts on R^m as with the tangent bundle, and it acts on the dual via the transpose of the inverse, i.e., if $v \in R^{m*}$ = the dual of R^m, $x \in R^m$, $g \in$ GL(m,R), then $gv(x) = v(g^{-1}x)$. Clearly, TM is a special case of a tensor bundle; this is similar to that a tangent vector is a special tensor, a contravariant tensor of rank one.

Vector bundles, which we shall encounter later, in which the fiber is a vector space and they are frequently defined with no explicit mention made of the structural group (although often it is a subgroup of the general linear group of the vector space). It is usually defined as the union of vector spaces, all of the same dimension, each associated to an element of the base space and defining the manifold structure via smooth, linearly independent cross-sections over a covering system of coordinate neighborhoods. In fact, we did this for TM, a special case of a vector bundle.

The quotient space bundle of an imbedding, sometimes considered as a normal bundle for Riemannian manifold, may be defined as follows: Let i: N → M be the imbedding of the submanifold N in M. The fiber over $q \in N$ is the quotient space $M_{i(q)}/di(N_q)$, and the bundle space is the union of these fibers, so the bundle space can be considered as the collection of pairs $(q, t+di(N_q))$, where $t \in M_{i(q)}$.

The Whitney (or direct) sum of two bundles ξ = (B,M,G,π,F) and $\zeta = (B',M,G',\pi',F')$ over the same base space M is the bundle $\xi \oplus \zeta$ whose fiber over $x \in M$ is $F \oplus F'$. If

ϕ, ψ are charts for ξ, ζ over U respectively, a chart ρ for $\xi \oplus \zeta$ over U is $\rho_x = \phi_x \oplus \psi_x : F + F' \to R^m \oplus R^n$. Thus $\dim(\xi \oplus \zeta) = \dim B + \dim B'$.

Let two bundles ξ and ζ over the same base space M with $B \subset B'$, then ξ is a <u>sub-bundle</u> of ζ if each fiber F is a sub-vector-space of the corresponding F'.

<u>Lemma 2.7.3</u> Let ξ and ζ be sub-bundle of η such that each vector space $F(\eta)$, fiber of η, is equal to the direct sum of the subspaces F amd F'. Then η is isomorphic to the Whitney sum $\xi \oplus \zeta$.

Then the question arises, given a sub-bundle $\xi \subset \eta$ does there exist a complementary sub-bundle so that η splits as a Whitney sum? If η is provided with a Euclidean metric (provided the base space is paracompact) then such a complementary summand can be constructed by letting $F(\xi^\perp)$ be the subspace of $F(\eta)$ consisting of all vectors v such that $v \cdot \omega = 0$ for all $\omega \in F(\xi)$. Let $B(\xi^\perp) \subset B(\eta)$ be the union of all $F(\xi^\perp)$. One can show that $B(\xi^\perp)$ is the total space of a sub-bundle $\xi^\perp \subset \eta$. Moreover, η is isomorphic to the Whitney sum $\xi \oplus \zeta^\perp$ [Milnor and Stasheff 1974]. Here $\xi^\perp$ is called the <u>orthogonal complement of ξ</u> in η.

Suppose $N \subset M$ are smooth manifolds and M is provided with a Riemannian matric. Then the tangent bundle TN is a sub-bundle of the restriction TM|N. Then the orthogonal complement $TN^\perp \subset TM|N$ is the <u>normal bundle of N in M</u>, i.e.,

$\perp(N) = \{(q,t) \in TM | t \in M_q$ for some $q \in N$ and $t \perp N_q\}$. We would like to mention that the notion of a normal bundle is not only useful in differential geometry (such as in discussing geodesics and completeness of the Riemannian manifold) but also very useful in algebraic and differential topology (such as using the Whitney duality theorem to relate the immersibility of a n-dim manifold in R^{n+k}). Interested readers may want to consult the following books: Steenrod [1951], Milnor and Stasheff [1974]. Now let us get back to some properties of tangent bundles.

<u>Theorem 2.7.4</u> [Steenrod 1951] The tangent bundle to a differentiable manifold admits a nonzero cross-section and

is equivalent to the existence of a nowhere zero vector field on M iff the Euler characteristic of M is zero.

As we have defined earlier, the Euler characteristic of a manifold M is defined in Section 2.5. For example, we have pointed out that (i) for a noncompact manifold, its Euler characteristic vanishes; (ii) for compact manifolds, only odd dimensional ones have a vanishing Euler characteristic. Thus, only odd dimensional spheres have a nowhere zero vector field. One can convince oneself that there does not exist a nowhere zero tangent vector on 2-dim sphere. Milnor [1965] gives an interesting and illuminating view on this.

When TM^m is homeomorphic to $M^m \times R^m$, it is a trivial bundle. In such a case, one says that the manifold has a trivial tangent bundle, or the manifold is <u>parallelizable</u>.

Although all odd-dimensional spheres have a nowhere zero vector field, but it is a deep result [Bott and Milnor 1958, Adams 1962] that only S^1, S^3, and S^7 have trivial tangent bundle (i.e., they are the only parallelizable n-spheres).

Normally, some heavy machinery in algebraic topology such as characteristic classes [e.g., Milnor and Stasheff 1974, Steenrod 1951, Husemoller 1975] and obstructions [e.g., Milnor and Stasheff 1974, Steenrod 1951] are needed to prove the following theorem. We shall relate the historical origin of the term parallelizable manifold with a non-zero vector field.

Let us assume M is parallelizable, i.e., $TM = M \times R^m$. Thus M has m linearly independent tangent vector fields $t_i(p) = (p, (0, .., 1, .., 0))$ with 1 in the ith place, for any $p \in M$. In other words, at any point the m vectors $t_i(p)$ are a basis for $\pi^{-1}(p) \subset TM$. Thus, a nonzero vector $v \in \pi^{-1}(p)$ can be expressed by $v = \Sigma \alpha_i t_i(p)$, then one can transport it parallel to itself over the entire manifold to obtain a nowhere zero vector field by setting $v(p) = \Sigma \alpha_i t_i(p)$ for all $p \in M$. This is the global notion of parallel transport in such a manifold.

From the above demonstration, it is almost trivial that if M is parallelizable, it has a global C^∞ base field.

Theorem 2.7.5   A manifold M^m is said to be
parallelizable, (i.e., has a trivial tangent bundle TM = M x
R^m) iff M admits a global C^∞ base vector field [e.g., Milnor
and Stasheff 1974].

Theorem 2.7.6 (generalization of a theorem due to
Cartan) Any connected Lie group G is topologically a product
space H X E where H is a compact subgroup of G and E is a
Euclidean space.

E.g., In physics, the proper Lorentz group L_o, or
SO(3,1), $\approx$ SO(3) x R^3, and its universal covering group
SL(2,C) $\approx$ S^3 X R^3.

If (B,M,G) is a principal bundle, H a subgroup of G,
then G is reducible to H iff there exists a principal bundle
(B',M,H) which admits a bundle map f: (B',M,H) → (B,M,G)
such that f_M is the identity map on M, f_B is one-to-one, and
f_G is the inclusion map H $\hookrightarrow$ G.

Theorem 2.7.7 [Steenrod 1951] If (B,M,G) is a principal
bundle, H a maximal compact subgroup of G, then G can be
reduced to a bundle with structure group H.

Corollary 2.7.8 Every principal bundle with GL(m,R) as
the structure group, e.g., bundle of bases B(M), can be
reduced to a bundle with the structural group being the
orthogonal group O(m).

The reduced bundle with O(m) as structural group is
called the bundle of orthonormal bases and denoted by O(M).

Many modern differential geometry books have at least
one or two chapters covering fiber bundles. For instance,
Bishop and Crittenden [1964], Helgason [1962], Kobayashi and
Nomizu [1963]. For specific details in tangent and cotangent
bundles, see Yano and Ishihara [1973]. For more advanced
readers, the classic by Steenrod [1951] is highly
recommended. For a more modern and broader treatment,
Husemoller [1975] is also recommended.

2.8   Differential forms and exterior algebra
Tensor analysis is part of the usual mathematical

repertoire of a physicist or engineer. Differential forms
are special types of tensors. Yet, its utility and
conceptual implications are far beyond the capabilities of
tensors. Not only does it provide more compact formulations
of electrodynamics, Hamiltonian mechanics, etc. and simpler
mathematical manipulations, but it also provides topological
implications. In this section, we shall briefly define and
discuss some properties of differential forms and exterior
algebra, and illustrate its power. It is very tempting to
briefly discuss de Rham cohomology theorem. Once again, the
reader is urged to consult those differential geometry books
we have just mentioned earlier for further details.

For $p \in M$, the dual vector space M_p^* of M_p is called the
<u>cotangent space</u> (or the space of covectors at p). An
assignment of a covector at each p is called a <u>one-form</u>. If
$(u^1,..,u^n)$ is a local coordinate system in a neighborhood of
p, then $du^1,..,du^n$ form a basis for M_p^*, and they are the
dual basis of the basis $\partial/\partial u^1$, ..., $\partial/\partial u^n$ of M_p. So in a
coordinate neighborhood, a 1- form can be written as $\alpha = \Sigma_i$
$f_i du^i$. Clearly, α is C^∞, if f_i's are. Note, one-form can also
be defined as an $F(M)$ linear mapping of the $F(M)$-module $X(M)$
into $F(M)$. That is,
$$(\alpha(X))_p = \langle \alpha_p, X_p \rangle \text{ , where } X \in X(M), p \in M.$$
The <u>exterior product</u> is defined by $A \wedge B = (A \times B)_a$,
here a denotes that it is antisymmetrized, where A and B are
skew-symmetric, covariant tensors. It has the following
properties:

    (a) associativity: $(A \wedge B) \wedge C = A \wedge (B \wedge C)$,
    (b) distributivity: $(A + B) \wedge C = A \wedge C + B \wedge C$,
    (c) anticommutativity: If A is of degree p, and B is of
        degree q, then $A \wedge B = (-1)^{pq} B \wedge A$.

Of course, together with addition and scalar multiplication
operations they form the algebra.

An <u>r-form</u> can be defined as a skew-symmetric r-linear
mapping over $F(M)$ of $X(M) \times ... \times X(M)$ (r-times) into $F(M)$. If
$\alpha_1,..,\alpha_r$ are 1-forms, $X_1,..,X_r \in X(M)$, then
$$(\alpha_1 \wedge \alpha_2 \wedge ... \wedge \alpha_r)(X_1,..,X_r)$$

$$= \Sigma_{(i_1 \ldots i_1)} \; \text{sign}(i_1, \ldots, i_r)(\alpha_{i_1}(X_{im})) \ldots (\alpha_{ip}(X_{iq})).$$

For instance, if $r = 2$, α, $\beta \in \Lambda^1(M)$, X, $Y \in X(M)$ then

$$(\alpha \wedge \beta)(X, Y) = \begin{vmatrix} \alpha(X) & \beta(X) \\ \alpha(Y) & \beta(Y) \end{vmatrix}$$

In general, if α is a p-form, β is a q-form on M, and $X_1, \ldots, X_{p+q} \in X(M)$, then

$(\alpha \wedge \beta)(X_1, \ldots, X_{p+q})$

$= \Sigma_{(i \ldots i)} \; \text{sign}(i_1 \ldots i_{p+q}) \; \alpha(X_{i}, \ldots, X_{i}) \; \beta(X_{i}, \ldots X_{i}).$

We shall denote $\Lambda^p(M^*)$ to be the collection of all p-forms on M. When there is no danger of ambiguity, we usually will denote it by Λ^p. We shall now introduce a derivative called an exterior derivative.

Let M be C^∞. The <u>exterior derivative</u> is a map: $d: \Lambda \rightarrow \Lambda$ with the following properties:

    (i)   For $q = 0$, f is a C^∞ function, then df is a 1-form;

    (ii)   d is a linear map such that $d(\Lambda^q) \subset \Lambda^{q+1}$;

    (iii) For $\alpha \in \Lambda^q$, $\beta \in \Lambda^s$

        $d(\alpha \wedge \beta) = d\alpha \wedge \beta + (-1)^q \alpha \wedge d\beta$;

    (iv) $d(df) = 0$ and it follows that $d(d\alpha) = 0$ for any α.

Any form α is said to be <u>closed</u> if $d\alpha = 0$; if α is a p-form and there is a (p-1)-form β such that $\alpha = d\beta$, then we call α an <u>exact</u> p-form. From (iv) above, clearly every exact p-form is also closed. Let us denote Λ_e^p be the space of exact p-forms, Λ_c^p be the space of closed p-forms. Clearly, $\Lambda_e^p \subset \Lambda_c^p \subset \Lambda^p$.

$D^p(M) \equiv \Lambda_c^p / \Lambda_e^p$ is called the <u>p-dimensional de Rham cohomology group of M</u> obtained by using differential forms. Clearly D^p is an abelian group because Λ_e^p and Λ_c^p are themselves abelian.

    <u>Theorem 2.8.1 (de Rham)</u> For a paracompact and T_2 differentiable manifold M, $D^p(M) \approx H^p(M,R)$.

    Remark: Thus D^p is a topological invariant. It also follows immediately that from properties (iii) and (iv)

closed form Λ closed form = closed form, closed form Λ exact
form = exact.

Another way to define or to determine whether a manifold
is orientalble or not is the following useful theorem.

Theorem 2.8.2 A paracompact, T_2, C^∞ manifold M of dim m
is orientable iff M admits a continuous, nonvashing globally
defined n-form (for instance, a volume element).

In physics, a space-time M is a 4-dim, connected,
paracompact, C^∞-manifold with Lorentz signatures. The
orientability of a physical "space-time" is concerning the
consistent assignment of an "arrow of time" and of
"handedness" throughout the "space-time".

The strong principle of equivalence [Dicke 1959, 1962]
asserts that local experiments should be the same throughout
the space-time. From the strong principle of equivalence,
some local experiments in particle physics, C, P (β-decay),
and CP(K_L, K_s decays) noninvariance and T noninvariance, and
CPT invariance. We can conclude that the space-time must be
orientable.

Remark: Assuming space-time being orientable is also
convenient; otherwise we can always find its universal
covering space which is orientable.

Now we shall introduce a linear mapping which is also an
antiderivation.

Let X ϵ X(M) of a differentiable manifold M. A linear
mapping (inner product) i(X) : $\Lambda \to \Lambda$ such that

(i)   i(X) : $\Lambda^p \to \Lambda^{p-1}$, p $\geq$ 1; and i(X)(f) = 0;

(ii)   i(X) is an antiderivation, i.e., if $\theta \epsilon \Lambda^p$, $\phi \epsilon$
Λ^q, then i(X)($\theta \Lambda \phi$) = (i(X)θ) $\Lambda \phi$ + $(-1)^p\theta \Lambda$ (i(X)ϕ);

(iii)   if $\theta \epsilon \Lambda^1$, i(X)$\theta$ = θ(X).

Lie derivative is a geometric procedure which is very
useful in finding symmetries, such as Killing vector field :
$\mathcal{L}_X g = 0$. We shall define and state the properties of Lie
derivatives.

$\mathcal{L}_X Y \equiv [X,Y]$, also called the "dragging" of Y along X
(more appropriately, along the integral curve of X), as
indicated in the following diagram.

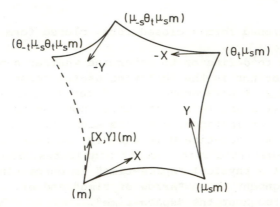

here $\{\mu_s\}$, $\{\theta_t\}$ are one parameter groups of X & Y.
If $\pounds_X Y = [X,Y] = 0$ then $(\theta_{-t}\ \mu_{-s}\ \theta_t\ \mu_s\ m) = m$.

Proposition 2.8.3 Let X, Y, Y_o, ..., $Y_p \in X(M)$, f $\in$
F(M), $\alpha \in \Lambda^p(M)$, then:

(i) $\pounds_X f\ =\ X(f)$

(ii) $\pounds_X\ =\ i(X)\cdot d\ +\ d\cdot i(X)$

(iii) $\pounds_X$ commutes with d

(iv) $\pounds_{Y_o}(\alpha(Y_1,..,Y_p))\ =\ (\pounds_{Y_o}\alpha)(Y_1,..,Y_p)$
       $+\ \Sigma^p_{i=1}\ \alpha(Y_1,..,Y_{i-1},\ \pounds_Y\ Y_i,Y_{i+1},..,Y_p)$.

(v)   d $\alpha(Y_o,..,Y_p)\ =\ \Sigma^p_{\ i=0}\ (-1)^i Y_i\ \alpha(Y_o,..,\hat{Y}_i,...,Y_p)$
       $+\ \Sigma_{i<j}\ (-1)^{i+j}\ \alpha([Y_i,Y_j],Y_o,..,\hat{Y}_i,...\hat{Y}_j,..Y_p)$.

Symplectic manifolds

Symplectic manifold is a very natural framework in which
to discuss classical mechanics, in particular, the
Hamiltonian systems. Here we shall very briefly define and
discuss symplectic manifolds and Hamiltonian systems.

Definition 2.8.4 A volume on n-dim M is an n-form $\Omega \in$
$\Lambda^n(M)$ such that $\Omega(p) \neq 0$ for all p $\in$ M. M is called
orientable iff there is a volume on M.

Definition 2.8.5 Let $\alpha \in \Lambda^2(M)$, α is non-degenerate iff
$\alpha(e_1,e_2) = 0$ for all $e_2 \in X(M)$, thus $e_1 = 0$.

Proposition 2.8.6 α is nondegenerate on M iff M is
even-dim (say 2n) and $\alpha^n = \alpha \wedge \alpha \wedge \ldots \wedge \alpha$ is a volume on

96

M. Thus if α is nondegenerate, then M is orientable.

Definition 2.8.7   A <u>symplectic form</u> (structure) on an even- dim M is a non-degenerate, closed 2-form α on M. A <u>symplectic manifold</u> (M, α) is an even-dim M together with a symplectic form α.

Theorem 2.8.8: (Darboux)   Suppose α is a non-degenerate 2-form on 2n-dim M. Then $d\alpha = 0$ iff there is a chart (U, ϕ) at each $p \in$ M such that $\phi(p) = 0$ and with $\phi(u) = (x^1(u),$ ..., $x^n(u), y^1(u), ..., y^n(u))$, we have $\alpha|U = \Sigma^n_{i=1}$ $dx^i \wedge dy^i$.

Remark: The charts guaranteed by the theorem are called symplectic charts and x^i, y^i are canonical coordinates. Thus, in a symplectic chart $\alpha = \Sigma$ $dx^i \wedge dy^i$ and $\Omega_\alpha = dx^1 \wedge$ ... $\wedge dx^n \wedge dy^1 \wedge ... \wedge dy^n$ is the volume element.

Remark: In most mechanics problems, the phase space is T^*M and the coordinates are $(q^1, ..., q^n, p_1, ..., p_n)$ then the canonical form on T^*M is
$\theta = \Sigma^n_{i=1}$ $p_i dq^i$ and $\alpha = \Sigma^n_{i=1}$ $dq^i \wedge dp^i$.

Proposition 2.8.9 T^*M of any manifold is orientable.

Theorem 2.8.10   Lie derivative of X, restricted to $\Lambda(M^*)$ is a derivation, i.e., ($\mathfrak{L}_x: \Lambda^P \to \Lambda^P$), $\mathfrak{L}_x = i(X)$ $d + d \cdot i(X)$.

Example: In electrodynamics, $f \in \Lambda^2(M)$ and in the symplectic chart, $f = 1/2$ F_{uv} $dx^u \wedge dx^v$. The Maxwell equations can be written as: $df = 0$, in terms of symplectic coordinate chart, it can be written as $F_{\mu v, \alpha} + F_{v\alpha, \mu} + F_{\alpha\mu, v} = 0$. In the usual vector notation, it is equivalent to the following set: div $\mathbf{B} = 0$ and curl $\mathbf{E} = -$ d$\mathbf{B}$/dt. Now define the star operator $* : \Lambda^P \to \Lambda^{n-p}$, where dim M = n and the Hodge's operator $\delta = d$ $*$, then the other set of Maxwell equations can be represented by: $\delta f \equiv d*f = 4\pi$ $*J$. It is equivalent to: div $\mathbf{E} = 4\pi q$ and curl $\mathbf{B} = $ d$\mathbf{E}$/dt $+ 4\pi \mathbf{J}$. For source-free, then $F^{\mu v}{}_{,v} = 0$, i.e., f is a harmonic 2-form (f is closed and coclosed). The existence of vector potential is equivalent to the non-existence of a magnetic monopole.

Definition 2.8.11   (M, α) and (N, β) be symplectic. A C^∞ map F : M $\to$ N is <u>symplectic</u> (or <u>canonical transformation</u>) iff $F_* \beta = \alpha$ where F_* is the pullback of forms, i.e., F_*:

$\Lambda^k(N) \to \Lambda^k(M)$.

   Theorem 2.8.12   If F: $M \to N$ is symplectic, then F is volume preserving, and F is a local diffeomorphism.

   Definition 2.8.13   (M, α) is symplectic. f, g ϵ F(M). X_f, X_g ϵ X(M). The <u>Poisson bracket</u> of f and g is the function $\{f,g\} \equiv -i_{X_f} i_{X_g} \alpha$.

   Proposition 2.8.14   As above:

   (i) $\{f,g\} = - \mathfrak{L}_{X_f} g = \mathfrak{L}_{X_g} f$,

   (ii) $d\{f,g\} = \{df,dg\}$

   (iii) $X_{\{f,g\}} = -[X_f, X_g]$

   (iv)   The real vector space F(M) and the composition { , } form a Lie algebra.

   (v)   (U, ϕ) be a symplectic chart, then the Poisson bracket has the usual form,

   $\{f,g\} = \Sigma^n_{i=1} [(\partial f/\partial x^i)(\partial g/\partial y^i) - (\partial f/\partial y^i)(\partial g/\partial x^i)]$.

See for instance, Goldstein [1950].

   Hamiltonian systems

   Definition 2.8.15   M be a manifold, X ϵ X(M). Let α ϵ $\Lambda^k(M)$. We call α an <u>invariant k-form of X</u> iff $\mathfrak{L}_X \alpha = 0$.

   Definition 2.8.16   Let (M, α) be symplectic and X ϵ X(M). We say that X is <u>locally Hamiltonian</u> iff α is an invariant 2-form of X, i.e., $\mathfrak{L}_X \alpha = 0$. The set of locally Hamiltonian vector fields on M is denoted by $X_{L.H.}(M)$.

   Proposition 2.8.17 (i)   Let X ϵ $X_{L.H.}(M)$ on 2n-dim (M, α). Then α, α^2, ..., α^n are invariant forms of X, (ii) $X_{L.H.}(M)$ is a Lie subalgebra of X(M).

   Proposition 2.8.18   The following are equivalent:

   (i)   X ϵ $X_{L.H.}(M)$,

   (ii) $i_X \alpha$ is closed,

   (iii)   U a neighborhood of p ϵ M and H ϵ F(U) such that
          $X|U = X_H$.

   Definition 2.8.19   Let X ϵ $X_{L.H.}(M)$. A function H as above is called a <u>local Hamiltonian of X</u>.

   Proposition 2.8.20   H a local Hamiltonian of X. Then H is constant along the integral curves of X in U (i.e., H is a constant of motion, or "energy" is conserved). This is because $\mathfrak{L}_X H = \mathfrak{L}_X H = \{H,H\} = 0$.

<u>Proposition 2.8.21</u> Let X ∈ $X_{L.H.}(M)$, H ∈ F(U) be a local Hamiltonian on a symplectic manifold (M, α). Let V ⊂ U, (V, φ) be a symplectic chart with φ(V) ⊂ R^{2n}, φ(v) = ($q^1(v),..,q^n(v)$, $p_1(v)..p_n(v)$). Then a curve c(t) on V is an integral curve of X iff
$$(dq^i/dt)c(t) = (\partial H/\partial P_i)c(t), \qquad\qquad i = 1,...,n$$
and $(dp_i/dt)c(t) = -(\partial H/\partial q^i)c(t), \qquad i = 1,...,n$
where $q^i(t) = q^i(c(t))$, $p_i(t) = p_i(c(t))$.

<u>Proposition 2.8.22</u> X, H, (M, α) as above. Then for f ∈ F(U), we have $f_X f = \{f,H\}$ on U.

Before ending this subsection, for our future use as well as for general interest, we are going to discuss briefly the Kolmogorov-Arnold-Moser theorem. This theorem was originally proved for Hamiltonian of two degrees of freedom, and it was later extended to n degrees of freedom systems with (2n - 2)- dimensional Poincaré maps. For simplicity, here we shall only state the theorem for two degrees of freedom. Let us consider the Hamiltonian H^ϵ is a small perturbation of an integrable Hamiltonian H^0. For simplicity, let us assume that the perturbed Hamiltonian has the following form:
$$H^\epsilon(q,p,\theta,I) = F(q,p) + G(I) + \epsilon H^1(q,p,\theta,I),$$
where H^1 is 2π periodic in θ and the unperturbed system $H^0(q,p,\theta) = F(q,p) + G(I)$ decouples directly into two independent systems with integrals F and G. Furthermore, we assume nondegeneracy, i.e., dG/dI ≠ 0. This implies that for ε small, H^ϵ is invertible and can be solved for I. By transforming (q,p) to a second set of action angle variables J and φ, and the unperturbed system H^0 becomes
$$dJ/dt = 0, \qquad d\phi/dt = 2\pi T,$$
and it associates with the Poincaré map P_0. And the perturbed system modifies the Poincaré map to P_ϵ. The KAM theorem asserts that for sufficiently small ε, most of the closed curves J = constant of P_0 are preseved for P_ϵ. In other words, the Poincaré map under consideration is an area preserving diffeomorphism.

<u>Theorem</u> (KAM) If an unperturbed Hamiltonian system is

nondegenerate and ϵ is sufficiently small, then the
perturbed map P_ϵ has a set of invariant closed curves of
positive Lebesgue measure $\mu(\epsilon)$ close to the original set $J = J^\alpha$; moreover, $\mu(\epsilon)/\mu(J) \to 1$ as $\epsilon \to 0$. The surviving
invariant closed curves are filled with dense irrational
orbits.

Instead of these technical versions, we can put the KAM
theorem in the following words:

<u>Theorem</u> (KAM) If an uperturbed system is nondegenerate,
then for sufficiently small conservative Hamiltonian
perturbations, most non-resonant invariant tori do not
vanish, but are only slightly deformed, so that in the phase
space of the perturbed system, there are invariant tori
densely filled with phase curves winding around them
conditionally periodically too, with the number of
independent frequencies equal to the number of degrees of
freedom. These invariant tori form a majority in the sense
that the measure of the complement of their union is small
when the perturbation is small.

As we shall see, the major part of these lectures will
be concerned with dissipative dynamical systems and with the
structure of the nonwandering and attracting sets occuring
in such systems. Nonetheless, it is often very useful to
consider such systems as perturbations of Hamiltonian
systems, this is because the existence of energy integrals
or other constants of motion enables us to obtain global
information on the structure of solutions. Thus, the KAM
theorem will be utilized many times in later discussions
either explicitly or implicitly. Recently, transport in
three-dimensional nonintegrable Hamiltonian flows is studied
and the destruction of KAM barriers in the presence of
stochastic perturbations are described [Gyorgyi and Tishby
1989]. They further extended the action principle to
Hamiltonians with small noise, which provided a framework to
determine universal scaling of characteristic times as a
function of the noise.

For the example of a Hamiltonian system with two degrees

of freedom, e.g., two coupled nonlinear oscillators, Greene [1979] developed a method for deciding the existence of any given KAM surface computationally. One finds, when given that KAM orbits exist, that the guiding hypothesis is that the disapearance of a KAM surface is associated with a sudden change from stability to instability of nearby periodic orbits. The relation between KAM surfaces and periodic orbits has been explored extensively in this paper by the numerical computation of a particular mapping.

Lagrangian system

We will be concerned with an alternative description of classical mechanics on another phase space, the tangent bundle of the configuration space M of the system, TM. We consider a function L on TM and solutions to a certain second-order equation. From L we can derive an energy function E on TM which, when translated to T^*M by means of the "fiber derivative" FL: TM → T^*M (usually called the Legendre transformation), the derivative of L in each fiber of TM yeilds a suitable Hamiltonian. Then the solution curves in T^*M (i.e., solutions of Hamiltonian equations) and in TM (i.e., solutions of Lagrangian equations) will coincide when projected to M. The following diagram will be helpful.

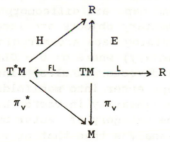

Definition 2.8.23 Let M be a manifold and L ϵ F(TM). Then the map FL: TM → T^*M : α_m → $DL_m(\alpha_m)$ ϵ L(T_mM,R) = T_m^*M is called the <u>fiber derivative</u> of L. L_m denotes the restriction of L to the fiber over m ϵ M.

101

<u>Proposition 2.8.24</u> FL: TM → T*M is a fiber preserving map.

It should be noted that FL is not necessarily a vector bundle mapping.

For further detail of "geometric theory of classical mechanics", Abraham and Marsden [1978] and Arnold [1978] are highly recommended.

2.9 Vector bundles and tubular neighborhoods

As we have mentioned earlier, a vector bundle can be thought of as a family of disjoint vector spaces $\{V_x\}_{x \in M}$ parameterized by the base space M. The union of these vector spaces is a space B, and the map π : B → M, $\pi(V_x) = x$ is continuous. Furthermore, π is locally trivial in the sense that locally B looks like a product of B with R^n; and there are open sets U covering M and homeomorphisms $\pi^{-1}(U) \approx U \times R^n$, mapping each fiber V_x linearly onto $\{x\} \times R^n$. A <u>morphism</u> from one vector bundle to another is a map taking fibers linearly into fibers.

A vector bundle is similar to a manifold in that both are building up from elementary objects glued together by special maps. For manifolds, the elementary objects are open subsets of R^n, the gluing maps are diffeomorphisms; for vector bundles the elementary objects are local "trivial bundles" $U \times R^n$, and the gluing maps are morphisms $U \times R^n \to U \times R^n$ of the form $(x,y) \to (x,g(x)y)$ where g: U → GL(n,R).

In both manifolds and vector bundles, linear maps play a central role. Linear maps enter into manifolds in a subtle way as derivatives, the linearity in vector bundle is more explicit. This makes the category of vector bundles more flexible and easier to analyze than that of manifolds. In fact, many natural constructions with vector bundles, such as direct sum, quotients, and pullbacks, are impossible to make for manifolds.

By introducing tubular neighborhoods, a new connection between vector bundles and the topology of manifolds is established. If $M \subset N$ is a submanifold and M has a certain neighborhood in N which looks like a normal vector bundle of M in N. Furthermore, such neighborhoods are essentially unique. Consequently, the study of the kinds of neighborhoods that M can have as a submanifold of a larger manifold is reduced to the classification of vector bundles over M. For example, the question whether the inclusion map $M \hookrightarrow N$ can be approximated by imbedding is tantamount to whether the normal bundle of M in N has a nonvanshing section.

Let F be a field (can be either real R, complex C or quaternion Q). A k-dimension vector bundle η over F is a bundle (B,M,π) together with the structure of a k-dimensional vector space over F on each fiber $\pi^{-1}(p)$ such that the local triviality condition is satisfied: For each point of M ther is an open neighborhood U and an isomorphism $i: UxF^k \rightarrow \pi^{-1}(U)$ such that the restriction $pxF^k \rightarrow \pi^{-1}(p)$ is a vector space isomorphism for each $p \in U$.

If F = R it is a real vector bundle, F = C a complex vector bundle, F = Q a quaternionic vector bundle. And the isomorphism $i: UxF^k \rightarrow \pi^{-1}(U)$ is a local coordinate chart of η.

One can also define vector bundles as a special case of a fiber bundle, or a special principal bundle. For instance, a real vector bundle is a fiber bundle where the fiber is a real vector space V^k and the structure group is GL(k,R).

It is helpful to introduce the notion of an exact sequence of vector bundles morphisms in just the way as an exact sequence of groups introduced earlier, i.e., a finite or infinite sequence

$$\ldots \rightarrow \eta_{i-1} \xrightarrow{f} \eta_i \xrightarrow{f} \eta_{i+1} \rightarrow \ldots$$

of morphisms such that for each $p \in M$ we have $\text{Im}(f_{i-1})_p = \text{Ker}(f_i)_p$ for all i. In particular, we are interested in the short exact sequence $0 \rightarrow \alpha \xrightarrow{f} \beta \xrightarrow{g} \mu \rightarrow 0$ where 0 denotes a 0-dim bundle over M. Such an exact sequence means that f is one-to-one, onto, and Im f = Ker g. One can show that the short sequence is unique up to isomorphism.

103

In the short exact sequence we call α the quotient bundle of the one-to-one map (monomorphism) f. Then every monomorphism has a quotient bundle and quotient bundle is unique up to isomorphism. If $\alpha \subset \beta$ is a subbundle, then the fibers of the quotient bundle are taken to be the vector spaces β_p/α_p and the quotient bundle is denoted by β/α.

The short exact sequence $0 \rightarrow \alpha \xrightarrow{f} \beta \xrightarrow{g} \mu \rightarrow 0$ is said to <u>split</u> if there is a monomorphism h : $\mu \rightarrow \beta$ such that gh = id_μ. Working through fibers, this is equivalent to the existence of an onto map (epimorphism) k: $\beta \rightarrow \alpha$ such that kf = id_α. Then the <u>Whitney sum</u> of bundles α, μ over M can be defined as the bundle $\alpha \oplus \mu$ whose fiber over p $\in$ M is $\alpha_p \oplus \mu_p$. If ϕ, τ are charts for α, μ respectively over U, a chart θ for $\alpha \oplus \mu$ over U is $\theta_p = \phi_p \oplus \tau_p : \alpha_p \oplus \mu_p \rightarrow R^m \oplus R^n$ as we have defined before.

The natural exact sequences of vector spaces
$$0 \rightarrow \alpha_p \xrightarrow{f_p} \alpha_p \oplus \mu_p \xrightarrow{g_p} \mu_p \rightarrow 0$$
can fit together to form a split exact sequence
$$0 \rightarrow \alpha \xrightarrow{f} \alpha \oplus \mu \xrightarrow{g} \mu \rightarrow 0.$$
Now let us apply this short exact sequence to the tangent bundle of a vector bundle. Let $\alpha = (B,M,\pi)$ be a C^{r+1} vector bundle, then each fiber α_p is a vector space with origin at p. Thus we can identify α_p with the tangent space of α_p, i.e., $(\alpha_p)_p$. Hence α is a subbundle of TB|M, the tangent bundle of the bundle space B restricted to M, in a natural way. Take note that TB|M is of C^r. Since M $\subset$ B is a submanifold, TM is a C^r subbundle of TB|M. Thus we have a short exact sequence $0 \rightarrow \alpha \rightarrow TB|M \xrightarrow{d\pi} TM \rightarrow 0$ which is split by di: TM $\rightarrow$ TB|M. Thus we have the following:

Theorem 2.9.1 Let $\alpha = (B,M,\pi)$ be a C^{r+1} vector bundle, $0 \leq r \leq \infty$. The short exact sequence of C^r vector bundles
$$0 \rightarrow \alpha \rightarrow TB|M \rightarrow TM \rightarrow 0$$
is naturally split by di:TM $\rightarrow$ TB|M. Thus there is a natural C^r isomorphism n_α : TB|M $\approx \alpha \oplus$ TM. And particularly, $\alpha \subset$ TB|M is a natural subbundle.

Here we provide the natural split of vector bundles. Recall the construction of orthogonal complements

and the normal bundles, one can see that it is easy to
establish the existence of such constructions. In fact, we
have the following useful result for a vector bundle:

   Theorem 2.9.2   Every short exact sequence of C^r vector
bundle splits, for $0 \leq r \leq \infty$, provided the base space is
paracompact. [Hirsch 1976].

   Whitney sums and restricting a bundle to a subset of the
base space are methods to construct new vector bundles out
of old. A more general construction is the induced bundle
method. Let $\alpha = (B, M, \pi)$ be a vector bundle. M_0 be an
arbitrary topological space, $f: M_0 \to M$ be any map, then one
can construct the _induced bundle_ (or _pullback_) $f^*\alpha$ over M_0.
The bundle space B_0 of $f^*\alpha$ is the subset $B_0 \subset M_0 \times B$ consisting
of all pairs (p, b) with $f(p) = \pi(b)$. The projection map π_0 :
$B_0 \to M_0$ is defined by $\pi_0(p, b) = p$. Thus one has the following
commutative diagram

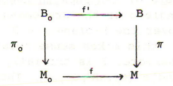

where $f'(p, b) = b$. The vector space structure in $\pi_0^{-1}(p)$ is
defined by $t_1(p, b_1) + t_2(p, b_2) = (p, t_1 b_1 + t_2 b_2)$. Thus f'
carries each vector space $(f^*\alpha)_p$ isomorphically onto the
vector space $(\alpha)_{f(p)}$. We leave it to the reader to show that
$f^*\alpha$ is a fiber bundle by showing that $f^*\alpha$ is locally
trivial.

   Now if α is a smooth vector bundle and f a smooth map,
then it can be shown that B_0 is a smooth submanifold of
$M_0 \times B$, and hence $f^*\alpha$ is also a smooth vector bundle.

   The above commutative diagram suggests the concept of a
bundle map. Let α and η be vector bundles, a _bundle map_ from
η to α is a continuous map $g: B(\eta) \to B(\alpha)$ which carries
each vector space η_p isomorphically onto one of the vector
space α_q.

105

<u>Lemma 2.9.3</u> If g: $B(\eta) \rightarrow B(\alpha)$ is a bundle map, and if g': $M(\eta) \rightarrow M(\alpha)$ is the cooresponding map of base spaces, then η is isomorphic to the induced bundle $g'^*\alpha$.

Before we state the classifying theorem of vector bundles, we must first introduce the concept of the universal bundle over a Grassmann manifold. Let $G_{n,k}$ be the set of k-dim linear subspaces of R^n (k-planes through the origin). Since any element of the orthogonal group O(n) carries k-plane into a k-plane, and in fact, O(n) is transitive on $G_{n,k}$. If R^k is a fixed k-plane and R^{n-k} is its orthogonal complement, then the subgroup of O(n) mapping R^k onto itself splits up into the direct product of O(k)xO(n-k) of two orthogonal subgroups whereby the first leaves R^{n-k} and the second leaves R^k pointwise fixed respectively. Thus we may identify $G_{n,k} = O(n)/O(k)xO(n-k)$. The set $G_{n,k}$ with this structure as an analytic manifold is called the <u>Grassmann manifold of k-planes in n-space</u>.

Let $\alpha_{n,k}$ be a vector bundle over the Grassmann manifold $G_{n,k}$, and the fiber of $\alpha_{n,k}$ over the k-plane $F \subset R^n$ is the set of pairs (F,p) where $p \in F$; this makes sense because F is a k-dim subspace of R^n. Furthermore, F is trivially a subbundle of the vector bundle $(G_{n,k}xR^n, G_{n,k}, \pi)$. Thus $\alpha_{n,k}$ can be made into an analytic k-dim vector bundle in a natural way. We call $\alpha_{n,k}$ <u>the universal bundle over $G_{n,k}$.</u> As we shall see, it is also called the <u>classifying bundle</u> for n-dim vector bundle.

The following two theorems will help us in understanding the construction of classifying bundles as well as preparing us for the classification theorem.

<u>Theorem 2.9.4</u> Let α be a k-dim C^r vector bundle over a manifold M, where $0 \leq r \leq \infty$. Let $U \subset M$ be a neighborhood of a closed set $A \subset M$. Assume f: $\alpha|U \rightarrow UxR^n$ to be a one-to-one C^r map of vector bundles over id_U. If $n \geq k + dim M$, then there is a one-to-one C^r map (C^r monomorphism) $\alpha \rightarrow MxR^n$ over id_M which agrees with f over some neighborhood of A in U.

<u>Theorem 2.9.5</u> Let α be a C^r k-plane bundle over a m-manifold M, $0 \leq r \leq \infty$. Then there is a C^r m-plane bundle η

over M such that $\alpha \oplus \eta \approx_r M \times R^n$ $(n > k + m)$.

Let us define a vector bundle map

$$\begin{array}{ccc} \alpha & \xrightarrow{\phi} & \beta_{n,k} \\ \downarrow & & \downarrow \\ M & \xrightarrow{g} & G_{n,k}. \end{array}$$

To $p \in M$, g assigns the k-plane $g(p) = f(\alpha_p) \in G_{n,k}$ where f: $\alpha \to M \times R^n$ is a monomorphism over id_M. The map g: $M \to G_{n,k}$ can be shown to have the property that $g^* \beta_{n,k} \approx \alpha$. The pullback is called the <u>classifying map for α</u>. From the previous two theorems, we are ready to state the classification theorem!

<u>Theorem 2.9.6</u> If $n \geq k+m$ then every C^r k-plane bundle α over a m-manifold M has a classifying map f_α: $M \to G_{n,k}$, when $n > k+m$, the homotopy class of f_α is unique, and if η is another k-plane bundle over M then $f_\alpha \approx f_\eta$ iff $\alpha \approx \eta$.

This theorem is of great importance because it converts the theory of vector bundles into a branch of homotopy theory. One can use what one knows about maps to study vector bundles. We shall not go into this, for those readers who are interested in this development, one can consult [Steenrod 1951, Husemoller 1975, Spanier 1966]. As an example, we have the following theorem:

<u>Theorem 2.9.7</u> Every C^r vector bundle α over a C^∞ manifold M has a compatible C^∞ bundle structure, and such a structure is unique up to C^∞ isomorphism.

Remark: This reminds us of the Whitney theorem about the C^∞ structure of manifolds. Thus from now on it is not necessary to specify the differentiability class of a vector bundle either. Although these last three theorems are stated for manifolds, they are also true (ignoring the differentiability) for vector bundles over simplicial or CW complexes of finite dimension [Steenrod 1951, Spanier 1966].

Now we are ready to introduce briefly the concept of a tubular neighborhood and its properties.

Let M be a submanifold of N. A <u>tubular neighborhood of M</u> is a pair (f, α) where $\alpha = (B,M,\pi)$ a vector bundle over M

107

and f: B → N is an imbedding such that : (i) f|M = id$_M$ where M is identified with the zero section of B; (ii) f(B) is an open neighborhood of M in N. Loosely speaking, we can refer the open set V = f(B) as a tubular neighborhood of M. It is understood that associated to V is a particular retraction g: V → M making (V,M,g) a vector bundle whose zero section is the inclusion M → V. A slightly more general concept is the partial tubular neighborhood of M. This is a triple (f, α, U) where α = (B,M,π) is a vector bundle over M, U ⊂ B is a neighborhood of the zero section and f: U → N is an imbedding such that f|M = id$_M$ and f(U) is open in N. A partial tubular neighborhood (f, α, U) contains a tubular neighborhood in the sense that there is a tubular neighborhood (h, α) of M in N such that h = f in a neighborhood of M.

   Theorem 2.9.8   Let M ⊂ R^n be a submanifold without boundary. Then M has a tubular neighborhood in R^n [Hirsch 1976, Golubitsky & Guillemin 1973].

   Theorem 2.9.9   Let M ⊂ N be a submanifold, and ∂M = ∂N = 0. Then M has a tubular neighborhood in N [Hirsch 1976].

   It is useful to be able to slide one tubular neighborhood of a manifold onto another one, and to map fibers linearly onto fibers. Such sliding is a special case of the concept of isotopy. Here we give a more restrictive version of isotopy.

   If M, N are manifolds, an isotopy of M in N is a homotopy f: M x I → N by f(p, t) = f$_t$(p) such that the related map f' : M x I → N x I, where (p, t) ℝ (f$_t$(p), t) is an imbedding. It is clear that "f is isotopic to g" is transitive.

   Theorem 2.9.10   Let M ⊂ N be a submanifold, and ∂M = ∂N = 0. Then any two tubular neighborhoods of M in N are isotopic [Hirsch 1976].

   It is clear that the boundary of a manifold cannot have a tubular neighborhood, nonetheless, it has a kind of "half-tubular" neighborhood called a collar. A collar on M is an imbedding f: ∂M x [0,∞) → M such that f(p,0) = p.

108

Theorem 2.9.11 ∂M has a collar. (Note, M can be a C°
manifold) [Brown 1962, Hirsch 1976].

When a submanifold M ⊂ N whose boundary is nicely
placed, N is called a neat submanifold. More precisely, we
call M a neat submanifold of N if ∂M = M ∩ ∂N and M is
covered by charts (ϕ_i, U_i) of N such that M ∩ U_i = $\phi_i^{-1}(R^m)$
where m = dim M. A neat imbedding is the one whose image is
a neat submanifold.

For example:

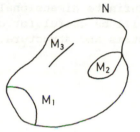

M_1 is neat
M_2, M_3 are not.

If M ⊂ N is a submanifold and ∂M = 0, then M is neat iff
M ∩ ∂N = 0. In general, M is neat iff ∂M = M ∩ ∂N and if
both M and N are at least C^1, M is not tangent to ∂N at any
point p ∈ ∂M; i.e., M_p ⊄ $(∂N)_p$.

Theorem 2.9.12   Let M ⊂ N be a closed neat submanifold,
then ∂N has a collar which restricts to a collar on ∂M in M
[Hirsch 1976].

Theorem 2.9.13   Let M ⊂ N be a neat submanifold, then M
has a tubular neighborhood in N. Moreover, every tubular
neighborhood of ∂M in ∂N is the intersection with ∂N of a
tubular neighborhood for M in N.

Finally, we have the important theorem on the existence
of tubular neighborhoods.

Theorem 2.9.14 Let M be a submanifold of N. Then there
exists a tubular neighborhood of M in N [Golubitsky and
Guillemin 1973].

In this section, we have illustrated the concept and
techniques to "thickening" a submanifold. This concept will
be very useful for the discussion of convergence of orbits
to a periodic orbit in stability analysis. Although in the

discussion we may not explicitly invoke the notion of
tubular neighborhoods, nonetheless, the reader can feel such
"construction". The simplest situation is discussing the
Poincaré return map. The mapping cylinder is the tubular
neighborhood.

So far in this chapter, we have discussed finite
dimensional manifolds and their topological and geometric
properties. In the next chapter, we shall give some brief
discussions of infinite dimensional manifolds and global
analysis which will be useful for our subsequent discussions
of dynamical systems and structural stabilities.

Chapter 3   Introduction to Global Analysis and Infinite Dimensional Manifolds

3.1   What is global analysis?

First recall that linear analysis is the study of
topological vector spaces, such as real, complex or vector
valued functions on R^n or on some domain in R^n, and linear
maps usually are differential (or integro-differential)
operators. This may be viewed as "local" linear analysis. To
generalize to "global" linear analysis, an arbitrary
differentiable manifold M replaces the domain in R^n, and
topological vector spaces of cross-sections of
differentiable vector bundles over M are considered. Again,
the linear maps are defined by linear differential or
integro- differential operators. Roughly speaking, the
questions here are relating analytic invariants of the
operators with topological invariants of M and the given
vector bundles, and it is the proper setting, for example,
for Hodge's theory of harmonic forms, the Atiyah-Singer
index theorem, and the Atiyah-Bott fixed point formula.

What about "global non-linear analysis"? Instead of
differentiable vector bundles over a differentiable manifold
M, we consider more general differentiable fiber bundles
over M; instead of topological vector spaces of sections of
the vector bundle, we consider differentiable manifolds of
sections of the general fiber bundle, and we take non-linear
differential operators which define differential maps
between such manifolds of sections. This seems to be the
proper arena for a variety of subjects, such as the theory
of non-linear differential operators and the calculus of
variations (in particular, Morse theory and
Lucternik-Schnirelman theory), and the general
transversality theorem, just to name a few. For an earlier
review, see [Eells 1966; Kahn 1980; Palais 1968; Berger
1977].

The underlying technique which runs through all the

nonlinear analyses is the idea of "linearization", i.e., approximating a nonlinear map "locally" by a linear map. Usually the sets where the map is defined and into which it maps have natural infinite dimensional manifold structures, and the map is differentiable with respect to such manifold structures. Moreover, the linearization of the maps near a given point is just its differential at that point. Thus abstract nonlinear analysis turns out to be the study of infinite dimensional manifolds and differentiable maps on them.

An analogous situation in differential geometry is that for a given differentail manifold, an open chart can be found such that in a neighborhood of a point, all the local properties of the manifold can be represented in a Cartesian coordinate system. For arbitrarily small neighborhoods, the manifold can be considered "Euclidean". Everything is "trivialized", which corresponds to "linearalization". Yet the different pieces at different points have different geometrical and local properties. The means to piece together is through the overlapping region $U_\alpha \cap U_\beta$. Here, the two pieces have to agree on all the local properties in the overlapping region. This reminds us that in calculus, the definition of continuity of a function at a point is that, not only limits on either side of the point exist, but they have to agree. The matching of the overlapping region over the entire manifold is the spirit of "linearization" without distroying the properties (or more plainly, losing some information) of the manifold.

3.2   Jet bundles

In the following, we shall define jet bundles and display their utility. But before we start the proper mathematical definition, it is helpful to remind ourselves an analogy in calculus. If $f(x)$ and $g(x)$ are C^∞ and analytic, one can easily show that $f(x) = g(x)$ up to k-th order iff the Taylor series of f and g are identical up to the k-th order at every point of x. The slightly more generalized case is the power series of a complex variable. Indeed, one could generalize to a function of multivariables. In the following, we go one step further, we generalize the idea of Taylor series to the setting of maps over manifolds.

Let M and N be C^∞ manifolds, and let $f, g : M \to N$ be C^∞ maps, f is said to agree with g up to order k at $p \in M$ if there are coordinate charts at $p \in M$ and $f(p) = g(p) \in N$ such that they have the same Taylor expansion up to and including order k. One can convince oneself that the agreement of f and g up to order k (denoted by $f \sim g$) is coordinate independent. In fact, $\sim_k$ is an equivalence relation, and the equivalence class of maps which agree with f to order k at p is called the k-jet of f at p and denoted by $j_p^k f$. Let x^a be local coordinates around $p \in M$ and y^μ be local coordinates around $f(p) \in N$, then $j_p^k f$ is specified by x^a, $y^\mu = f^\mu(p)$, $y_a^\mu = \partial_a f^\mu(p)$, $y_{ab}^\mu = \partial_{ab} f^\mu(p), \ldots, y_{a_1 \ldots a_k}^\mu = \partial_{a_1 \ldots a_k} f^\mu(p)$, where $f^\mu(p)$ is the coordinate presentation of f, and $\partial_a, \ldots, \partial_{a_1 \ldots a_k}$ denote partial derivatives

$$\partial_a = \partial / \partial x^a, \ldots, \quad \partial_{a_1 \ldots a_k} = \partial^k / \partial x^{a_1} \ldots \partial x^{a_k} ,$$

and Latin indices $a, b, \ldots a_1, \ldots, a_k$ range from 1 to dim M and Greek indices $\mu, \alpha, \ldots$ range from $1, \ldots, $ dim N. Conversely, any collection of numbers x^a, y^μ, $y_a^\mu, \ldots, y_{a_1 \ldots a_k}^\mu$, where x^a and y^μ are the corresponding coordinate charts and $y_{ab}^\mu, \ldots y_{a_1 \ldots a_k}^\mu$ are symmetric in their lower indices, determining a unique equivalence class. Putting this formally, one has the following:

Let M and N be smooth manifolds, $p \in M$. Suppose $f, g: M \to N$ are smooth maps with $f(p) = g(p) = q$. (i) f has first order contact (i.e., agrees to first order) with g at p if $(df)_p = (dg)_p$ as mapping of $M_p \to N_q$. (ii) By induction, f has k-th (k is a positive integer) order contact with g at p if $(df): TM \to TN$ has (k-1)-th order contact with (dg) at every point in M_p. This is written as $f \sim_k g$ at p. (iii) Let $J^k(M,N)_{p,q}$ denote the set of equivalence classes under "$\sim_k$ at p" of mappings $f: M \to N$ where $f(p) = q$. (iv) Let $J^k(M,N)$,

113

the <u>k-jet bundle of M and N</u>, be the disjoint union of
$J^k(M,N)_{p,q}$, i.e., $J^k(M,N) =$
$U_{(p,q)\in M\times N}$ $J^k(M,N)_{p,q}$. Any element h ϵ $J^k(M,N)$ is a <u>k-jet of</u>
<u>mappings</u> (or just <u>k-jet</u>) from M to N. (v) Let h ϵ $J^k(M,N)$ be
a k- jet, then there exist p in M and q in N such that h ϵ
$J^k(M,N)_{p,q}$. Then p is the <u>source of h</u> and q is <u>the target of</u>
<u>h</u>. The mapping α : $J^k(M,N) \to M$, defined by h $\to$ source of h,
is the <u>source map</u> and the mapping β : $J^k(M,N) \to N$, defined
by h $\to$ target of h, is <u>the target map</u>.

For a given smooth map f:M $\to$ N and for every p ϵ M,
there is a canonically defined mapping j^kf : M $\to$ $J^k(M,N)$
called <u>the k- jet of f</u> defined by $j^kf(p)$ = equivalence class
of f in $J^k(M,N)_{p,f(p)}$. Thus, j^kf at p is a local lifting from M
to $J^k(M,N)_{p,f(p)}$, i.e., $j^kf(p)$ is just an invariant way of
describing the Taylor expansion of f at p up to order k. One
can also show that j^kf is a smooth mapping.

Since $J^0(M,N) = M \times N$, so f has $-_0$ contact with g at p
iff f(p) = g(p), and $j^0f(p) = (p,f(p))$ is just the graph of
f!

Since any given map determines a k-jet at each point of
its domain, thus if f:M $\to$ N is smooth, the k-jet extension
of f is a map j^kf: U $\to$ $J^k(M,N)$ by x $\to$ j_x^kf, where U is the
domain of f. Clearly, the k-jet extension of f is a
cross-section of the source map α, i.e., $\alpha \cdot j^kf = id_U$.

A map from a jet bundle $J^k(M,N)$ to another smooth
manifold P induces maps of higher jet bundles and they are
called <u>prolongations</u>. More precisely, let M, N and P are
smooth manifolds and let ϕ: $J^k(M,N) \to P$ be smooth. The <u>1-th</u>
<u>prolongation of ϕ</u> is the unique map $p^l\phi$: $J^{k+l}(M,N) \to J^l(M,P)$
such that the following diagram commutes:

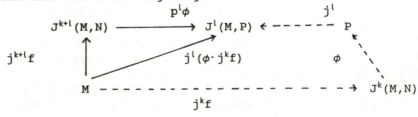

114

In appropriate local coordinates on M, N, and P, roughly speaking, the prolongations amount to the taking of total derivatives of ϕ. We shall not go into any details. For this, one may want to consult [Pirani, Robinson and Shadwick 1979]. Jet bundles not only are important in discussing global analysis and nonlinear differential operators on which we shall embark briefly, but they are also important in describing higher order vector bundles on smooth manifolds [see e.g., Yano and Ishihara 1973].

Over the past couple of decades, a class of transformations discovered by Bäcklund over a century ago has emerged as an important tool in the study of a wide range of nonlinear evolution equations in theoretical physics and continuum mechanics. Bäcklund transformations are significant because the invariance under Bäcklund transformations can be utilized to generate an infinite sequence of solutions of certain nonlinear evolution equations by purely algebraic procedures. Current work on Bäcklund transformations are mainly concerned with extending their applications to other nonlinear partial differential equations of physical interest and with encompassing the known results in a unified theory so that generalizations to higher dimensions can be made. Jet-bundle formulation of Bäcklund transformations seems to provide an appropriate geometric setting for the main concerns of Bäcklund transformations described above and their connection with the inverse scattering formalism, prolongation structures, and symmetries. For those readers who are interested in these areas of research, please consult Pirani, Robinson and Schadwick [1979], Rogers and Schadwick [1982], Hermann [1976, 1977], Estabrook and Walquist [1975].

<u>Lemma 3.2.1</u> Let $U \subset R^n$ be open, and $p \in U$. Let $f, g: U \to R^m$ be smooth mappings. Then $f \sim_k g$ at p iff $(D^\alpha f_i)(p) = (D^\alpha g_i)(p)$ for $|\alpha| \leq k$, $1 \leq i \leq m$ where $\alpha = (\alpha_1, .., \alpha_n)$ are n-tuples of non-negative integers, $|\alpha| = \alpha_1 + \alpha_2 + ... + \alpha_n$, $D^\alpha = \partial^{|\alpha|}/\partial x_1^{\alpha_1} ... \partial x_n^{\alpha_n}$, and f_i and g_i are the coordinate functions determined by f and g respectively.

115

This lemma sometimes is used to define the equivalence relation "$\sim_k$". As we have pointed out earlier, the following corollary follows immediately.

Corollary 3.2.2   f and g : U → R^m such that f $\sim_k$ g at p iff the Taylor expansions of f and g up to and including order k are identical at p.

The following lemma which concerns the equivalence relation between a pair of composite maps is intuitively clear and can be proven by induction.

Lemma 3.2.3   Let U ⊂ R^n and V ⊂ R^m are open subsets. Let f_1, f_2 : U → V and g_1, g_2: V → R^l be smooth maps such that $g_1 \cdot f_1$ and $g_2 \cdot f_2$ make sense. Let p ∈ U and suppose $f_1 \sim_k f_2$ at p and $g_1 \sim_k g_2$ at $f_1(p) = f_2(p) = q$. Then $g_1 \cdot f_1 \sim_k g_2 \cdot f_2$ at p.

Theorem 3.2.4   Let X, Y, Z and W be smooth manifolds. (i) Let h: Y → Z be smooth, then h induces a map h_*: $J^k(X,Y)$ → $J^k(X,Z)$ defined by: let σ ∈ $J^k(X,Y)_{p,q}$ and let f: X → Y represent σ. Then $h_*(\sigma)$ = the equivalence class of h·f in $J^k(X,Z)_{p,h(q)}$. (ii) Let a: Z → W be smooth. Then $a_* \cdot h_*$ = $(a \cdot h)_*$ as mappings of $J^k(X,Y)$ → $J^k(X,W)$ and $(id_Y)_*$ = $id_{J^k(X,Y)}$. Thus if h is a diffeomorphism, h_* is a bijection (one-to-one, onto). (iii) Let g : Z → X be a diffeomorphism, then g induces a map g^*: $J^k(X,Y)$ → $J^k(Z,Y)$ defined as follows: let σ ∈ $J^k(X,Y)_{p,q}$ and let f:X → Y represent σ. Then $g^*(\sigma)$= the equivalence class of f·g in $J^k(Z,Y)_{g^{-1}(p),q}$. (iv) Let b: W → Z be a diffeomorphism. Then the induced composite mappings $b^* \cdot g^*$ = $(g \cdot b)^*$ are mappings of $J^k(X,Y)$ → $J^k(W,Y)$ and $(id_X)^*$ = $id_{J^k(X,Y)}$ so that g^* is also a bijection.

Let A_n^k be the vector space of polynomials in n-variables of degree ≤ k with their constant term equal to zero. Choose the coefficients of the polynomials as coordinates of A_n^k, then A_n^k is isomorphic to some Euclidean space and thus a smooth manifold. Let $B_{n,m}^k$ = $\oplus_{i=1}^m A_n^k$. Here ⊕ is the Whitney sum. Then $B_{n,m}^k$ is also a smooth manifold.

Theorem 3.2.5   Let X and Y be smooth manifolds and dim X = n, dim Y = m.
(i) $J^k(X,Y)$ is a smooth manifold, and dim $J^k(X,Y)$ = m + n + dim($B_{n,m}^k$).

(ii) The source map α : $J^k(X,Y) \to X$ and the target map β : $J^k(X,Y) \to Y$, and $\alpha \times \beta$: $J^k(X,Y) \to X \times Y$ are submersions.
(iii) If h : Y $\to$ Z is smooth, then h_* : $J^k(X,Y) \to J^k(X,Z)$ is smooth. If g : Z $\to$ X is a diffeomorphism, then g^* : $J^k(X,Y) \to J^k(Z,Y)$ is a diffeomorphism.
(iv) If h : X $\to$ Y is smooth, then $J^k h$: X $\to$ $J^k(X,Y)$ is smooth. [Golubitsky & Guillemin 1973, p.40].

Note that in general $J^k(X,Y)$ is not a vector bundle because there is no natural addition in $J^k(X,Y)_{p,q}$. But if Y $= R^m$, then $J^k(X,R^m)$ is indeed a vector bundle over X $\times$ R^m where the addition of jets is given by the addition of functions representing these jets in $J^k(X,R^m)_{p,q}$.

Although $J^k(X,Y)$ is not a vector bundle in general, it is more than just a manifold, it is a fiber bundle. In fact, if k > l and ignoring all derivatives above the l-th yields the natural projection $\pi_{k,l}$: $J^k(X,Y) \to J^l(X,Y)$ by $j^k_x f \to j^l_x f$. Thus, $J^k(X,Y)$ is a fiber bundle over $J^l(X,Y)$ with projection $\pi_{k,l}$. In particular, $J^0(X,Y)$ may be identified with X $\times$ Y and $J^1(X,Y)$ may be thought of as a vector bundle over X $\times$ Y. Furthermore, $\pi_{l,l}$ is understood as the identity map of $J^l(X,Y)$.

As we have seen in this section, jet bundle is the proper language to describe relations between smooth maps among smooth manifolds. Thus, it will be very important in discussing stable maps, their singularities, and bifurcations. Of course, it will also be very important to describe global stability properties. Indeed, it is one of the mathematical techniques frequently used in much current literature on the theoretical dynamical systems. There is no single volume which discusses jet bundles and their appliations. Jet bundles can be found in many books discussing global analysis. The following books cover some discussions of jet bundles: Pirani, Robinson and Schadwick [1979], Golubitsky and Guillemin [1973], and of course, Palais [1968].

3.3 Whitney C^∞ topology

As we have mentioned earlier, in order that any physical
fields, be they scalar, vector, or tensor fields, have
physical significance, they must have some form of stability
under "perturbation". In other words, "nearby" fields should
have the same properties and physical significance. But in
order to give a precise meaning of "nearby", one has to
define a topology on the set of fields. There are various
ways in which this can be defined depending on whether one
requires a "nearby" field to be nearby in just its values
(C^0 topology) or also in its derivatives up to the kth order
(C^k topology) and whether one requires it to be nearby
everywhere (open topology) or only on compact sets (compact
open topology).

Let M and N be smooth manifolds and $C^\infty(M,N)$ be the set
of all smooth mappings from M to N. For a fixed non-negative
integer k, let U be a subset of $J^k(M,N)$. Let $\theta(U)$ denote the
set $\{f \in C^\infty(M,N) \mid j^k f(M) \subset U\}$. Also note that $\theta(U) \cap \theta(V) =
\theta(U \cap V)$. This family of sets $\{\theta(U)\}$ forms a basis for the
Whitney C^k topology. Let W_k denote the set of open subsets
of $C^\infty(M,N)$ in the Whitney C^k topology. Thus <u>the Whitney C^∞
topology</u> on $C^\infty(M,N)$ is the topology whose basis is the union
of all W_k's where k= $0,..,\infty$, i.e., $W = U_{k=0}^\infty W_k$. This basis is
a well-defined one because $W_k \subset W_l$ for $k \leq l$.

In usual literature, the Whitney C^k topology is also
called the <u>strong</u> or fine topology on $C^k(M,N)$ sometimes
denoted by $C_s^k(M,N)$, and the compact-open topology on $C^k(M,N)$
is also called the weak topology on $C^k(M,N)$, denoted by
$C_w^k(M,N)$. If M is compact, the weak topology is the same as
the strong topology. Thus the strong topology is more
useful. But it is only fair to point out that weak topology
has very nice features: it has a countable basis (we shall
point this out for strong topology with compact M), a
complete metric, and for compact M, $C_w^k(M,R^n)$ is a Banach
space. Moreover, the Whitney imbedding theorem (Sect.2.4)
can be stated in relation to the weak topology: Let M be a
compact C^r manifold, $2 \leq r \leq \infty$. Then imbeddings are dense

in $C_w^r(M,R^q)$ if $q > 2m$, while immersions are dense if $q \geq 2m$, $m = \dim M$.

Let us describe a neighborhood basis in the Whitney C^k topology for a function f in $C^\infty(M,N)$. Since all manifolds are metrizable [Kelley 1955], one can choose a metric on $J^k(M,N)$ compatible with the C^k topology. Let $B_\delta(f) = \{g \in C^\infty(M,N) \mid$ for all $p \in M$, $d(j^kf(p),j^kg(p)) < \delta(p)\}$, where $\delta: M \to R^+$ is C^0. Thus one can consider $B_\delta(f)$ as the set of smooth mappings of $M \to N$ whose partial derivatives up to the k-th order are δ-close to f's. Let U be an open neighborhood of f in $C^\infty(M,N)$, one can show that $B_\delta(f) \subset U$, and indeed, $B_\delta(f)$ is open for every δ.

If M is compact, $C^\infty(M,N)$ satisfies the first axiom of countability. This is because one can find a countable neighborhood basis of f by $B_\delta(f)$ where $\delta_n(p) = 1/n$ for all $p \in M$, thus δ is bounded below for large n. One can easily prove that: a sequence of functions f_n in $C^\infty(M,N)$ converges to f (in the C^k topology) iff j^kf_n converges uniformly to j^kf. Clearly, in the local situation, f_n and all of the partial derivatives of f_n of order $\leq k$ converge uniformly to f's.

For noncompact manifolds, the concept of the convergence of f_n to f is stronger than uniform convergence and the weak topology does not control the behavior of a map at "infinity" very well. Thus the statement for compact manifolds has to be modified. Since the manifolds are paracompact, then it is locally compact (Sect.2.4). Thus one can extend the result of a compact manifold to a compact subset of the noncompact manifold. More precisely; the sequence of mappings f_n converge to f (in the C^k topology) iff there is a compact subset C of M such that j^kf_n converges uniformly to j^kf on C and all but a finite number of the f_n's equal to f outside C. In fact, for a noncompact M, $C^\infty(M,N)$ in the Whitney C^k topology does not satisfy the first axiom of countability. Nonetheless, the Whitney (strong) C^k- topology is very important for differential topology for the fact that many interesting and important

119

subsets are open. For example: let $Imm^r(M,N)$ be the set of C^r immersions of M in N, $Imb^r(M,N)$ be the set of C^r imbeddings of M in N, $Diff^r(M,N)$ be the set of C^r diffeomorphisms from M onto N, and $Prop^r(M,N)$ be the set of proper C^r maps $M \to N$ where $f:M \to N$ is proper if f^{-1} takes compact sets to compact sets. Then we have the following important theorem.

Theorem 3.3.1 $Imm^r(M,N)$ and $Imb^r(M,N)$ are open in $C_s^r(M,N)$ (Whitney C^k toplogy) for $r \geq 1$; $Prop^r(M,N)$ is open in $C_s^r(M,N)$ for $r \geq 0$; if M and N are C^r manifolds without boundary, then $Diff^r(M,N)$ is open in $C_s^r(M,N)$ for $r \geq 1$; and if M and N are manifolds without boundary and $f:M \to N$ is a homeomorphism, then f has a neighborhood of onto maps in $C_s^0(M,N)$.

A very basic approximation theorem for a manifold without boundary is:

Theorem 3.3.2 Let M and N be C^s manifold without boundary, $1 \leq s \leq \infty$. Then $C^s(M,N)$ is dense in $C_s^r(M,N)$, $0 \leq r \leq s$.

For its many applications in differential topology, see e.g., [Hirsch 1976, Chapter 2, section 2]. For example, from the above theorem and the openness theorem and a theorem on diffeomorphisms, one has:

Theorem 3.3.3 Let $1 \leq r \leq \infty$, every C^r manifold is C^r diffeomorphic to a C^∞ manifold. (This is the theorem due to Whitney we quoted at the begining of Chapter 2).

The following definitions and propositions are of technical nature and they are useful in describing the generalized transversality theorem, or in proving some results. The reader may skip the following five propositions without loss of continuity.

Let X be a topological space, then: (i) A subset Y of X is <u>residual</u> if it is the countable intersection of open dense subsets of X. (ii) X is a <u>Baire space</u> if every residual set is dense.

Proposition 3.3.4 Let M and N be smooth manifolds, then $C^\infty(M,N)$ is a Baire space in the Whitney C^∞ topology.

120

[Goluoitsky & Guillemin 1973, p. 44].

Let us list some properties of $C^\infty(M,N)$ which will be used later.

Proposition 3.3.5 Let M and N be smooth manifolds. Then the mapping j^k: $C^\infty(M,N) \to C^\infty(M,J^k(M,N))$ defined by $f \to j^k f$ is continuous in the Whitney C^∞ topology.

Proposition 3.3.6 Let M, N and P be smooth manifolds and let ϕ: $N \to P$ be smooth. Then the induced map ϕ_*: $C^\infty(M,N) \to C^\infty(M,P)$ defined by $f \to \phi \cdot f$ is a continuous mapping in the Whitney C^∞ topology.

Proposition 3.3.7 Let M, N and P be smooth manifolds. Then $C^\infty(M,N) \times C^\infty(M,P)$ is homeomorphic to $C^\infty(M,N \times P)$ in the C^∞ topology by the standard identification $(f,g) \to f \times g$ where $f \in C^\infty(M,N)$, $g \in C^\infty(M,P)$, $p \in M$ and $(f \times g)(p) = (f(p),g(p))$.

Proposition 3.3.8 Let M, N and P be smooth manifolds, and in addition, M is compact, then the mapping of $C^\infty(M,N) \times C^\infty(N,P) \to C^\infty(M,P)$ defined by $(f,g) \to g \cdot f$ is continuous, where $f \in C^\infty(M,N)$, $g \in C^\infty(N,P)$.

Note that this proposition will not be true if M is not compact. Nonetheless, if we replace $C^\infty(M,N)$ by its open subset of proper mappings f of M into N (recall, f is proper if f^{-1} maps compact sets to compact sets), then the same conclusion is valid. Note also that the induced mapping f^*: $C^\infty(N,P) \to C^\infty(M,P)$ defined by $g \to g \cdot f$ is (not) continuous if f is (not) a proper mapping. In particular, if M is an open subset of N and f is just the inclusion map, then the restriction map of $C^\infty(N,P) \to C^\infty(M,P)$ given by $g \to g|M$ is not continuous.

We now state Thom's transversality theorem in a more general form and show that Theorem 2.4.27 is a corollary.

Theorem 3.3.9 (Thom's transversality theorem) Let M and N be smooth manifolds and W a submanifold of $J^k(M,N)$. Let $T_W = \{f \in C^\infty(M,N) \mid j^k f \pitchfork W\}$. Then T_W is a residual subset of $C^\infty(M,N)$ in the Whitney C^∞ topology.

Since $J^0(M,N) = M \times N$ and $j^0 f(p) = (p,f(p))$, and the target map β: $M \times N \to N$ is onto, so $\beta^{-1}(W)$ is a submanifold of M x N. If $j^o f \pitchfork \beta^{-1}(W)$ at p then clearly $f \pitchfork W$ at p. [If $j^o f(p) \notin \beta^{-1}(W)$ then, $f(p) \in W$. But if $j^o f(p) \in \beta^{-1}(W)$ and $(\beta^{-1}(W))_{(p,f(p))} + (dj^o f)M_p = (MxN)_{(p,f(p))}$. Then apply $d\beta$ one obtains $W_{f(p)} + (df)M_p = N_{f(p)}$. Thus $f \pitchfork W$ at p]. Since the set of all maps which are transverse to W contains the set $\{f \in C^\infty(M,N) | j^o f \pitchfork \beta^{-1}(W)\}$, which is dense by Theorem 3.3.9. Thus it is established that Theorem 2.4.27 is a special case of Theorem 3.3.9. Note that usually Thom's transversality theorem is stated as a combination of Theorems 2.4.26 and 2.4.27. Sometimes Theorems 2.4.26 and 2.4.27 are called the elementary transversality theorem.

Let us describe an "obvious" generalization. Let M and N be smooth manifolds and define $M^s = Mx...xM$ (s times) and $X^{(s)} = \{(x_1,..x_s) \in M^s \mid x_i \neq x_j$ for $1 \leq i \leq j \leq s\}$. Let α be the source map, i.e., $\alpha: J^k(M,N) \to M$. Now define $\alpha^s: J^k(M,N)^s \to M^s$ in an obvious way. Then $J_s^k(M,N) = (\alpha^s)^{-1}(X^{(s)})$ is the s-fold k-jet bundle. A multijet bundle is some s-fold k-jet bundle. And $X^{(s)}$ is a manifold since it is an open subset of M^s. Thus $J_s^k(M,N)$ is an open subset of $J^k(M,N)^s$ and is also a smooth manifold. Now let $f: M \to N$ be smooth. Then define $j_s^k f: X^{(s)} \to J_s^k(M,N)$ in the natural way, i.e., $j_s^k f(x_1,..,x_s) = (j^k f(x_1),..,j^k f(x_s))$.

Theorem 3.3.10 (multijet transversality theorem) Let M and N be smooth manifolds and W a submanifold of $J_s^k(M,N)$. Let $T_w = \{f \in C^\infty(M,N) | j_s^k f \pitchfork W\}$. Then T_w is a residual subset of $C^\infty(M,N)$. Furthermore, if W is compact, then T_w is open [G & G 1973, p.57].

3.4   Infinite dimensional manifolds

Differentiable manifolds are natural generalization of Euclidean spaces by building up from Euclidean spaces and gluing together by some overlapping maps and charts. Infinite dimensional manifolds can also be viewed in a similar way. For instance, starting with some fixed, infinite dimensional linear vector spaces, such as a Hilbert space or a Banach space, one can form a manifold by gluing

122

together open sets by nice overlapping functions. The theory of infinite dimensional manifolds can provide the proper framework for certain analyses and various connections between geometry and analysis, particularly the operator theory. But the need for infinite dimensional manifolds also arises in another vein. For instance, let M and N be two differentiable manifolds, there is great interest in the space of continuous maps from M to N and its subspaces of various degrees of differentiability for these maps. We shall state that under certain conditions, this space of continuous maps is an infinite dimensional manifold.

First, we shall review some definitions and properties of Banach and Hilbert spaces, then an infinite dimensional manifold on a given topological vector space will be defined and facts recalled. The spaces of maps between manifolds will be defined and some results stated.

Let V be a vector space over the field F (either R or C). V is a topological vector space if there is a topology on V such that ϕ: V x V → V;   ($\phi(v_1,v_2) = v_1 + v_2$)   and ϕ: F x V → V;   ($\phi(a,v) = av$) are continuous.

Let V be a vector space, a norm on V is a function V → R, $\|x\|$, for each x ϵ V, such that:

(i) $\|x\| \geq 0$, $\|x\| = 0$ iff x = 0;

(ii) $\|\alpha x\| = |\alpha|\ \|x\|$, for x ϵ V, α ϵ F;

(iii) $\|x + y\| \leq \|x\| + \|y\|$, for any x, y ϵ V.

With the given norm, one can define a metric on V by e(x,y) = $\|x - y\|$. Then one can define a topology on V by taking the open ball, $B_\epsilon(x) = \{y \epsilon V | e(x,y) < \epsilon \}$, as a basis.

A topological vector space whose topology arises from the norm, such as the one given above, is a Banach space provided that it is complete with respect to the norm or the metric (this is the Cauchy completion).

Let B_1 and B_2 be Banach spaces and let f be a linear map f: B_1 → B_2. Suppose f is bounded, i.e., there is k > 0 such that $\|f(x)\| \leq k\|x\|$ for all x ϵ B_1. Then one can show that f is continuous.

If we define $\|f\| = \text{g.l.b.}\{k|\ \|f(x)\| \le k\|x\|$ for all $x \in B_1\}$, then this norm makes the space of linear maps of Banach spaces, into a Banach space $\text{Lin}(B_1, B_2)$.

Let $U_1 \subseteq B_1$, $U_2 \subseteq B_2$ are open subsets of Banach spaces, and let $g: U_1 \to U_2$ be continuous, g is _differentiable_ at $x \in U_1$ with derivative $g' \in \text{Lin}(B_1, B_2)$ if for all $h \in B_1$ of sufficiently small norm $\epsilon(h) = g(x + h) - g(x) - g'(x)h$ satisfies $\lim_{|h|\to 0} \|\ \epsilon(h)/\|h\|\ \| = 0$. It is not difficult to check the usual linearity and the chain rule.

Next we shall introduce the notion of an infinite dimensional manifold modeled after some given topological vector space, rather than just usual R^n, and discuss some basic properties.

Let V be a topological vector space (usually a Banach space). A T_2 space M is called a _manifold modeled on V_, if for every $p \in M$ there is some open subset O_p of M, with $p \in O_p$, and a homeomorphism ϕ_p of O_p onto an open subset of V.

Of course there are many O_p and ϕ_p for each $p \in M$. This definition is very similar to an earlier definition of a finite dimensional manifold except that ϕ_p is only required to be a homeomorphism onto an open subset of V, not onto all of V. In fact, in general we shall not be able to specify the range of ϕ_p. It is clear that every finite dimensional manifold is a manifold modeled on R^n for some n.

If V is a topological vector space, then any open subset of V is a manifold modeled on V. If M^n is a finite dimensional manifold and B a Banach space, then $M^n \times B$ is a manifold modeled on the (Banach) space $R^n \times B$.

It is more convenient to work with norms in order to define an infinite dimensional manifold. For the time being, we shall restrict our attention to manifolds modeled on a Banach space and it will be called a _Banach manifold_.

Let B be a given Banach space, M is a _smooth manifold modeled on B_ if M is a manifold modeled on B such that if $O_p \cap O_q \ne 0$, then the composite map
$$\phi_p(O_p \cap O_q) \xrightarrow{(\phi^{-1}/\phi_?(O_?\cap O_q))} O_p \cap O_q \xrightarrow{\phi_q} \phi_q(O_p \cap O_q)$$
is a smooth map in the sense that it has continuous

derivatives of any order (obtained from iterating the definition of the differentiable at a given point).

Let M and N are manifolds modeled on Banach spaces. Let $f: M \to N$ be a continuous map. f is <u>smooth (or differentiable)</u> if for each $p \in O_p \subseteq M$ with homeomorphism $\phi_p: O_p \to \phi_p(O_p)$ and $q \in O_q \subseteq N$ with $\phi_q: O_q \to \phi_q(O_q)$, where $f(p) = q$, then in a sufficiently small neighborhood of p, $\phi_p(O_p) \xrightarrow{\phi_p^{-1}} M \xrightarrow{f} N \xrightarrow{\phi_q} \phi_q(O_q)$ is smooth (or differentiable).

Some of the notions for finite dimensional manifolds can easily be generalized to our current interest, nonetheless, occasionally some care has to be exercised.

<u>Submanifold</u> is defined by requiring that each point has a neighborhood homeomorphic to $O_1 \times O_2$ where O_1 and O_2 are open in B_1 and B_2 respectively. And the big manifold is modeled on $B_1 \times B_2$, and the manifold is described locally in terms of $O_1 \times \{pt\}$. One can easily check that such a submanifold is a manifold.

A <u>diffeomorphism</u> of infinite dimensional manifolds is a smooth map with a smooth, two sided inverse. An <u>imbedding</u> ϕ: $M \to N$ is a smooth one-to-one map that is a diffeomorphism of M onto a submanifold of N. An <u>immersion</u> is a smooth map which is locally an imbedding.

For a given smooth manifold M modeled on a Banach space B, let us choose a coordinate neighborhood O_p, for each $p \in M$, endowed with a homeomorphic ϕ_p onto an open set in B, i.e., $\phi_p: O_p \to \phi_p(O_p) \subseteq B$. Consider the sets $O_p \times B$ and define an equivalence relation in their union by specifying that if $(u, v_1) \in O_p \times B$, $(u, v_2) \in O_q \times B$, then (u, v_1) is equivalent to (u, v_2) iff $(\phi_q \cdot \phi_p^{-1})'(v_1) = v_2$ where the prime refers to the derivative taken at u. The quotient space is defined to be the <u>tangent bundle</u> TM and the projection map $\pi : TM \to M$ is defined by $\pi\{(u, v)\} = u$. And as before, the fiber is the tangent space at p, i.e., $\pi^{-1}(p) = M_p$.

As before, the differential of a smooth map between smooth manifolds can be defined. Then it is possible to characterize immersion in terms of the tangent space by requiring that at every point the differentials of the

125

immersion $(df)_p$: $M_p \to N_{f(p)}$ have a left inverse. By
generalizing the inverse function theorem to Banach spaces
one can prove the equivalence of the two notions of
immersion.

Since a separable Hilbert manifold, a manifold modeled
on L^2, is a fortiori a paracompact space, hence partition of
unity exists from a theorem in point-set topology. It can be
shown that such a manifold has a smooth partition of unity
[Kahn 1980, p.216].

<u>Theorem 3.4.1</u> Let M be a separable manifold modeled on
L^2, and let $\{B_\alpha\}$ be an open covering, where $B_\alpha(x)$ is an open
ball of radius $\alpha > 0$ about $x \in L^2$. Then there is a
countable, locally finite open covering $\{O_n\}$, a refinement
of $\{B_\alpha\}$, and there exists a smooth partition of unity
subordinate to this covering $\{O_n\}$.

Let M be a smooth m-dim Riemannian manifold, i.e., for
any $p \in M$, and vectors u, v $\in M_p$ we have a continuous,
symmetric, positive definite inner product defined by $(u,v)_p$
$= u^t(g(p))v$ and the norm is defined by $\|u\|_p = \sqrt{(u,u)_p}$.

Let X be any compact Hausdorff space, M be a Riemannian
manifold. We set $F(X,M) = \{f|$ f: $X \to M$, f continuous$\}$. A
metric on $F(X,M)$ can be introduced by
$$d(f,g) = l.u.b._{x \in X} \sigma(f(x),g(x)), \quad f,g \in F(X,M) \quad \text{where}$$
$\sigma(a,b)$ is the greatest lower bound of the length of all
smooth curves in M from a to b. It is easy to show that d
makes $F(X,M)$ into a metric space.

For a given f $\in F(X,M)$, one may define a tangent space
to $F(X,M)$ at that given map, denoted by $F(X,M)_f$, by letting
ϕ: $TM \to M$ be the projection map of the tangent bundle of M.
Then $F(X,M)_f$ is the set of all f': $X \to TM$ such that $\pi \cdot f' =$
f. Obviously one can make $F(X,M)_f$ into a linear space
because if we set $(f_1' + f_2')(x) = f_1'(x) + f_2'(x)$, where f_1',
f_2': $X \to TM$ and $x \in X$, then $\pi \cdot (f_1' + f_2')(x) = f(x)$ and
likewise for $\alpha f'$. One can also introduce a norm into $F(X,M)_f$
by setting $\|f'\|_f = l.u.b._{x \in X} \|f'(x)\|_{f(x)}$. Note that $\|f'(x)\|_{f(x)}$
is the norm in $M_{f(x)}$ defined by the Riemannian metric. It is
then straight forward to show that $F(X,M)_f$ is a complete

126

metric space with respect to the norm. That is, $F(X,M)_f$ is a real Banach space.

Theorem 3.4.2 If X is a compact Hausdorff space and M is a smooth Riemannian manifold, then $F(X,M)$ is a smooth manifold modeled on a real Banach space - $F(X,M)_f$, independent of the choice of f [Eells 1958, see also Eells 1966, Kahn 1980, p.218].

If here X and M are compact manifolds, as before, $F(X,M)_f$ be the vector space of all smooth liftings of $f \in F(X,M)$, i.e., $f' \in F(X,M)_f$ is smooth and $f': X \to TM$ such that $\pi \cdot f' = f$ where $\pi: TM \to M$ is the canonical projection. Earlier, we have pointed out that such space $F(X,M)_f$ is a complete linear vector space, thus a Fréchet space. (A Fréchet space is a topological vector space which is metrizable and complete). Thus it is not surprising to note that:

Theorem 3.4.3 The group of diffeomorphisms of M, Diff(M), is a locally Fréchet C^∞ group [Leslie 1967].

For the relations between the homeomorphism group and the diffeomorphism group of a smooth manifold and their homotopic type, one should consult [Burghelea and Lashof 1974a,b].

In addition to the manifold and differentiable structures for spaces of differentiable maps, one can also construct such structures to a more general class, the spaces of sections of fiber bundles.

Let α be a smooth vector bundle over a compact smooth manifold M, then define $S(\alpha)$ to be the set of all sections s of α such that $s \in S(\eta)$ for some open vector subbundle η of α, i.e., $S(\alpha) = U\eta \, S(\eta)$ where the union is over all open vector subbundles η of α. Then one can show that $S(\alpha)$ not only is a Banach manifold, but also has a unique differentiable structure [see Palais 1968].

We have only scratched the surface of this evolving and very interesting area. Interested readers are urged to consult Lang [1962], Eells [1966], Burghelea and Kuiper [1969], Eells [1958], Eells and Elworthy [1970], Marsden,

127

Ebin, and Fischer [1972], and some papers in Anderson [1972], Palais [1965, 1966a,b, 1971].

Although most results in finite dimensional manifolds can easily be extended to infinite dimensional manifolds, nonetheless, there are a few surprises. It has been established that every separable, metrizable C^0-manifold can be C^0-imbedded as an subset of its model [Henderson 1969]. In other words, any reasonable Hilbert manifold is equivalent to an open subset of L^2 space. This is contrary to the case of finite dimensional manifold!

Let M be a C^∞ manfold modeled on any infinite dimensional Hilbert space E, Kuiper [1965] asserts that M is parallelizable. Eells and Elworthy [1970] observe that there is a diffeomorphism of M onto MxE. They also observe that if M and N are two C^∞ manifolds modeled on E and if there is a homotopy equivalence ϕ: M → N, then ϕ is homotopic to a diffeomorphism of M onto N. As a corollary, the differentiable structure on M is unique. Again, these results are different from the finite dimensional cases as we have pointed out earlier in Sect.2.4 [see also Milnor 1956].

3.5 Differential operators

Most of the basics of linear differential operators may be generalized to manifolds by piecing together differential operators on Euclidean spaces, but it turns out much more elegant and convenient to give a general definition in terms of vector bundles. We shall adopt this approach to avoid cumbersome details involving coordinate charts at the beginning. After that, we shall discuss various important cases and examples of differential operators. Finally, we shall briefly discuss the important notions of ellipticity, the symbol of an operator, linearization of nonlinear operators, and the analytic index of operators. References for further reading will be provided.

Let (B_1,M,π_1) and (B_2,M,π_2) be two smooth vector bundles over an m-dim compact smooth manifold M, and let $C^\infty(M,B_i)$ be

128

the vector space of sections of the bundle (B_i, M, π_i), here i = 1, 2. A _linear differential operator_ is a linear map of vector spaces $P: C^\infty(M, B_1) \to C^\infty(M, B_2)$ such that supp $P(s) \subseteq$ supp s, where supp s = closure of $\{p \in M \mid s(p) \neq 0\}$.

Note that this definition is very elegant and simple, nonetheless, some work is needed so that we can "visualize" that these operators are locally generated by differentiation. In the following, we shall specify the smooth manifold M and its vector bundles, the linear map, and sections to make the definition "visualizable".

If $M = R^m$ and B_1, B_2 are trivial 1-dim vector bundles (line bundles) over M, and if $\alpha = (\alpha_1, \ldots, \alpha_m)$ is an index set, then for a smooth f one can define a linear differential operator D^α by $D^\alpha f = \partial^{|\alpha|} f / \partial x_1^{\alpha_1} \ldots \partial x_m^{\alpha_m}$. If P is any polynomial over the ring $C^\infty(M, R)$ in m variables $z_1, \ldots, z_m$, then if we substitute $\partial/\partial x_i$ for z_i, then the resulting polynomial gives a linear differential operator P : $C^\infty(M, R) \to C^\infty(M, R)$ by $P(f) = P(\partial/\partial x_1, \ldots, \partial/\partial x_m)(f)$. If s vanishes in an open set, every term of $P(s)$ and $P(s)$ itself vanishes on that open set. Clearly supp $P(s) \subseteq$ supp s.

Let B_1, B_2 and M be as before, and let $P: C^\infty(M, B_1) \to C^\infty(M, B_2)$ be a linear differential operator. Then we say that _P has order k at the point $p \in M$_ if k is the largest non-negative integer such that there is some $s \in C^\infty(M, B_1)$ and some smooth function f defined in an open neighborhood of p and vanishing at p such that $P(f^k s)(p) \neq 0$. The _order of P_ is the maximum of the orders of P at all points of M.

It is easy to check that this notion of order for P agrees with the usual definition of order for a linear differential operator defined on Euclidean space. Here a word of caution is called for. Recall in our definition of differential operator and its order, we assume that M being compact. If it happens that M is noncompact, then the order of a linear operator may not exist. For example, $M = R^1$ and choose $\phi_i \in C^\infty(R^1) = C^\infty(R^1, R^1)$ with support supp$\phi_i \subseteq [i, i+1]$ and $\phi_i(i+1/2) > 0$. Set $P(f)(p) = \Sigma_i \phi_i(p) d^i f(p)/dx^i$. Clearly, with such construction, P is a linear differential operator, but the order of P is not defined.

Fortunately, the next theorem asserts that a linear differential operator locally will have an order, and consequently a linear differential operator defined on a compact manifold will have a finite order.

Theorem 3.5.1 (local theorem)   Let $O \subseteq R^n$ be open and let P: $C^\infty(O,\Phi) \rightarrow C^\infty(O,\eta)$ be a linear differential operator from the sections of the trivial s-dim vector bundle Φ over O to those of the trivial t-dim vector bundle η over O. Let $O_1 \subseteq O$ have a compact closure contained in O. Let $V(\Phi,\eta)$ be the vector space of linear maps from R^s to R^t. Then there is an $m \geq 0$ such that for every multi-index α, $|\alpha| \leq m$, there are C^∞ maps $g_\alpha: O_1 \rightarrow V(\Phi,\eta)$ such that for any $f \in C^\infty(O_1,\Phi)$, $p \in O_1$, $(Pf)(p) = \Sigma_{|\alpha| \leq m} g_\alpha(p)(D^\alpha f)(p)$ [Kahn 1980, p.194; Peetre 1960].

The next theorem has the essential idea and it is a global theorem for differential operators defined on a manifold.

Theorem 3.5.2 (global theorem)   Let M be a smooth manifold, and let B_1 and B_2 be two vector bundles over M, with dim $B_i = a_i$. Let P: $C^\infty(M,B_1) \rightarrow C^\infty(M,B_2)$ be a linear differential operator. Take $p \in M$, then there is a coordinate neighborhood of p, U, over which both bundles are trivial and a positive integer m so that in U, $(Pf)(p) = \Sigma_{|\alpha| \leq m} g_\alpha(p)(D^\alpha f)(p)$ for smooth maps $g_\alpha: U \rightarrow V(B_1,B_2)$ [Kahn 1980, p.196].

In the last section we have mentioned that the space of sections of a vector bundle is a Banach manifold and has a unique differentiable structure. We shall say a little bit more and state the theorem due to Hormander [1964].

Let (B,M,π) be a smooth vector bundle, $s_i \in C^\infty(M,B)$ be a sequence of C^∞ sections of B, and a given fixed section s. We say that s_i converges to s locally uniformly if for any $p \in M$ there is a coordinate neighborhood U of p over which B is trivial such that in U, s_i and $D^\alpha s_i$ converge uniformly to s and $D^\alpha s$, respectively. For two given smooth vector bundles B_1 and B_2, a linear map L: $C^\infty(M,B_1) \rightarrow C^\infty(M,B_2)$ is weakly

<u>continuous</u> if whenever s_i converges to s locally uniformly, then $L(s_i)$ converges to $L(s)$ uniformly over $K \subseteq M$.

The term weakly refers to the fact that we have not yet topologized the vector spaces $C^\infty(M, B_i)$, so it does not make sense to ask whether L is continuous. Nonetheless, clearly any classically defined linear differential operator is weakly continuous.

<u>Theorem 3.5.3</u> Let B_1 and B_2 be two smooth vector bundles over M, let L: $C^\infty(M, B_1) \to C^\infty(M, B_2)$ be a weakly continuous linear map. The necessary and sufficient condition for L to be a linear differential operator of order $\leq$ m is that: for any $s \in C^\infty(M, B_1)$, $p \in M$, and f a smooth function of M, then there exist a function g in x from R to $R^{\dim B_1}$, $g(x)(p) = e^{-ixf(p)}[L(e^{ixf}s)(p)]$ such that in each coordinate it is a polynomial of degree $\leq$ m.

For the proof we refer the reader to [Hormander 1964; Kahn 1980, p.198].

We wish to define the symbol and ellipticity of a linear differential operator before we discuss nonlinear differential operators. Locally, the symbol of a linear partial differential operator of order m defined over Euclidean space R^n can be thought of by: (a) ignoring the terms of order less than m, and (b) in each term, replacing $\partial/\partial x_i$ by Φ_i thus obtaining a form in the variables Φ_i with smooth functions as coefficients. For higher derivatives, they are written as a power of Φ_i, i.e., replace $\partial^k/\partial x_i^k$ by Φ_i^k. If the symbol is definite, i.e., all the coefficients are positive, or in other words, the symbol vanishes only when all the variables are set equal to zero, then we say that the linear differential operator is elliptic.

For instance, the 3-dim Laplacian $\equiv \partial^2/\partial x^2 + \partial^2/\partial y^2 + \partial^2/\partial z^2$ has symbol $\Phi_1^2 + \Phi_2^2 + \Phi_3^2$ and clearly it is elliptic. The linear operator L = $a\partial^2/\partial x^2 + b\partial^2/\partial y^2 - c\partial/\partial x + d\partial/\partial y$ has symbol $a\Phi_1^2 + b\Phi_2^2$ and it is elliptic. But the wave operator $\partial^2/\partial x^2 + \partial^2/\partial y^2 + \partial^2/\partial z^2 - c^{-2}\partial^2/\partial t^2$ has symbol $\Phi_1^2 + \Phi_2^2 + \Phi_3^2 - c^{-2}\Phi_4^2$ is not elliptic. The term elliptic comes from the theory of quadratic forms as it relates to conic sections. Thus, there are also linear differential operators of the types parabolic and hyperbolic. For instance, the wave equation is hyperbolic, and the diffusion equation is parabolic.

In order to treat the symbol in a global setting, one cannot just consider a single form, we need the following machinery:

Given a smooth m-dim manifold M and two smooth vector bundles over M, (B_i, M, π_i), $(i = 1, 2)$. Let T^*M be the cotangent bundle of M with projection p. Let P: $C^\infty(M, B_1) \rightarrow C^\infty(M, B_2)$ be a linear differential operator of order k. Let us now consider the induced bundle $p^{-1}(B_1) = \{(u,v) \in T^*MxB_1 | \pi_1(v) = p(u)\}$. Let $a \in M$, $\Omega \in M_a^*$. Let f be a smooth function in a neighborhood of a where $f(a) = 0$ and $df(a) = \Omega$. For $e \in \pi_1^{-1}(a)$, let s be a smooth section of (B_1, M, π_1) such that $s(a) = e$. Now set $\sigma_p(\Omega, e) = P(f^k s)(a)$. Since $\Omega \in M_a^*$ and $e \in \pi_1^{-1}(a)$, clearly $(\Omega, e) \in p^{-1}(B_1)$.

We call the map σ_p: $p^{-1}(B_1) \rightarrow B_2$ the symbol of the differential operator P. For each $a \in M$ and $\Omega \in M_a^*$, σ_p is a map from $\pi_1^{-1}(a)$ to $\pi_2^{-1}(a)$. A linear differential operator P is elliptic if for each nonzero $\Omega \in M_a^*$, $a \in M$, the map σ_p is one- to-one.

As a simple example, consider the two-dimensional Laplacian in the plane $\Delta = \partial^2/\partial x^2 + \partial^2/\partial y^2$ in a trivial one-dimensional vector bundle over R^2. Then $\pi^{-1}(B)$ is a trivial one-dimensional vector bundle over $R^2 x R^2 = R^4$. For a given 1-form α in R^2, $\alpha = adx + bdy$, choose $f(x,y) = ax + by$. Let the section be $s(x,y) = ((x,y), m)$, where m is a variable. Then
$$(\partial^2/\partial x^2 + \partial^2/\partial y^2)((ax + by)^2 m) = 2a^2 m + 2b^2 m.$$
If a and b are not both zero, this is clearly a one-to-one map in terms of the variable m, thus Δ is elliptic.

As another example, let M be a smooth n-dim manifold and let $\Lambda^k(T^*M)$ be the bundle of k-forms. We have a linear differential operator of order 1, d : $D^k(M) \rightarrow D^{k+1}(M)$ where $D^k(M) = C^\infty(M, \Lambda^k(T^*M))$ are smooth differential k-forms. When i = 0, it is easy to show that the symbol σ_d of d is elliptic.

132

Let f be a smooth function of M, $\alpha \in M_a^*$ and $(df)(a) = \alpha$.
Let $s \in D^0(M) = C^\infty(M)$ and $s(a) = e$. Then in a local
coordinate chart,

$\sigma_d(\alpha,e) = d(fs)(a) = \Sigma_{i=1} (\partial fs(a)/\partial x_i) dX_i$
$= e \cdot \alpha + \Sigma_{i=1} f(a)(\partial s(a)/\partial x_i) dX_i$.

Since we may choose any s such that $s(a) = e$, we choose $s(x)$
= e identically, i.e., $\partial s(a)/\partial x_i = 0$. Thus $\sigma_d(\alpha,e) = e\alpha$, and
$d : D^0(M) \to D^1(M)$ is elliptic.

    With the same information as provided earlier, it is
known   [Ch.1 of Palais 1965] that the vector spaces for an
elliptic differential operator P of positive order on a
compact manifold

    Ker P = $\{f \in C^\infty(M,B_1) | P(f) = 0\}$ and

    Coker P = $C^\infty(M,B_2)/ \{g \in C^\infty(M,B_2) | g = Pf\}$ are both
finite dimensional. The <u>analytic index of P</u> is an integer

        $i_a(P) = dim (Ker P) - dim (Coker P)$.

It is known that this index is invariant under deformation
of P, and thus suggests that there might be a topological
description of i_a. This has led to the theory of the
topological index and the theorem of Atiyah and Singer which
asserts that these two indices are the same.

    For the definition of the topological index see [Ch. 1,
3, 4 of Palais 1965], analytic index see [Ch. 1, 5 of Palais
1965], both analytic and topological indeces on unit
ball-bundle or unit sphere bundle of M see [Ch.15 of Palais
1965], and the index theorem see [App.I of Palais 1965],
applications see [Ch.19 of Palais 1965] and the topological
index of elliptic operators see [App.2 of Palais 1965]. A
very good review of differential operators on vector bundles
is Ch. 4 of [Palais 1965]. A good source of reference on the
subject of Atiyah-Singer index theorem is the book edited by
Palais [1965]. For those readers who want to go to the
source, the series of papers by Atiyah and Singer [1963,
1968] are recommended.

    So far we have defined linear differential operators,
the symbol of a linear operator, the ellipticity and
analytic index of a linear differential operator. We shall

133

briefly introduce the notion of nonlinear differential operators, the linearization and the symbol of such operators, and the index of a nonlinear elliptic operator.

Recall that, let $\pi: B \to M$ be a C^∞ fiber bundle over M. Given $b_0 \in B$ with $\pi(b_0) = p_0$ and local sections s_1, s_2 of B defined near p_0 with $s_1(p_0) = s_2(p_0) = b_0$. By choosing a chart at $p_0 \in M$, a local trivialization of B near p_0, and a chart near b_0 in the fiber $\pi^{-1}(p_0)$, we can define the kth order Taylor expansions of s_1 and s_2 at p_0. If the Taylor expansions are the same for one set of choices, they will be the same for any other (i.e., all in the same equivalence class), then we say that s_1 and s_2 have the same k-jet at p_0. Recall again, this set of equivalence classes is denoted by $J^k(B)_b$ and the equivalence class of s is denoted by $j_k(s)_p$. Let $J_0^k(B) = U_{b \in B} J^k(B)_b$ and let $\pi_0^k: J_0^k(B) \to B$ maps $J_0^k(B)_b$ to b. It can be shown that π_0^k has the structure of a C^∞ fiber bundle over B whose fiber at b_0 is, $\oplus^k_{m=1} L_s^m(M_{p_0}, (B_{p_0})_{b_0})$, all polynomial maps of degree less then or equal to k from M_{p_0} into $(B_{p_0})_{b_0}$. We define a C^∞ fiber bundle $\pi^k: J^k(B) \to M$ whose total space is $J_0^k(B)$ and the projection is $\pi^k = \pi \cdot \pi_0^k$, i.e., $\pi^k(j_k(s)_{p_0}) = p_0$. If $0 \leq n \leq k$, we can generalize the fiber bundle by defining $\pi_n^k : J_n^k(B) \to J^n(B)$ and whose total space is $J^k(B)$ and the projection is given by $\pi_n^k(j_k(s)_{p_0}) = j_n(s)_{p_0}$. We then have a natural map, the k-jet extension map, j_k: $C^\infty(B) \to C^\infty(J^k(B))$, defined by $j_k(s)(p) = j_k(s)_p$. If Φ is a vector bundle neighborhood of $s \in C^\infty(B)$ then $J^k(\Phi)$ is a vector bundle, and in fact it is a vector bundle neighborhood of $j_k(s) \in C^\infty(J^k(B))$. Furthermore, j_k: $C^\infty(B) \to C^\infty(J^k(B))$ restricts to a linear map of $C^\infty(\Phi)$ to $C^\infty(J^k(\Phi))$.

Equipped with the knowledge of proper relationships, we recall that if Φ and η are vector bundles over M, then a linear differential operator of order k from $C^\infty(\Phi)$ to $C^\infty(\eta)$ is a linear map P: $C^\infty(\Phi) \to C^\infty(\eta)$ and it can be factored as a composition $C^\infty(\Phi) \xrightarrow{j_k} C^\infty(J^k(\Phi)) \xrightarrow{f_*} C^\infty(\eta)$ where f: $J^k(\Phi) \to \eta$ is a C^∞ vector bundle morphism over M. By analogy, it is clear how to define a non-linear differential operator of order k.

Let B_1 and B_2 be C^∞ fiber bundles over M. A mapping D:

$C^\infty(B_1) \to C^\infty(B_2)$ is called a <u>non-linear differential operator</u> <u>of order k</u> from $C^\infty(B_1)$ to $C^\infty(B_2$ if it can be factored as $C^\infty(B_i) \xrightarrow{j_k} C^\infty(J^k(B_1)) \xrightarrow{f_*} C^\infty(B_2)$ where $f: J^k(B_1) \to B_2$ is a C^∞ fiber bundle morphism over M. Let us denote the set of all differential operators of order k from $C^\infty(B_1)$ to $C^\infty(B_2)$ (or from B_1 to B_2 for simplicity) by $D_k(B_1,B_2)$.

Theorem 3.5.4 If $l \le k$, $D_l(B_1,B_2) \subseteq D_k(B_1,B_2)$ we can construct higher order differential operators from the lower order ones by repeating the jet extension map (or by induction method through jet extension map) as indicated in the following lemma.

Lemma 3.5.5 Let $j_k: C^\infty(B) \to C^\infty(J^k(B))$ and $j_l': C^\infty(J^k(B)) \to C^\infty(J^l(J^k(B)))$ be jet extension maps. Then $j_l' \cdot j_k$ is a differential operator of order k+l from B to $J^l(J^k(B))$.

Using the definition of $D_k(B_1,B_2)$ and the above lemma, we can prove the following:

Theorem 3.5.6 If $P_1 \in D_k(B_1,B_2)$, and $P_2 \in D_l(B_2,B_3)$, then $P_2P_1 \in D_{k+l}(B_1,B_3)$.

Now we shall see what a nonlinear differential operator really looks like in local coordinates. Let us choose coordinates $x_1,\ldots,x_m$ in a neighborhood U of p_0 in M, coordinates $y_1,\ldots,y_n$ in a neighborhood V of b_0 in B_p and using a local trivialization to identify a neighborhood of b_0 in B with U x V so that π restricted to U x V is a projection on the first component. A section s of B over U with $s(p) \in V$ for $p \in U$ is given by a map $p \to (p,s(p))$ of U into U x V. Let $s_i(p) = y_i(s(p))$, then $j_k(s)$ in coordinate form is given by a map $p \to (p,s(p),y_i^\alpha(j_k(s)(p)))$ where $y_i^\alpha(j_k(s)(p)) = D^\alpha s_i(p)$, and as usual,
$$D^\alpha \equiv \partial^{|\alpha|}/\partial x_1^{\alpha_1} \ldots \partial x_m^{\alpha_m}, \text{ where } 0 \le |\alpha| \le k.$$

Now let $\pi': \eta \to M$ be a C^∞ vector bundle over M and let P $\in D_k(B,\eta)$, say $P = f_*j_k$ where $f: J^k(B) \to \eta$ is a C^∞ fiber bundle morphism. Let $v_1,\ldots,v_r$ be a basis of C^∞ sections of η over U. If we restrict f to the part of $J_0^k(B)$ over U x V, it is given by f_j (j = 1,..,r) with coordinates $(x_1,\ldots,x_m,y_1\ldots,y_n,y_i^\alpha)$ by $(x,y,y_i^\alpha) \to \Sigma^r_{j=1} f(x,y,y_i^\alpha)v_j(x)$. Then the explicit expression for Ps is

135

$Ps(x) = \Sigma^r_{j=1} f_j(x, s_1(x), \ldots, s_n(x), D_s^\alpha(x) v_j(x)$.

We shall call the ordered r-tuple of functions $f_j(x,y,y_i)$ the <u>parametric expressions for the operator P near b_0</u> relative to the choices made.

By utilizing the linear structure in η, it is possible to single out certain vector subspaces of the vector space $D_k(B,\eta)$ of kth order differential operators from B to η which are polynomial functions of certain derivatives. These classes play an important role in nonlinear analysis, particularly in calculus of variations.

Let B be a C^∞ fiber bundle over M, Φ a C^∞ vector bundle over M and $P \in D_k(B,\Phi)$. Let w and l are integers with $w > 0$, and $0 \le u < k$. We say that <u>P is a polynomial of weight $\le$ w with respect to derivatives of order $> u$</u> (denoted by $P \in D_k^{w;u}(B,\Phi)$) if for each parametric representation of P as above, and each of the functions $f_j(x,y,y_i^\alpha)$ can be written as a sum of functions of the form $G(x,y,y_i^\tau) y_{l_i}^{\beta_l} \ldots y_{l_q}^{\beta_q}$ where all $|\tau| \le u$, all $|\beta_i| > u$, and $|\beta_1| + \ldots + |\beta_q| \le w$. We abbreviate $D_k^{w;o}(B,\Phi)$ to $D_k^w(B,\Phi)$ and elements of this space are referred as polynomial differential operators of order k and weight $\le$ w.

As a special case, elements of $D_k^{k;k-1}(B,\Phi)$, i.e., the f_j of a parametric representation are linear in derivatives of order k, are called <u>quasi-linear differential operators of order k from B to Φ</u>.

It is evident from the definition that $D_o(B,\Phi) \subseteq D_k^{o;u}(B,\Phi)$ for all k and $u < k$. Nonetheless, it is not immediately clear that $D_k^{w;u}(B,\Phi)$ form a subspace of $D_k(B,\Phi)$ with positive dimension if $w > 0$. The next theorem shows how to construct lots of operators in $D_k^{w;u}(B,\Phi)$.

<u>Theorem 3.5.7</u> Let B be a C^∞ fiber bundle over M, Φ a C^∞ vector bundle over M and $P \in D_k(B,\Phi)$. In order that $P \in D_k^{w;u}(B,\Phi)$, it is sufficient that for each $b_0 \in B$ there exist at least one parametric expression for P near b_0 satisfying the conditions in the above definition of polynomial differential operators of weight $\le$ w.

It should be noted that in order to prove the theorem,

136

it suffices to assume that B is a vector bundle. This is not surprising. After all, Theorem 3.5.7 is a local statement. With a choice of coordinates, there are many formulas and results of polynomial differential operators. Since some of them will involve some results in functional analysis, which we have neither the time nor the space to discuss, we shall only recommend a few references for further reading [Palais 1965, Ch.4; Palais 1968; Berger 1977].

Let V be a function which associates to each C^∞ vector bundle Φ over a compact n-dim C^∞ manifold a complete, normable, topological vector space $V(\Phi)$ which includes $C^\infty(\Phi)$.

Axiom A (for V)   Let M and N be compact n-dim C^∞ manifolds and let $\phi\colon M \to N$ be a diffeomorphism of M into N. If Φ is a vector bundle over N, then $s \to s\cdot\Phi$ is a continuous linear map of $V(\Phi)$ into $V(\phi^*\Phi)$.

In usual examples, this map is onto. If $M \subseteq N$ and ϕ is inclusion, then the Axiom says that restriction is continuous from $V(\Phi)$ to $V(\Phi/M)$ and the onto-ness expresses the possibility of extending $s \in V(\Phi/M)$ to an element of $V(\Phi)$.

Axiom B   If Φ is a vector bundle over a compact C^∞ m-dim manifold M then $V(\Phi) \subseteq C^0(\Phi)$ and the inclusion map is continuous. Furthermore, if η is another vector bundle over M and f$\colon \Phi \to \eta$ is a C^∞ fiber preserving map, then $f_*\colon C^0(\Phi) \to C^0(\eta)$ restricts to a continuous map $V(f)\colon V(\Phi) \to V(\eta)$.

Let Φ be a C^∞ vector bundle over a compact C^∞ m-manifold M, and let $V_{(k)}(\Phi) = \{s \in C^k(\Phi) \mid j_k(s) \in V(J^k(\Phi))\}$. Thus j_k is a continuous linear injection of $V_{(k)}(\Phi)$ into the Banach space $V(J^k(\Phi))$ and $V_{(k)}(\Phi)$ become a normable topological vector space if we topologize it by requiring that j_k be a homeomorphism into. Let $V_k(\Phi)$ to be the completion of $V_{(k)}(\Phi)$ so that j_k extends to a continuous linear isomorphism of $V_k(\Phi)$ onto a closed linear subspace of $V(J^k(\Phi))$.

With these axioms stated and notations defined, we are ready to discuss briefly the linearization and the symbol of a differential operator.

Let B_1 and B_2 be C^∞ fiber bundles over a compact n-dim C^∞ manifold M and let P: $C^\infty(B_1) \to C^\infty(B_2)$ be an element of $D_k(B_1, B_2)$, let us say $P = f_* j_k$ where f: $J^k(B_1) \to B_2$ is a fiber bundle morphism over M. Let $\mathbf{V}$ satisfy Axiom A and B so that P extends to a C^∞ map of $\mathbf{V}_k(B_1)$ into $\mathbf{V}(B_2)$. If $s \in C^\infty(B_1)$, then dP_s is a linear map of $(\mathbf{V}_k(B_1))_s$ into $(\mathbf{V}(B_2))_{Ps}$. It can be shown that there exists $\Lambda(P)_s \in \text{Diff}_k((B_1)_s, (B_2)_{Ps})$, the "<u>linearization of P at s</u>", such that $\Lambda(P)_s$: $C^\infty((B_1)_s) \to C^\infty((B_2)_{Ps})$ extends to dP_s. Furthermore, $\Lambda(P)_s$ depends only on P but not on $\mathbf{V}$. The linear differential operator $\Lambda(P)_s$ has a <u>symbol</u> $\sigma_k(\Lambda(P)_s)$. This symbol is a function on the cotangent bundle of M, $T^*(M)$, and for $(v,x) \in T^*(M)$, $\sigma(\Lambda(P)_s)(v,x)$ is a linear map of the fiber of the tangent space of $(B_1)_x$ at $s(x)_1$, $((B_1)_x)_{s(x)}$, into the the fiber of $((B_2)_x)_{Ps(x)}$. Furthermore, $\sigma_k(\Lambda(P)_s)(v,x)$ depends only on $j_k(s)(x)$ and v, thus we can denote it by $\sigma_k(P)(j_k(s)(x),(v,x))$. In other words, $\sigma_k(P)$ is a function defined on the total space of the fiber product $J^k(B_1) \times_M T^*(M)$, which we denote by $T^*_k(B_1)$. There are also two natural maps π: $T^*_k(B_1) \to B_1$ and f': $T^*_k(B_1) \to B_2$ defined by $\pi(j_k(s)(x),(v,x)) = s(x)$ and $f'(j_k(s)(x),(v,x)) = fj_k(s)(x) = Ps(x)$. But these give rise to two C^∞ vector bundles over $T^*_k(B_1)$, i.e., $\pi^*(Tf(B_1))$ and $f'^*(Tf(B_2))$ and their fibers at $(j_k(s)(x),(v,x))$ are $((B_1)_x)_{s(x)}$ and $((B_2)_x)_{Ps(x)}$ respectively. Thus $\sigma(P)$ is an element of $\text{Hom}(\pi^*(Tf(B_1)), f'^*(Tf(B_2)))$, and this vector bundle homomorphism over $T^*_k(B_1)$ is called the <u>(k-th order) symbol of the differential operator P</u>.

Theorem 3.5.8 Given $s \in C^\infty(B_1)$, $\delta_{j_k(s)}f$ is a C^∞ vector bundle homomorphism of $J^k(B_1)_s$) into $(B_2)_{Ps}$ and hence defines an element $\Lambda(P)_s \in \text{Diff}_k((B_1)_s, (B_2)_{Ps})$ called the <u>linearization of P at s</u>. If Φ is a vector bundle neighborhood of s in B_1 (so we may identify $(B_1)_s$ with Φ), then for $\sigma \in C^\infty(\Phi)$, $\Lambda(P)_s\sigma(x)$ depends only on $j_k(s)(x)$ and $j_k(\sigma)(x)$. Moreover, $\Lambda(P)_s(\sigma)(x) = d/dt|_{t=0}(P(s+t\sigma)(x))$.

Corollary 3.5.9 Let $s \in C^\infty(B_1)$, Φ_1 a vector bundle neighborhood of s in B_1 and Φ_2 a vector bundle neighborhood of Ps in B_2. Then for $\sigma \in C^\infty(\Phi_1)$, $(P(s+t\sigma) - Ps)/t$ converges in the C^∞ topology to $\Lambda(P)_s(\sigma)$ as $t \to 0$.

We shall now derive an expression for $\Lambda(P)_s$ in local coordinates. We can suppose $M = D^n$, and we can replace B_1 and B_2 with vector bundle neighborhoods of s and Ps, which we can identify with $M \times R^m$ and $M \times R^q$ respectively. Then a section s of B_1 is given by n real functions of $x = (x_1, \ldots, x_n) \in D^n$, $s(x) = (s_1(x), \ldots, s_m(x))$ and similarly Ps is given by q-real functions of x, $Ps(x) = ((Ps)_1(x), \ldots, (Ps)_q(x))$ and when we say $P \in D_k(B_1, B_2)$ we mean that there exist q of C^∞ functions $f_j(x, y_i, y_i^\alpha)$ ($i = 1, \ldots, m$, and α range over n-multi-indices with $|\alpha| \leq k$) such that $(Ps)_j(x) = f_j(x, s_i(x), P^\alpha s_i(x))$. Recall that the f_is defined earlier are called the parametric representation of P. Now $\Lambda(P)_s \in \text{Diff}_k(B_1, B_2)$. Similarly, there are q C^∞-functions $g_j(x, y_i, y_i^\alpha)$ such that if $\sigma = (\sigma_1, \ldots, \sigma_m)$ is a C^∞-section of B, then $(P)_s(\sigma)(x) = (g_1(x, \sigma_i(x), P^\alpha\sigma_i(x)), \ldots, g_q(x, \sigma_i(x), P^\alpha\sigma_i(x)))$. Since $\Lambda(P)_s$ is a linear differential operator, $g_j(x, y_i, y_i^\alpha)$ are functions linear in (y_i, y_i^α). That is, there exists C^∞-functions of x, A_i^j and $A^j_{\alpha,i}$ such that
$$g_j(x, y_i, y_i^\alpha) = \Sigma_i\, A_i^j(x)y_i + \Sigma_{\alpha,i}\, A^j_{\alpha,i}(x)y_i^\alpha.$$
The problem is to express these A_i^j and $A^j_{\alpha,i}$ in terms of the f_j and s. As we shall see in the following theorem, the answer is $A_i^j(x) = (\partial f_j/\partial y_i)(x, s_i(x), P^\alpha s_i(x))$ and
$$A^j_{\alpha,i} = (\partial f_j/\partial y_i^\alpha)(x, s_i(x), P^\alpha s_i(x)).$$
More precisely, we have the following theorem.

<u>Theorem 3.5.10</u>   Let $P \in \text{Diff}_k(B_1, B_2)$ be given parametrically by $P(s) = (f_1(x, s_i(x), P^\alpha s_i(x)), \ldots, f_q(x, s_i(x), P^\alpha s_i(x)))$, then $\Lambda(P)_s$ is given parametrically by $\Lambda(P)_s(\sigma)(x) = (g_1(x, \sigma_i(x), P^\alpha\sigma_i(x)), \ldots, g_q(x, \sigma_i(x), P^\alpha\sigma_i(x)))$, where $g_q(x, y_i, y_i^\alpha) = \Sigma_i(\partial f_j/\partial y_i)(x, s_i(x), P^\alpha s_i(x))y_i$
$$+ \Sigma_{\alpha,i}(\partial f_j/\partial y_i^\alpha)(x, s_i(x), P^\alpha s_i(x))y_i^\alpha.$$

Let us recall the definition of the symbol of a linear differential operator. Let Φ_1 and Φ_2 be C^∞ vector bundles over M and let $P \in \text{Diff}_k(\Phi_1, \Phi_2)$. If v is a cotangent vector of M at x, the symbol of P at (v, x) is a linear map $\sigma_k(P)(v, x)$ of $(\Phi_1)_x$ into $(\Phi_2)_x$ defined as: choosing any C^∞ function g on M such that $g(x) = 0$, and $dg_x = v$ and given e

ϵ $(\Phi_1)_x$ choose any C^∞ section f of Φ_1 such that f(x) = e. Then $\sigma_k(P)(v,x)e = (k!)^{-1}P(g^k f)(x)$. It is immediate from the definition of the symbol that if $P = F_* j_k$ where F is a vector bundle homomorphism of $J^k(\Phi_1)$ into Φ_2, then $\sigma_k(P)(v,x)$ depends on F only through its value at x. Indeed, $\sigma_k(P)(v,x)e = F_x(j_k(g^k f)(x)/k!)$ where g and f as before. Let us now restate the definition of the symbol of $P \epsilon D_k(B_1,B_2)$ in a more compact form.

Let B_1 and B_2 be C^∞ fiber bundle over M and let $P \epsilon$ $D_k(B_1,B_2)$, say $P = f_* \cdot j_k$ where f: $J^k(B_1) \to B_2$ is a C^∞ fiber bundle morphism. Let $T_k^*(B_1)$ be the fiber bundle over M which is the fiber product of $J^k(B_1)$ and $T^*(M)$ and one defines the maps π and f' of $T_k^*(B_1)$ into B_1 and B_2 respectively by $\pi(j_k(s)_x,(v,x)) = s(x)$ and $f'(j_k(s),(v,x)) = f(j_k(s)_x) = Ps(x)$. The symbol of P, $\sigma_k(P)$, is then defined to be the element of $\text{Hom}(\pi^* Tf(B_1), f'^* Tf(B_2))$ given by $\sigma_k(P)(j_k(s)_x,(v,x)) = \sigma_k(\Lambda_k(P)_s)(v,x)$ where $\Lambda_k(P)_s$ is the linearization of P at s. (See Theorem 3.5.8 for the definition of $\Lambda(P)_s$).

The next theorem describes how to compute $\sigma_k(P)$ in local coordinates.

<u>Theorem 3.5.11</u> Let M = D^n, B_1 = $M \times R^m$, B_2 = $M \times R^q$ and P ϵ $D_k(B_1,B_2)$ be given by
Ps(x) = $(f_1(x,s_1(x),P^\alpha s_i(x)),\ldots,f_q(x,s_i(x),P^\alpha s_i(x)))$
where $f_j(x,y_i,y_i^\alpha)$ are C^∞-function of x = $(x_1,\ldots,x_n)$, Y_i = $(y_1,\ldots,y_m)$ and y_i^α (i = 1,..,m and α ranges over all n-multi- indices with $1 \le |\alpha| \le k$). Since each fiber of $Tf(B_1)$ and $Tf(B_2)$ is canonically isomorphic to R^m and R^q respectively, $\sigma_k(P)(j_k(s)_x,(v,x))$ is given by a (q x m)-matrix $\sigma_{ij}(x,j_k(s)_x,(v,x))$. If v = Σ $v_i dx_i$ then this matrix is given explicitly by:
$\sigma_{ij}(x,j_k(s)_x,(v,x)) = \Sigma_{|\alpha|=k} v^\alpha (\partial f_j/\partial y_i^\alpha)(x,s_i(x),P^\alpha s_i(x))$
where $v^\alpha = v_1^\alpha \ldots v_n^\alpha$.

Let Φ_1 and Φ_2 be C^∞ vector bundles over M and let P = $f_* \cdot j_k$ ϵ $\text{Diff}_k(\Phi_1,\Phi_2)$ where f: $J^k(\Phi_1) \to \Phi_2$ is a C^∞ vector bundle morphism. It is natural to ask what is the connection between $\sigma_k(P)$ ϵ $\text{Hom}(p^*\Phi_1,p^*\Phi_2)$, where p: $T^*(M) \to M$ is the natural projection, and $\sigma_k(P)$ ϵ $\text{Hom}(\pi^* Tf(\Phi_1),\pi^* Tf(\Phi_2))$, where

$\pi: T_k^*(\Phi_1) \to \Phi_1$ is the natural map.

Now if g: $T_k^*(\Phi_1) = J^k(\Phi_1) \times_M T^*(M) \to T^*(M)$ is the natural projection, then $g^* p^*(\Phi_1) = \pi^* Tf(\Phi_1)$ and $g^* p^*(\Phi_2) = f^* Tf(\Phi_2)$, so we can regard $\sigma_k(P)$ as an element of $\text{Hom}(g^* p^*(\Phi_1), g^* p^*(\Phi_2))$, when P is considered as a nonlinear operator. The composition with g maps $\text{Hom}(p^* \Phi_1, p^* \Phi_2)$ into $\text{Hom}(g^* p^*(\Phi_1), g^* p^*(\Phi_2))$ where "$\sigma_k(P) = \sigma_k(P) \cdot g$" expresses the relation between the nonlinear $\sigma_k(P)$, the left hand side, and the linear $\sigma_k(P)$, the right hand side, through the composition of g. The whole point is that if P is linear, then $\Lambda(P)_s = P$ for all $s \in C^\infty(\Phi_1)$, so $\Lambda(P)_s$ is independent of s, and $\sigma_k(P)(j_r(s),(v,x)) = \sigma_k(\Lambda(P)_s)(v,x) = \sigma_k(P)(v,x)$ does not depend on $j_k(s)$, i.e., $\sigma_k(P)$ is lifted from $T^*(M)$.

Let us now introduce the notion of an elliptic differential operator.

An element $P \in D_k(B_1,B_2)$ is called an elliptic differential operator of order k from B_1 to B_2 if for all $(j_k(s),(v,x)) \in T_k^*(B_1)$ with $v \neq o$, $\sigma_k(P)(j_k(s),(v,x))$: $T((B_1)_x)_{s(x)} \to T((B_2)_x)_{Ps(x)}$ is a linear isomorphism. We will denote the set of all such elliptic differential operators by $\text{Elp}_k(B_1,B_2)$. Note that $\text{Elp}_k(B_1,B_2)$ is nonempty only if B_1 and B_2 have the same fiber dimension.

Recall that if Φ_1 and Φ_2 are C^∞ vector bundles over M then the set $\text{Ell}_k(\Phi_1,\Phi_2)$ of k-th order linear elliptic differential operators from Φ_1 to Φ_2 is the subset of $P \in \text{Diff}_k(\Phi_1,\Phi_2)$ such that $\sigma_k(P)(v,x)$ is a linear isomorphism of $(\Phi_1)_x$ onto $(\Phi_2)_x$ for all $(v,x) \in M_x^*$ with $v \neq 0$. Obviously, $\text{Ell}_k(\Phi_1,\Phi_2) \subseteq \text{Elp}_k(\Phi_1,\Phi_2)$.

<u>Theorem 3.5.12</u> If $P \in \text{Elp}_k(B_1,B_2)$ then $\Lambda(P)_s \in \text{Ell}_k((B_1)_s,(B_2)_{Ps})$ for all $s \in C^\infty(B_1)$. Conversely, if each element of $J^k(B_1)$ is the k-jet of a global section of B_1, then $P \in \text{Elp}_k(B_1,B_2)$ provided each linearization of P is elliptic.

This theorem gives some conditions for the linearization of nonlinear elliptic differential operators. In the remainder of this section, we shall give a definition and some properties of the analytic index of a nonlinear elliptic operator.

For a given C^∞ fiber bundles B_1 and B_2 over a compact C^∞ manifold M without boundary, we will associate to each $P \in$ $\text{Elp}_r(B_1, B_2)$ an element $i_a(P) \in K(C^\infty(B_1))$ called <u>the analytic index of P</u>. If we choose a "base point" s_0 for $C^\infty(B_1)$ then we have a canonical augmentation map dim: $K(C^\infty(B_1)) \to Z$ and $\dim(i_a(P))$ turns out to be the usual analytic index of the linearized elliptic operator $\Lambda(P)_s : C^\infty((B_1)_s) \to C^\infty((B_2)_{Ps})$, i.e., $\dim(\text{Ker}(\Lambda(P)_s)) - \dim(\text{coKer}(\Lambda(P)_s))$. Note that, in general, when $C^\infty(B_1)$ is not homotopically trivial, $i_a(P)$ can carry more information about P than does its "dimension".

Let X and Y be C^1 Hilbert manifolds and let f: X $\to$ Y be a C^1 map. We say f is a <u>Fredholm map</u> if df: TX $\to$ f*TY is a Fredholm bundle morphism over X and in this case we define ind(f) $\in$ K(X) by ind(f) = ind(df). More generally if j: $\Omega \to$ X is a continuous map we say f is a <u>Fredholm map relative to j</u> if df$\cdot$j: j*TX $\to$ j*f*TY is a Fredholm bundle morphism (i.e., if $df_{j(w)}$: $TX_{j(w)} \to TY_{fj(w)}$ is a Fredholm operator for each w $\in$ Ω), and we define ind(f,j) $\in$ K(Ω) by ind(f,j) $\geq$ ind(df$\cdot$j). Note that if f: X $\to$ Y is a Fredholm map then clearly f is also a Fredholm map relative to any j: $\Omega \to$ X and then ind(f,j) = j*ind(f) where j*: K(X) $\to$ K(Ω) is the functorial K(j).

In order to define K for an arbitrary space X, the basic requirement is that K should be a functor from spaces and homotopic classes of maps to abelian groups. For compact spaces, such a functor is naturally isomorphic to the Grothendieck group of vector bundles and is representable, i.e., K(X) should be naturally equivalent to [X,C], the homotopy classes of maps of X into a homotopy abelian H-space C. This uniquely determines C up to homotopy type, i.e., C = Z x B_G where B_G means the classifying space of G, and G is either $\underrightarrow{\lim}$ O(n) or $\underrightarrow{\lim}$ U(n) depending on whether we mean K_o or K_u. By a theorem proved by Janich [1965], and Atiyah independently, there is a nice choice of C for our purpose. Let H be a separable, infinite dimensional Hilbert space and let Fred(H) be the space of Fredholm operators on

H, i.e., bounded linear maps. Let T: H → H be such a map
such that Ker T and coker T are finite dimensional. T(M) is
then automatically closed in H so we can take cokerT = T(H)$^{\perp}$
= Ker(T*). Fred(H) is topologized as a subspace of the space
of all bounded operators on H with the usual norm $\|T\|$ =
sup{$\|Tx\|$ | $\|x\|$ = 1}. If X is a topological space, then a
continuous map f: X → Fred(H) is called <u>admissible</u> if Ker(f)
= {(v,x) ∈ H x X| v ∈ Ker f(x)} and coker(f) = {(v,x) ∈ H x
X| v ∈ Im f(x)$^{\perp}$} are vector bundles over X under the obvious
projection. Then it is known that if X is a compact space
and r ∈ [X,Fred(H)] then r has an admissible representative
g and the element ind(r) = [Ker(g)] - [Coker(g)] of the
Grothendieck group K(X) of vector bundles over X is well
defined and independent of the choice of admissible g, and
ind: [X,Fred(H)] → K(X) is a bijection. Furthermore, Fred(H)
is homotopy abelian H-space under usual operator
composition, making [X,Fred(H)] an abelian group, and ind is
a group isomorphism.

  Now let X be a paracompact space and let B$_1$ and B$_2$ be
Hilbert space bundles over X with GL(H), the general linear
group of Hilbert space with the norm topology as structural
group. A Hilbert bundle morphism f: B$_1$ → B$_2$ is a Fredholm
bundle morphism if f$_x$: (B$_1$)$_x$ → (B$_2$)$_x$ is a Fredholm map for
each x ∈ X. In this case, by a theorem of Kuiper which
states that GL(H) is constructable, thus there exist bundle
isomorphisms g: X x H ≈ B$_1$ and h: B$_2$ ≈ X x H. Then x → h$_x$f$_x$g$_x$
is a map hfg: X → Fred(H). By the contractability of GL(H),
it follows that g and h are well determined up to homotopy
and hence the homotopy class of hfg is a well determined
element of [X,Fred(H)] = K(X) and we denote by ind(f).

  Now we are ready to define the index of a nonlinear
elliptic differential operator.

  <u>Definition and Theorem 3.5.13</u> Let M be a compact n-dim
C$^{\infty}$ manifold without boundary, let B$_1$ and B$_2$ are C$^{\infty}$ fiber
bundles over M, and let P ∈ Elp$_r$(B$_1$,B$_2$). Let k > n/2+r, then
P: C$^{\infty}$(B$_1$) → C$^{\infty}$(B$_2$) extends to a C$^{\infty}$ map of Hilbert manifolds
P$_{(k)}$: L^2$_k$(B$_1$) → L^2$_{k-r}$(B$_2$). Then P$_{(k)}$ is a Fredholm map relative

to the inclusion map $\Gamma_k\colon C^\infty(B_1) \to L^2{}_k(B_1)$ and moreover $\mathrm{ind}(P_{(k)}, \Gamma_k) \in K(C^\infty(B_1))$ is independent of k and hence defines an element $i_a(P) \in K(C^\infty(B_1))$ called the <u>analytic index of P</u>.

For more detailed definitions of topological and analytic index of operators, the index theorem for manifolds with boundary and other applications of the index theorem, see articles in Palais [1965].

This chapter, in particular, discusses the subjects which are not extensively utilized in the subsequent chapters, at least not explicitly. Nonetheless, some of the concepts and even terminology do find their way to our later discussions. This chapter is included, and indeed is lectured, to prepare the students with some concepts and understanding about global analysis in general, and some techniques important to the global theory of dynamical systems. In particular, the reader may find it useful when they venture to theoretically oriented research literature.

We have reviewed a broad range of mathematical concepts and techniques useful for our discussion, we can now turn to our main interest, namely, the nonlinear dynamical systems and their structural stability.

Chapter 4   General Theory of Dynamical Systems

4.1   Introduction

In order to place our discussion of dynamical systems in proper perspective, let us discuss briefly three aspects of the theory of dynamical systems. For lack of a better description, we shall refer to them as the local, the global, and the abstract theories.

The local theory is concerned with the application of geometrical and topological methods to the qualitative study of differential equations. The general setting is a set of differential equations in R^n and one is interested in asking: "What does the omega-limit set look like?"; "What happens in the neighborhood of a fixed point?"; "Is it stable?", etc.

The object for the study of the global theory is the set of vector fields on a manifold. One is interested in characterizing the structurally stable vector fields and in studying the "orbit picture" of the flow associated with a given vector field.

The setting of the abstract theory is a general transformation group but the notions studied are those arising in the qualititive study of differential equations. One can show that many of the results for differential equations are valid in a much broader domain.

In current literature, the abstract theory is known as <u>topological dynamics</u>, the global theory is known as <u>smooth dynamical systems</u>, and the local theory is known as <u>qualitative theory of diffferential equations</u>. Topological dynamics deals with continuous actions of any topological group G on a topological space X. Smooth dynamical systems are smooth actions of the group R or Z on a differentiable manifold M. We shall begin by illustrating a few fundamental definitions with some simple examples. Most of these definitions and examples are also common to the qualitative theory of differential equations. Indeed, the latter theory provides the proper intuition and phenomena for the development of dynamical systems.

Let G be either the additive topological group R of real numbers or the additive topological group Z of integers. A <u>dynamical system</u> on a topological space X is a continuous map ϕ: G x X → X such that for all x ϵ X, for all g, h ϵ G, $\phi(g+h,x) = \phi(g, \phi(h,x))$, and $\phi(0,x) = x$. (4.1-1) The space X is called the <u>phase space</u> of ϕ. If X is a differentiable manifold and ϕ is a C^r map, $r \geq 0$, then we call ϕ a C^r dynamical system.

For instance, for any X the trivial dynamical system is defined by $\phi(t,x) = x$. For X = R, $\phi(t,x) = e^t x$ defines a C^w dynamical system on X.

Let ϕ be a dynamical system on X. Given t ϵ G, we define the partial map ϕ^t: X → X by $\phi^t(x) = \phi(t,x)$. If G = R, ϕ^t is sometimes called the <u>time map of ϕ</u>. Likewise, given x ϵ X, we define the partial map ϕ_x: G → X by $\phi_x(t) = \phi(t,x)$. Note that if ϕ is C^r, then so are ϕ^t and ϕ_x. Then Eq.(4.1-1) can be written as $\phi^{g+h} = \phi^g \phi^h$, and $\phi^0 = $ id. (4.1-2) Sometimes for brevity we denote $\phi(t,x)$ by t·x, when under the context there is no confusion. With this convention, Eqs. (4.1-1) and (4.1-2) become:

$\qquad$ (g + h)·x = g·(h·x), and o·x = x. (4.1-3)

<u>Proposition 4.1.1</u> For all t ϵ G, ϕ^t is a homeomorphism. If ϕ is C^r then ϕ is a C^r diffeomorphism.

<u>Exercise</u> Prove this by the definition of homeomorphism and diffeomorphism.

Note that, if G = **R**, then the dynamical system ϕ is called <u>a flow</u> on X, or an one-parameter group of homeomorphism of X.

Let ϕ be a dynamical system on X. We define a relation ⁻ on X by putting x ⁻ y iff there exists t ϵ G such that $\phi^t(x)$ = y.

<u>Proposition 4.1.2</u> The relation ⁻ is an equivalence relation.

The equivalence classes of ⁻ are called <u>orbits</u> of ϕ (or of the homeomorphism ϕ' in the case G = Z). For each x ϵ X, the equivalence class containing x is called <u>the orbit</u>

through x. It is the image of the partial map ϕ_x: G → X. We
sometimes denote it by G·x. Thus Prop. 4.1.2 implies that
two orbits either coincide or are disjoint. We denote the
quotient space X/~ by X/ϕ and call it the <u>orbit space</u> of ϕ.
The quotient map, which takes x to its equivalence class, is
denoted by γ_ϕ: X → X/ϕ or just γ: X → X/ϕ. As usual, we give
X/ϕ the finest topology with respect to which γ is
continuous. (That is, a subset U of X/ϕ is open in X/ϕ iff
γ^{-1}(U) is open in X). Now let us look at some examples of
dynamical systems.

Earlier we have pointed out that if G = **R**, the trivial
dynamical system on any X is the point x ϵ X, and the orbit
is the set of single element {x}. For the nontrivial flow
t·x = e^tx, there are three orbits, namely, the origin, the
positive and negative half lines.

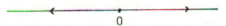

$$0$$

The arrows on the orbits indicate the orientations induced
by the flow. If, however, for all t,x ϵ **R** we put t·x = x +
t, then this flow has only one orbit, **R** itself.

For any θ ϵ **R**, put t·[x] = [x + θt] where [x] is a fixed
value of S^1. If θ = 0, we have the trivial flow on S^1,
otherwise we have the single orbit S^1. If we imbed S^1 in the
plane by the standard imbedding [x] → (cos 2πx, sin 2πx),
then the rotation is counter-clockwise if θ is positive, and
clockwise if θ is negative. We call θ the angular speed of
the flow.

For flows on R^2, it is sometimes more convenient to
identify R^2 with the complex line C because the two are
topologically indistinguishable. The simple non-trivial flow
is for all t ϵ **R**, and for all (x,y) ϵ R^2, put t·(x,y) =
(xe^t, ye^t). The origin is the only point orbit, and all other
orbits are open rays radiating from the origin.
Fig.4.1-1(a). If we change the formula slightly to t·(x,y) =
(xe^t, ye^{-t}), the phase portrait is radically changed, this is
because the new flow has only two orbits beginning at the

147

origin. It is associated with a <u>saddle point</u> of the flow.
Fig.4.1-1(b).

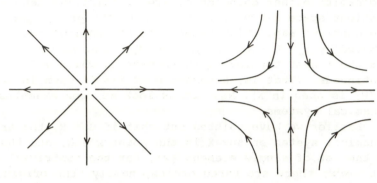

Fig.4.1-1(a)   Fig.4.1-1(b)

For all $t \in \mathbf{R}$ and $z = x + iy \in \mathbf{C}$, put $t \cdot z = ze^{it}$. Then the
origin is a point orbit and the other orbits are all circles
with center at the origin. But if $t \cdot z = ze^{(i-1)t}$, then the
origin is a point orbit, and all other orbits spiral in
towards it. See Fig.4.1-2.

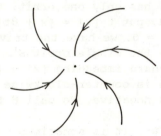

Fig.4.1-2

There are several ways to construct new dynamical
systems from the given ones. The most simplest and direct
one is the product. Let $\phi: G \times X \to X$ and $\mu: G \times Y \to Y$ be
dynamical systems. The product of the two systems is a
dynamical system on X x Y defined for all $g \in G$ and $(x,y) \in$
X x Y by $g \cdot (x,y) = (g \cdot x, g \cdot y)$. For example, let ϕ be the flow
$t \cdot x = e^t x$ and μ be the rotation flow $t \cdot [x] = [x + \theta t]$ for $\theta \in$

148

R. Then $\phi \times \mu$ is a flow on the circular cylinder $R \times S^1$. The circle $\{o\} \times S^1$ is an orbit, and all the other orbits spiral away from it.

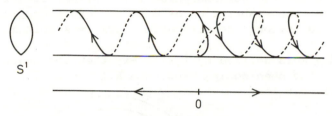

Fig.4.1-3

Let ϕ: $R \times X \to X$ be any flow. For any $t \in R$, ϕ^t: $X \to X$ is a homeomorphism which generates a discrete dynamical system μ: $Z \times X \to X$, then we say ϕ^t (or μ) is <u>imbedded</u> in ϕ. For instance, any rotation of S^1 is imbedded in any non-trivial rotation flow on S^1. It is not true that every homeomorphism of every topological space X can be imbedded in a flow on X.

In general, every flow on X yields a homeomorphism of X (in fact, many homeomorphisms), but the reverse is not usually true. Nonetheless, we can associate with a given homeomorphism f a flow with similar properties provided we allow the flow to be on a larger space Y. Thus we have the space X imbedded in Y and the homeomorphism of X imbedded in a flow ϕ on Y. The way to construct the larger space Y and the flow ϕ is known as <u>suspension</u>. It should be noted that this is different from the suspension discussed in algebraic topology (see, for instance, Bourgin [1963]; Greenberg [1967]; Eilenberg and Steenrod [1952]). We shall not get into this any further, except to say that when reading the literature, be aware of the difference of meaning of terms. It is fairly easy to distinguish them under very different context.

Let f: $X \to X$ be a homeomorphism generating a discrete dynamical system μ. Let ~ be the equivalence relation defined on $R \times X$ by (u,x) ~ (v,y) iff $u = v + m$ for some $m \in$

149

Z and y = f^m(x). Then there is a flow ϕ: R x Y → Y on Y = (R x X)/~ defined by ϕ(t,[u,x]) = [u+t,x], where [u,x] denotes the equivalence class of (u,x) ϵ R x X. The flow ϕ is called the <u>suspension</u> of the homeomorphism f (or of μ). Clearly, for any u ϵ R, the restriction of ϕ to any <u>cross-section</u> [u,X] with the obvious identification [u,X] = X coincides with f.

If f: R → R is defined by f(x) = -x, then the suspension is a flow on the open Möbius band, and all its orbits are topologically circles.

$$Y = R^2/\sim$$

We shall discuss rational and irrational flows on a torus $T^2 = S^1$ x S^1 and we shall show that all rational flows are essentially the same, but there are infinitely many different types of irrational flow. It should also be pointed out here that the phase portrait of a product flow is not uniquely determined by the phase portraits of its factors. This is because the phase portraits of rational and irrational flows are completely different topologically, but they both come from factors with identical phase portraits. We shall discuss this point in detail later.

Let ϕ: G x X → X be a dynamical system on X, let α: G → G be a continuous automorphism of the additive group G, and let h: X → Y be a homeomorphism. Then μ: G x Y → Y, μ = h·ϕ·$(\alpha$xh$)^{-1}$, is a dynamical system on Y. Then μ is the dynamical system <u>induced from ϕ by the pair (α,h)</u> (or by h, if α = id).

The simplest example is that if h = id: X → X, and α = -id: R → R, we then obtain ϕ^-, the <u>reverse flow</u> of ϕ, by

$\phi^{-1}(t,x) = \phi(-t,x)$. That is, in ϕ^{-}, points moving along the orbits of ϕ at the same speed but in opposite direction.

The construction of induced systems is not particularly interesting, but the generalization to quotient systems does produce new systems.

Let ϕ: G x X → X be a dynamical system on X and let ~ be an equivalence relation on X such that for all t ϵ G and all x, y ϵ X, $\phi(t,x)$ ~ $\phi(t,y)$ iff x ~ y. Then ϕ induces a dynamical system μ, called the quotient system on the quotient space X/~ by $\mu(t,[x]) = [\phi(t,x)]$ where t ϵ **R** and [x] is the equivalence class of x ϵ X.

If f and g are commuting homeomorphisms of X, i.e., fg = gf, then f takes orbits of g onto orbits of g, thus induces a homeomorphism of the orbit space of g. An example of a quotient system ϕ, which is the discrete dynamical system generated by f and ~ is the equivalence relation giving orbits of g as equivalence classes. Similarly, if ϕ and μ are commuting flow on X, i.e., $\phi^s\mu^t = \mu^t\phi^s$ for all s, t ϵ **R**, then ϕ induces a quotient flow on the orbit space X/μ.

Let us define an equivalence relation ~ on R^n by x ~ y iff x- y ϵ Z^n. Then the quotient space R^n/~ is the n-dim torus $T^n = S^1 \times S^1 \times ... \times S^1$, the Cartesian product of n copies of the circle S^1. Let f be a linear automorphism of R^n whose matrix A, with respect to the standard basis of R^n, is in $GL_n(Z)$. That is, A has integer entries and detA = ±1. Then f maps Z^n onto itself and thus f and f^{-1} preserve the equivalence relation. Thus f induces a homeomorphism (in fact, a diffeomorphism) of T^n. The induced homeomorphism of T^n is called a hyperbolic toral automorphism. For any hyperbolic toral automorphism g: T^n → T^n, its periodic and non-periodic point sets are dense in T^n. Note that, for a point x ϵ T^n which is periodic if $g^r(x) = x$ for some r > 0. We shall see that hyperbolic toral automorphisms are the simplest examples of Anosov diffeomorphisms on compact manifolds. Originally, the hyperbolic toral automorphisms were counter-examples to the conjecture that structurally stable diffeomorphisms have finite periodic sets.

151

Theorem 4.1.3   The periodic set of the hyperbolic toral automorphism $g: T^n \to T^n$ is precisely $Q^n/\sim$, where Q is the rational numbers.

Remark: One can verify that induced and quotient systems are dynamical systems.

Let ϕ be a dynamical system on a topological space X. For each $x \in X$, the subset $G_x = \{g \in G, \phi(g,x) = x\}$ is a subgroup of G and is called an <u>isotropy subgroup (or stabilizer) of x (or of ϕ at x)</u>.

Proposition 4.1.4   If X is a T_1 space, then for all $x \in$ X, G_x is a closed subgroup of G.

Proposition 4.1.5   If X is T_2 and G/G_x is compact, then the orbit $G \cdot x$ is homeomorphic to G/G_x.

Proposition 4.1.6   Every orbit of every flow is connected.

The definitions and propositions in this section are common to the theory of dynamical systems in general. The next section will introduce various equivalence relations and conjugacy, which are essential to the introduction of limiting sets.

4.2   Equivalence relations

To classify dynamical systems is one of the center themes and is of special interest to the subject. One begins by placing certain equivalence relations upon the set of all dynamical systems. Such equivalence relations should be natural in the sense that the systems have qualitative resemblance. Equipped with the equivalence relation, one can form the equivalence classes and be able to distinguish them by means of algebraic or topological invariants (quantities that are associated with all systems and are equal for all systems in the same equivalence class). A good classification scheme requires a careful choice of "intrinsic" equivalence relations with tractable invariants. We shall consider several "obvious" equivalence relations.

In order to appreciate the difficulties involved in the

152

classification problem, one only has to look at the situation when the manifold is the circle S^1. See, for instance, [Irwin 1980, Section 2-II; Nitecki 1971, Chapter 1].

Let f: X → X, g: Y → Y be homeomorphisms of topological spaces X and Y. A <u>topological conjugacy</u> from f to g is a homeomorphism h: X → Y such that h·f = g·h. That is, the following diagram commutes:

$$
\begin{array}{ccc}
X & \xrightarrow{\ f\ } & X \\
h \downarrow & & \downarrow h \\
Y & \xrightarrow[g]{} & Y
\end{array}
$$

If such a homeomorphism h exists, then the homeomorphisms f and g are said to be <u>topologically conjugate</u>. Clearly, topological conjugacy is an equivalence relation. It is easy to show that a topological conjugacy maps orbits onto orbits, periodic points to periodic points, and it also preserves periods.

In our discussion, we are mainly concerned with differentiable manifolds M and N, and f and g are diffeomorphisms. It seems natural to require the map h to be a diffeomorphism. This gives the notion of <u>differentiable conjugacy</u>. It is a stronger relation than topological conjugacy, and in general, there are many more equivalence classes with respect to it. Nonetheless, with differentiable conjugacy, we do find stable diffeomorphisms, ones which stay in the same equivalence class when slightly perturbed, which are very rare. Moreover, we also have to classify as non-equivalent diffeomorphisms which most people would feel are qualitatively the same such as the contractions x → x/2, and x → x/3 of the real line. For these reasons, topological conjugacy remains as the basic equivalence relation even when we are dealing with a differentiable category.

Let ϕ and μ be flows on topological spaces X and Y respectively. h: X → Y is a <u>flow map from ϕ to μ</u> if it is continuous and if there exists an increasing continuous

153

homomorphism α: R → R such that the diagram

$$\begin{array}{ccc} R \times X & \xrightarrow{\phi} & X \\ \alpha \times h \downarrow & & \downarrow h \\ R \times Y & \xrightarrow{\mu} & Y \end{array}$$

commutes. Recall that α is just multiplication by a positive constant. If h is a homeomorphism, we then call (α,h) or h, in the case $\alpha = $ id, a <u>flow equivalence from ϕ to μ</u>. We say that <u>ϕ is flow equivalent to μ</u> if such a pair of (α,h) exists. In this case, μ is the flow induced on Y from ϕ by (α,h).

The definition of flow equivalence seems to be very natural. But it is rather too strong for the qualitative theory of flows. For instance, it preserves the ratios of periods of closed orbits, but flows may differ in this respact and yet have a very similar appearance. We now define a weaker equivalence relation which is regarded as a basic notion. We call h: X → Y a <u>topological equivalence from ϕ to μ</u> if it is a homeomorphism, which maps each orbit of ϕ onto an orbit of μ, and it preserves the orientation of orbits. For instance,

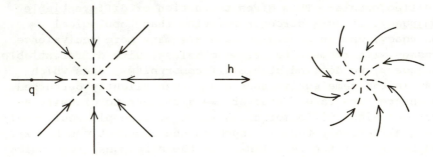

We have remarked earlier that it is undesirable to strengthen "topological" to "smooth" in the definition of conjugcy of homeomorphisms. The arguments apply equally well to the definition of equivalence of flows.

We say that homeomorphisms f and g of a topological

space X are topologically equivalent if there is a
homeomorphism h of X which maps orbits of f onto orbits of
g. This relation is different from topological conjugacy,
but oftenly they are the same. One can also show that flow
equivalence implies topological equivalence and both are
equivalence relations.

In some situations, we need to deal with topological
conjugacy, topological equivalence and flow equivalence in a
local form. It is often possible to modify the definitions
of such relations by paraphrasing them as "in some
neighborhood of a given point". Even if the "local"
definition does not make perfect sense, it often points
toward a sensible concept.

Let U and V are open subsets of X and Y respectively and
let f: X → X and g: Y → Y be homeomorphisms. Somewhat
abusing the notation, we say that f|U is topologically
conjugate to g|V if there is a homeomorphism h: U ∪ f(U) → V
∪ g(V) such that h(U) = V and, for all x ∈ U, hf(x) = gh(x).
If p ∈ X and q ∈ Y, we say that f is topologically conjugate
at p to g at q if there exists open neighborhood U of p and
V of q such that f|U is topologically conjugate to g|V and
by a conjugacy h which takes p to q.

Let ϕ: D → X be a continuous map where X is a
topological space and D is a neighborhood of {0} x X in R x
X. We write $t \cdot x$, as before, for $\phi(t,x)$ and D_x = {t ∈ R:
(t,x) ∈ D}. We say that is a partial flow on X if, for all x
∈ X,

    (i) D_x is an interval,

    (ii) $0 \cdot x = x$

    (iii) for all t ∈ D_x with s ∈ $D_{t \cdot x}$, $(s+t) \cdot x = s \cdot (t \cdot x)$,

    (iv) for all t ∈ D_x, $D_{t \cdot x}$ = {s-t : s ∈ D_x}.

That is, ϕ is a flow not defined for all time. (iv) implies
ϕ is maximal and it cannot be extended.

Proposition 4.2.1 Let ϕ: D → X be a partial flow on X.
Then (i) D is open in R x X, (ii) if D_x = (a,b) with b < ∞ ,
ϕ cannot be extended to a continuous map of D ∪ {(b,x)} into
X, (iii) if x is a fixed point or periodic point of ϕ, then

155

$D_x = R$.

It is clear that flow equivalence and topological equivalence for partial flows can be defined in a straightforward way. If μ is a partial flow on a topological space Y, then a _flow equivalence from ϕ to μ_ is a pair (α,h) where h: X → Y is a homeomorphism, α: R → R is a multiplication by a positive constant, and $h(t \cdot x) = a(t) \cdot h(x)$ for all $(t,x) \in D$. And a _topological equivalence from ϕ to μ is a homeomorphism_ h: X → Y that maps all orbits $D_x \cdot x$ of ϕ onto orbits of μ and preserves their orientation.

Proposition 4.2.2 Flow equivalence and topological equivalence are equivalence relations on the set of all partial flow on topological spaces.

Let ϕ be any flow (or partial flow) on X and let U be an open subset of X. Once again, we are abusing the notation and defining a map $\phi|U$: D → U by $(\phi|U)(t,x) = \phi(t,x)$ where D $= \cup_{x \in U} D_x \times \{x\}$ of R x U, and call $\phi|U$ _the restriction of the flow ϕ to the subset U_. Notice again, this is an abuse of notation, since U is not even in the domain of ϕ. One can easily check that $\phi|U$ is a partial flow on U. If μ is a flow on a topological space Y and V is an open subspace of Y, then we say that $\phi|U$ is _flow_ or _topologically equilvalent to $\mu|V$_ if they are equivalent as partial flows. If p $\in$ X and g $\in$ Y, we say that ϕ is equivalent at p to μ at q, if there exists open neighborhoods U of p and V of q and an equivalence from $\phi|U$ to $\mu|V$ taking p to q.

Proposition 4.2.3 Flow equivalence and topological equivalence are equivalent relations on $\{(\phi,p): \phi$ is a flow, p $\in$ phase space of $\phi\}$. If $\phi|p$ is flow equivalent to $\mu|q$, then $\phi|p$ is topologically equivalent to $\mu|q$.

4.3   Limit sets and non-wandering sets

Let ϕ and μ be flows on topological spaces X and Y. Suppose h is a topological equivalence from ϕ to μ, then h maps the closure $\bar{\Gamma}$ in X of each orbit Γ of ϕ onto the closure $\overline{h(\Gamma)}$ of h(Γ) in Y. Thus, h maps the set $\bar{\Gamma}/\Gamma$ onto the set $\overline{h(\Gamma)}/h(\Gamma)$.

We want to study the orbit with very large positive and negative t. Let $I_t = [t,\infty)$. The ω-set $\Omega(x)$ of $x \in X$ with respect to the flow ϕ is defined by $\Omega(x) = \cap_{t \in R} \overline{\phi_x(I_t)}$. Intuitively, $\Omega(x)$ is the subset of X that $t \cdot x$ approaches as $t \to \infty$. It follows from the definition of flow that for all $t \in R$, $\Omega(t,x) = \Omega(x)$. Thus we may define the ω-set, $\Omega(\Gamma)$ of any orbit Γ of ϕ by $\Omega(\Gamma) = \Omega(x)$ for any $x \in \Gamma$. Note that if Γ is a fixed point or periodic orbit then $\phi_x(I_t) = \Gamma$ for all $x \in \Gamma$ and $t \in R$, and $\Omega(\Gamma) = \Gamma$. Thus $\Omega(\Gamma)$ is not necessarily part of $\overline{\Gamma}/\Gamma$. Similarly, the α-set of a point $x \in X$ is defined by $\alpha(x) = \cap_{t \in R} \overline{\phi_x(J_t)}$, where $J_t = (-\infty,t]$, and the α-set $\alpha(\Gamma)$ of an orbit Γ is denoted by $\alpha(\Gamma) = \alpha(x)$ for any $x \in \Gamma$. Again, intuitively, the α-set is the subset that $t \cdot x$ approaches as $t \to -\infty$. Since the ω-set and α-set are "symmetric" in time, we shall confine our attention to ω-set, and the corresponding results for α-sets are exactly analogous.

Let a vector field $X \in X(M)$ and let ϕ be the flow of X. The orbit of X through $p \in M$ is the set $\Gamma(p) = \{\phi(t,p) \mid t \in R, p \in M\}$. If $X(p) = 0$, the orbit reduces to p, then we say that p is a singularity of X. Otherwise, the map $\alpha: R \to M$, $\alpha(t) = \phi(t,p)$ is an immersion. If α is not injective, there exists $\beta > 0$ such that $\alpha(\beta) = \alpha(0) = p$ and $\alpha(t) \neq p$ for $0 < t < \beta$. In this case, the orbit of p is diffeomorphic to S^1 and we say that it is a closed orbit with period β. If the orbit is not singular or closed, it is called regular. Thus a regular orbit is the image of an injective immersion of the line.

The ω-limit set of $p \in M$, $\Omega(p) = \{q \in M$, for which there exists a sequence $t_n \to \infty$ with $\phi(t_n,p) \to q\}$. Simiarly, we define the α-limit set of p, $\alpha(p) = \{q \in M \mid \phi(t_n,p) \to q$ for some sequence $t_n \to -\infty \}$. Thus, α-limit of p for the vector field X is ω-limit of p for the vector field -X. Also, $\Omega(p) = \Omega(p')$ if p' belongs to the orbit of p. Intuitively, $\alpha(p)$ is where the orbit of p is "born" and $\Omega(p)$ is where it "dies".

Example 1. Let us consider the system of differential equations defined in R^2 (in polar coordinates):

$$dr/dt = r(1 - r), \qquad d\theta/dt = 1.$$

It can easily be shown that the solutions are unique and all solutions are defined on R^2. Thus, the system defines a dynamical system. The orbits are shown in the following. They consist of (i) a critical point, namely the origin 0, (ii) a periodic orbit Γ coinciding with the unit circle, (iii) spiraling orbits through each point $p = (r,\theta)$ with $r \neq 0$, $r \neq 1$. For points p with $0 < r < 1$, the Ω-limit set of p is the unit circle and the α-limit set of p is the {0}. For points p with $r > 1$, $\Omega(p)$ is the unit circle and $\alpha(p)$ is an empty set.

Example 2. Consider the unit sphere $S^2 \quad R^3$ with center at the origin and use the standard coordinates (x,y,z) in R^3. Let $p_n = (0,0,1)$ and $p_s = (0,0,-1)$ be north and south poles of S^2 respectively. Let the vector field X on S^2 be X $= (-xz, -yz, x^2+y^2)$. Clearly, X is C^∞, and the singularities of X are p_n and p_s. Since X is a tangent to the meridians of S^2 and points upwards, $\Omega(p) = p_n$ and $\alpha(p) = p_s$ if $p \in S^2 - \{p_n, p_s\}$.

158

Theorem 4.3.1 (Poincaré-Bendixson) Let X ∈ X(S²) with a
finite number of singularities, and let p ∈ S² and Ω(p) be
the Ω- limit set of p. Then one of the following
possibilities holds:
    (i)   Ω(p) is a singularity,
    (ii)   Ω(p) is a closed orbit,
    (iii) Ω(p) consists of singularities $p_1,\ldots,p_n$ and
regular orbits such that if Γ ⊂ Ω(p), then α(Γ) = p_i, and
Ω(Γ) = p_j.
    Example Let X be a vector field on S² as in the
following:

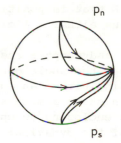

p_n

p_s

Both p_n and p_s are singularities and the equator is a closed
orbit. The other orbits are born at a pole and die at the
equator.
    Proposition 4.3.2 Let Γ be an orbit of φ and let α⁻(Γ)
and Ω⁻(Γ) denote the α-set and Ω-set of Γ as an orbit of φ⁻.
Then
α⁻(Γ) = Ω(Γ) and Ω⁻(Γ) = α(Γ).
    Proposition 4.3.3 Let h: X → Y be a topologcial
equivalence from φ to μ. Then for each orbit Γ of φ, h maps
Ω(Γ) onto Ω(h(Γ)), the Ω-set of the orbit h(Γ) of μ.
    Proposition 4.3.4 Let Γ be an orbit of φ. Then Ω(Γ) is
a closed subset of the topological space X, and Ω(Γ) ⊂ Γ̄. In
fact, Γ̄ = Γ ∪ α(Γ) ∪ Ω(Γ). Moreover, if δ is another orbit
of φ such that Γ ⊂ Ω(δ), then Ω(Γ) ⊂ Ω(δ).
    A Ω-limit cycle (or α-limit cycle) is a closed orbit Γ
such that Γ belongs to Ω-set (or α-set) of x for some x

159

which does not belong to Γ. Note that, limit cycles enjoy certain properties not shared by other closed orbits, namely, if Γ is an Ω-limit cycle, then there exists x not on Γ such that $\lim_{t \to \infty} d(\phi_t(x), \Gamma) = 0$. Geometrically, this means that some trajectory spirals toward Γ as t → ∞. For an α-limit cycle, replace ∞ by -∞.

We list some other results of limit cycles for future reference.

(A) Let Γ be an Ω-limit cycle. If Γ = Ω(x), x is not on Γ, then x has a neighborhood V such that Γ = Ω(y) for all y ∈ V. In other words, the set $\{y| \Gamma = \Omega(y)\}$ - Γ is open.

(B) A nonempty compact set K that is positively or negatively invariant contains either a limit cycle or an equilibrium.

(C) Let Γ be a closed orbit and suppose that the domain W of the dynamical system includes the whole open region U enclosed by Γ. Then U contains either an equilibrium or a limit cycle. Indeed, one can state a much sharper one.

(D) Let Γ be a closed orbit enclosing an open set U contained in the domain W of the dynamical system. Then U contains an equilibrium.

(E) Let H be a first integral of a planar C^1 dynamical system. If H is not constant on any open set, then there are no limit cycle.

Any union of orbits of a dynamical system is called an invariant set of the system. We then have the following:

Proposition 4.3.5   Any Ω-set of φ is an invariant set of φ.

Consequently, if Ω(Γ) is a single point p, then p is a fixed point of φ.

We may have noticed in linear examples that an orbit may have an empty Ω-set, and this seems to be associated with the orbit "going to infinity". Thus, it is reasonable to suppose that if we introduce some compactness conditions, we can ensure non- emptiness of the Ω-sets. Indeed, we have the following propositions:

Proposition 4.3.6   If X is compact and Hausdorff, then,

for any orbit Γ, $\Omega(\Gamma)$ is non-empty, compact and connected.

A complimentary statement is the following theorem: If X is locally compact, then a Ω-limit set is connected whenever it is compact. Furthermore, whenever a Ω-limit set is not compact, then none of its components is compact.

We have noticed earlier that if Γ is a fixed point or a periodic orbit, then $\Omega(\Gamma) = \Gamma$. In fact, one can show that under very general conditions, some Γ are periodic. We now want to investigate this relation more closely.

Proposition 4.3.7   Let X be compact and Hausdorff, then an orbit Γ of ϕ is closed in X iff $\Omega(\Gamma) = \Gamma$. (That is, closed orbits are periodic or fixed points).

Theorem 4.3.8   Let X be Hausdorff. Then an orbit Γ of ϕ is compact iff it is a fixed point or a periodic orbit.

Corollary 4.3.9   Let X be compact and Hausdorff, then the following three statements on an orbit Γ of ϕ are equivalent:

(i) Γ is a closed subset of X,

(ii) Γ is a fixed point or a periodic orbit,

(iii) $\Omega(\Gamma) = \Gamma$.

This Corollary can be generalized to non-compact but locally compact X.

A minimal set of a dynamical system is a non-empty, closed invariant set that does not contain any closed invariant proper subset. By using the Zorn's lemma, one can prove that if, for any orbit Γ of ϕ, $\overline{\Gamma}$ is compact, then it contains a minimal set.

For discrete dynamical systems, the theory of α- and Ω-limit sets can also be developed similarly. If f is a homeomorphism of a topological space X, then the Ω-set $\Omega(x)$ of $x \in X$ with respect to f is defined by $\Omega(x) = \cap_{n \in N} \overline{\{f^r(x):}$ $\overline{r \geq n\}}$. Again, the α-set $\alpha(x)$ of x is the Ω-set of x with respect to f^{-1}. All results of α- and Ω-sets have analogues form except that Ω-sets of homeomorphisms need not be connected.

When classification of dynamical systems becomes difficult, there are new equivalence relations with respect

161

to which classification can be made easier. One of the most
important concepts is concerned with an invariant set called
the non-wandering set which was defined by Birkhoff [1927].
The intuitive idea is the following. If one compares phase
portraits of dynamical systems, such as the following two:

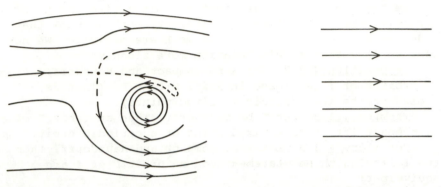

It seems that certain parts are qualitatively more important
(or attract more attention) than others. For instance, if we
were asked to pick out the more significant features of the
left hand picture, we would inevitably begin by mentioning
the fixed points and closed orbits. In general, qualitative
features in a phase portrait of a dynamical system can be
traced to sets of points that exhibit some form of
recurrence. Of course, the strongest form of recurrence is
periodicity, where a point resumes its original position
arbitrarily often, but there are weaker forms which are also
important. We shall define such recursiveness and some of
its properties.

Note that, for two-dimensional flows, all the possible
nonwandering sets fall into three classes: (i) fixed points,
(ii) closed orbits, and (iii) the unions of fixed points and
the orbits connecting them [Andronov et al 1966]. The third
sets are referred to as heteroclinic orbits when they
connect distinct points, and homoclinic orbits when they
connect a point to itself. Closed paths formed of
heteroclinic orbits are called homoclinic cycles. It is

162

worthwhile to note that the fixed points contained in such
cycles must all be saddle points, if they are hyperbolic,
this is because sinks and sources necessarily have wandering
points in their neighborhoods.

A set $A \subset X$ is said to be positively recursive with
respect to a set $B \subset X$ if for $T \in R$ there is a $t > T$ and an
$x \in B$ such that $t \cdot x \in A$. Negative recursiveness can be
defined by using $t < T$. A set A is self positively recursive
if it is positively recursive with respect to itself.

A point $x \in X$ is positively Poisson stable if every
neighborhood of x is positively recursive with respect to
$\{x\}$.

Theorem 4.3.10   Let $x \in X$. The following statements are
equivalent:

(i) x is positively Poisson stable,

(ii)   given a neighborhood U of x and a $T > 0$, $t \cdot x \in U$
for some $t > T$,

(iii) $x \in \Omega(x)$,

(iv) $\overline{\Gamma^+(x)} = \Omega(x)$, where $\Gamma^+(x) = \{\phi(t,x) \mid 0 \leq t < a\}$,

(v) $\Gamma(x) \subset \Omega(x)$,

(vi)   for every $\epsilon > 0$, there is a $t \geq 1$ such that $t \cdot x \in$
$B(x,\epsilon)$, where $B(x,\epsilon)$ is an open ball centered at x with
radius ϵ. Note also that if x is positively Poisson stable,
then so is $t \cdot x$ for every $t \in R$.

The following alternative definition of Poisson
stability is customary in the literature and is clearly
suggested by the above theorem.

A point $x \in X$ is positively or negatively Poisson stable
whenever $x \in \Omega(x)$ or $x \in \alpha(x)$ respectively. It is Poisson
stable if it is both positively and negatively Poisson
stable.

Theorem 4.3.11 $\Gamma^+(x) = \Omega(x)$ iff s is a periodic point.

Remark: Note that, if $\Gamma^+(x) = \Omega(x)$ then the point x is
indeed Poisson stable. The following example shows that
there exist points which are Poisson stable but not
periodic.

Example: Consider a dynamical system defined on a torus

163

by the differential system
$$dx/dt = f(x,y), \qquad dy/dt = \alpha f(x,y)$$
where $f(x,y) = f(x+1,y+1) = f(x+1,y) = f(x,y+1)$, and $f(x,y) > 0$ if x and y are not both zero, $f(0,0) = 0$. Let $\alpha > 0$ be irrational. It is clear that there is a fixed point $p = (0,0)$. There is only one orbit r_1 such that $\alpha(r_1) = \{p\}$, and exactly one orbit r_2 such that $\Omega(r_2) = \{p\}$. For any other orbit r, $\alpha(r) = \Omega(r) =$ the torus. Moreover, $\Omega(r_1) = \alpha(r_2) =$ the torus. It is also clear that points on r_1 are positively, but not negatively, Poisson stable. Likewise, points on r_2 are negatively, but not positively, Poisson stable. All other points are Poisson stable. But, no point except the fixed point p is periodic.

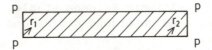

The following theorem sheds some light on positively Poisson stable points x when $\Gamma^+(x) \neq \Omega(x)$.

Theorem 4.3.12   Let X be a complete metric space. Let x ϵ X be positively Poisson stable, and let it not be a periodic point. Then the set $\Omega(x) - \Gamma(x)$ is dense in $\Omega(x)$, i.e.,
$$\overline{\Omega(x) - \Gamma(x)} = \Omega(x) = \overline{\Gamma(x)}.$$

Corollary 4.3.13   If X is complete, then $\Gamma(x)$ is periodic iff $\Gamma(x) = \Omega(x)$.

This corollary is closely related to Corollary 4.3.9.

Now let us introduce the definition of a non-wandering point. A point x ϵ X is non-wandering if every neighborhood U of x is self-positively recursive.

We shall state a few theorems showing the connection between Poisson stable points and non-wandering points.

Lemma 4.3.14   Let x ϵ X. Every y ϵ $\Omega(x)$ is non-wandering.

Theorem 4.3.15   Let P $\subset$ X such that every x ϵ P is either positively or negatively Poisson stable. Then every x

164

ϵ P is non- wandering.

Theorem 4.3.16   Let X be complete. Let every x ϵ X be non- wandering, then the set of Poisson stable points P is dense in X.

Again, a closely related theorem is the following:

Theorem 4.3.17   For any dynamical system ϕ on X, the set of all non-wandering points of ϕ, $\Omega(\phi)$, is a closed invariant subset of X, and is non-empty if X is compact. Furthermore, topological conjugacies and equivalences preserve the set of non-wandering points.

Before we introduced the notion of non-wandering points, we were seeking some new (or additional) equivalence relations with respect to which classification might be made easier. The new equivalence relation are called Ω-equivalence (for flow) and Ω-conjugacy (for homeomorphisms). They are just the old ones, topological equivalence and conjugacy, restricted to Ω-sets. Thus if $\phi|\Omega(\phi)$ denotes the restriction of the flow ϕ to $\Omega(\phi)$, defined by $(\phi|\Omega(\phi))(t,x) = \phi(t,x)$ for all (t,x) ϵ R x $\Omega(\phi)$, then ϕ is Ω- equivalent to μ iff $\phi|$ $\Omega(\phi)$ is topologically equivalent to $\mu|\Omega(\mu)$. Similarly, homeomorphisms f and g are Ω-conjugate iff their restrictions $f|\Omega(f)$ and $g|\Omega(g)$ are topologically conjugate. From the last theorem, topological equivalence (or conjugacy) is stronger than Ω-equivalence (or conjugacy).

Earlier, we touched upon the concept of a minimal set. In the following, we shall characterize a minimal set and its existence theorems.

Theorem 4.3.18   A non-empty set A $\subset$ X is minimal iff $\Gamma(x)$ = A for every x ϵ A.

Theorem 4.3.19   If A $\subset$ X is minimal and the interior of A is non-empty, then A = Int(A).

Theorem 4.3.20   Let A $\subset$ X be non-empty and compact. Then the following statements are equivalent:

(i)   A is minimal,
(ii) $\overline{\Gamma(x)}$ = A for every x ϵ A,
(iii) $\Gamma^{\pm}(x)$ = A for every x ϵ A,

165

(iv) $\Omega(x)$ = A for every x $\in$ A,

(v) $\alpha(x)$ = A for every x $\in$ A.

A fixed point and a periodic orbit are examples of compact minimal sets. From the above theorem, every point in a compact minimal set is Poisson stable. The example for an irrational flow on T^2 indicates that the closure of a Poisson stable orbit need not be a minimal set. This is because the closure of every Poisson stable orbit except the fixed point is the whole torus, which is not minimal, for it contains a fixed point.

Birkhoff [1927] discovered an intrinsic property of motions in a compact minimal set, which is usually called the property of recurrence.

For any x $\in$ X, the motion ϕ_x is <u>recurrent</u> if for each ϵ > 0 there exists a T = T(ϵ) > 0, such that $\Gamma(x)$ $\subset$ B([t-T,t+T]·x,ϵ) for all t $\in$ R.

Since every motion ϕ_y with y $\in$ $\Gamma(x)$ is also recurrent if ϕ_x is recurrent, thus we shall speak of the orbit $\Gamma(x)$ being <u>recurrent</u>. Moreover, a point x $\in$ X is recurrent if ϕ_x is recurrent. Note also that every recurrent motion is Poisson stable.

<u>Theorem 4.3.21</u> Every orbit in a compact minimal set is recurrent. Thus every compact minimal set is the closure of a recurrent orbit.

<u>Theorem 4.3.22</u> If $\Gamma(x)$ is recurrent and $\overline{\Gamma(x)}$ is compact, then $\Gamma(x)$ is also minimal.

<u>Corollary 4.3.23</u> If X is complete, then $\overline{\Gamma(x)}$ of any recurrent orbit is a compact minimal set.

So far our discussions were centered on compact minimal sets, not much is known about the properties of non-compact minimal sets. It has been established that all minimal sets in R^2 consist of single orbit with empty limit sets [Bhatia & Szego 1967]. Nonetheless, usually compact minimal sets contain more than one orbit.

<u>Lemma 4.3.24</u> There exists non-compact minimal sets which contain more than one orbit.

Consider the dynamical system of irrational flow on a

166

torus discussed earlier restricting the system to the complement of the fixed point in that example. The resulting space X is non-compact, but for each $x \in X$, $\overline{\Gamma(x)} = X$, so that X is minimal. This proves the lemma. Note that, in the above construction, ϕ_x are not recurrent. This shows that Theorem 4.3.21 is not necessarily true for non-compact minimal sets.

For any $x \in X$, the motion ϕ_x is <u>positively Lagrange stable</u> if $\overline{\Gamma^+(x)}$ is compact. If $\overline{\Gamma^-(x)}$ is compact, then the notion ϕ_x is <u>negatively Lagrange stable</u>. It is <u>Lagrange stable</u> if $\overline{\Gamma(x)}$ is compact.

Remark: If $X = R^n$, then the above statements are equivalent to the sets $\Gamma^{\pm}(x)$, $\Gamma(x)$ being bounded. One can also show that (i) If X is locally compact, then a motion ϕ_x is positively Lagrange stable iff $\Omega(x)$ is a non-empty compact set; (ii) If ϕ_x is positively Lagrange stable, then $\Omega(x)$ is compact and connected; (iii) If ϕ_x is positively Lagrange stable, then $d(t \cdot x, \Omega(x)) \to 0$ as $t \to \infty$.

Theorem 4.3.25 Every non-empty compact invariant set contains a compact minimal set.

Theorem 4.3.26 The space X contains a <u>compact minimal</u> set iff there is an $x \in X$ such that either $\overline{\Gamma^+(x)}$ or $\overline{\Gamma^-(x)}$ is compact.

It is worthwhile to note that the only recurrent motins in R^2 are the periodic ones. As a consequence, all compact minimal sets in R^2 are the orbits of periodic points. Indeed, Hajek [1968] shows that all positively Poisson stable points in R^2 are periodic. Moreover, the only noncompact minimal sets in R^2 consist of a single orbit with empty Ω-limit and α-limit sets [Bhatia and Szegö 1967]. Then Theorem 4.3.18 implies that all minimal sets in R^2 have empty interiors. This theorem also poses an important and interesting problem, i.e., which phase spaces (or manifolds) can be minimal.

A special case of recurrence, namely almost periodicity, is deferred to next chapter because it is intimately connected to the notion of stability of motion. It is

167

worthwhile to note that the concepts of recursiveness can be generalized to non-metric topological spaces, whereas almost periodicity requires a uniformity on the space.

4.4 Velocity fields, integrals, and ordinary differential equations

In this section, we shall discuss the existence and uniqueness of a flow whose velocity is a given vector field. By using a chart, the local problem is equivalent to the existence and uniqueness of solutions (integral curves) of a system of ordinary differential equations.

Recall that, if M is a differentiable manifold, a vector field on M is a map $X: M \to TM$ associated with each point $p \in M$ a vector $X(p)$ in the tangent space M_p. We can think of M_p as the space of all possible velocities of a particle moving along paths on M at p. Also recall that if X is a given vector field on M, we call any flow ϕ on M an _integral flow_ of X if X is the velocity vector field of ϕ. We also say that X is _integrable_ if such a flow exist.

Theorem 4.4.1 Let ϕ be a flow on M such that, for all $p \in M$, the map $\phi_p: R \to M$ is differentiable. Then the velocity of ϕ at any point is independent of time. Thus ϕ has a well defined velocity vector field.

An _integral curve_ of X is at least a C^1 map $\Gamma: I \to M$, where I is any real interval such that $\Gamma'(t) = X\Gamma(t)$ for all $t \in I$. A _local integral_ of X is a map $\phi: I \times U \to M$, where I is an interval neighborhood of O and U is a non-empty open subset of M, such that for all $p \in U$, $\phi_p: I \to U$ is an integral curve of X at p. For all $p \in U$, we call ϕ a local integral at p, and say that X is _integrable at p_ if such a local integral exists. If it does and is at least C^1, then the diagram

$$
\begin{array}{ccc}
T(I \times U) & \xrightarrow{T\phi} & TM \\
Y \uparrow & & \uparrow X \\
I \times U & \xrightarrow{\phi} & M
\end{array}
$$

168

commutes. If I can be extended to R, and U to M, then the local integral is a flow on M. We shall come back to this shortly.

Theorem 4.4.2 Let X be a vector field on M, and let h: M → N be a C^1 diffeomorphism. If Γ is an integral curve of X, then h is an integral curve of the induced vector field $h_*(X) = (Th)Xh^{-1}$ on N from X by h. If ϕ: IxU → M is a local integral of X at p, then $h\phi(id \times (h|U)^{-1})$ is a local integral of $h_*(X)$ at h(X).

It is intuitively clear that the following theorem follows.

Theorem 4.4.3 Any integral curve of a C^k vector field is C^{k+1}.

Let V be an open subset of a Banach space B, and let X be a C^k vector field on V (k ≥ 0). Also suppose that ϕ: IxU → V is a local integral of X. We now express the condition that Γ be an integral curve of X in terms of local representatives with respect to natural charts. Let (U, μ) be a chart of M and suppose the image of Γ is contained in U. Then the local representative of Γ with respect to the identity of R and (U, μ) is $\Gamma_\mu = \mu \cdot \Gamma$, while the local representative of the curve Γ' with respect to the identity of R and natural chart (TM|U,Tμ) is given by $(\Gamma')_\mu(t) = T\mu \cdot \Gamma'(t) = T\mu \cdot T\Gamma(t) = T(\mu \cdot \Gamma)(t) = (\Gamma_\mu)'(t)$ by the composite mapping theorem. Also, the local representative of X·Γ with respect to the identity of R and the natural chart Tμ is $T\mu \cdot X \cdot \Gamma = T\mu \cdot X \cdot \mu^{-1} \cdot \mu \cdot \Gamma = X_\mu \cdot \Gamma_\mu$ where X_μ is the local representative of X. Thus Γ is an integral curve of X iff X·Γ = Γ', iff $X_\mu \cdot \Gamma_\mu = \Gamma_\mu'$, iff Γ_μ is an integral curve of X_μ. This condition takes a simple and usual form if $\mu(U) \subset R^n$. Then we have $X_\mu(p) = (p;X_1(p),...,X_n(p))$ where p ∈ $\mu(U) \subset R^n$, {$X_i(p)$} are the components of X_μ, $\Gamma_\mu(t) = (\Gamma_1(t),...,\Gamma_n(t))$, $\Gamma_\mu'(t) = (\Gamma(t); \Gamma_1'(t),...,\Gamma_n'(t))$, and $X_\mu \cdot \Gamma_\mu = \Gamma_\mu'$ iff $\Gamma_\mu'(t) = X_i(\Gamma(t))$ for i = 1,...,n and all t ∈ I. Thus, Γ is an integral curve of X iff the local representatives satisfy the system of first-order ordinary diffeerential equations

$$\Gamma_1'(t) = X_1(\Gamma_1(t),...,\Gamma_n(t)),$$

169

$$\Gamma_n'(t) = X_n(\Gamma_1(t),\ldots,\Gamma_n(t)).$$

Note that t does not appear explicitly on the right. Such a system of equations (a local dynamical system) is called an <u>autonomous</u> system. It includes regular equations of higher order and the Hamiltonian equations of motion as special cases.

We should also point out that any m-th order ordinary differential equation in standard form

$$d^m y/dt^m = h(t,y,dy/dt,\ldots,d^{m-1}y/dt^{m-1}),$$

can be reduced to the form $dx/dt = g(t,x)$, where $x = (x_1,\ldots,x_m) \in R^m$, and g is a vector valued function. By substituting $x_1 = y$, $x_2 = dy/dt$, $\ldots$, $x_m = d^{m-1}y/dt^{m-1}$, and $g(t,x) = (x_2,x_3,\ldots,x_m, h(t,x_1,\ldots,x_m))$.

Example: In Chapter 1 we introduced the pendulum equation $d^2\theta/dt^2 = -g\sin\theta$ and it was reduced to $d\theta/dt = \Omega$, $d\Omega/dt = -g\sin\theta$ by the substitution $\Omega = d\theta/dt$.

Example: In classical mechanics, with a conservative force field, Lagrange's equations of motion can be obtained from the Euler-Lagrange equation

$$d/dt(\partial L/\partial q_i) - \partial L/\partial q_i = 0,$$

where q_i $(1 \le i \le n)$ are the "generalized coordinates", the Lagrangian $L = T - V$, where T is the kinetic energy and V is the potential energy. In terms of the generalized coordinates, $T = 1/2\ \Sigma_{i=1}^n\ m_{i(dq_{i/dt})}{}^2$, then the equations of motion becomes

$$m_i\ dq_i^2/dt^2 = - \partial V/\partial q_i.$$

With the substitution $m_i dq_i/dt = p_i$, the "generalized momentum", converts the equations of motion to $dp_i/dt = -\partial V/\partial q_i$. The generalized coordinates q_i and its time derivatives dq_i/dt form the coordinates of the tangent bundle of the configuration space M; while the generalized coordinates and generalized momenta form the coordinates of the cotangent bundle T^*M of the configuration space. In the T^*M, the equations of motion are the Hamilton's equations

$$dq_i/dt = \partial H/\partial p_i, \text{ and } dp_i/dt = - \partial H/\partial q_i,$$

where the Hamiltonian H = T + V is also the total energy of the system.

Example: (Van der Pol's equation)   This equation models the electronic oscillators in electric engineering.

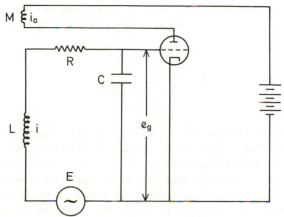

Consider the vacuum tube circuit represented by the integro-differential equation

$L\ di/dt + Ri + C^{-1}\int_0^t idt - M\ di_a/dt = E = E_o\sin\Omega_1 t.$

Assume $i_a = ke_g(1 - e_g^2/3V_s^2)$ where k is the transconductance of the tube, V_s is the saturation voltage, i.e., a sufficiently high grid voltage beyond which the current i_a does not change appreciably. Neglecting the grid current and anode reaction, let

$x = e_g/V_s = (V_sC)^{-1}\int_0^t idt,$ $a = Mk/LC - R/L,$
$b = Mk/3LC,$ $\Omega_o^2 = 1/LC,$ and $A = E_o/V_sLC = B\Omega_o^2.$

Then the integro-differential equation becomes a differential equation of the form:

$x" - ax' + b(dx^3/dt) + \Omega_o^2 x = A\sin\Omega_1 t.$

If the driving force function $A\sin\Omega_1 t$ is set equal to zero (i.e., A = 0), then it simplifies to:

$x" - (a - 3bx^2)x' + \Omega_o^2 x = 0.$

For simplicity, one can write the van der Pol equation in the following "normalized" form:

$x" - \alpha x'(1 - x^2) + x = 0,$

where α is a positive constant. We can transform this second order equation into a pair of first order equations as the following,
$$x' = y, \qquad y' = -x + \alpha y(1 - x^2).$$
The phase portrait of this vector field on R^2 is topologically equivalent to the following figure.

It has a unique closed orbit which is the Ω-set of all the orbits except the fixed point [Hirsch & Smale 1974]. The system is said to be auto-oscillatory since all (except one) solutions tend to become periodic as time increases.

Let p be a point of a Banach space B, let X be a vector field on some neighborhood U of p. We want to find out whether for some neighborhood V of p there are unique integral curves at each point of V. In order to prove uniqueness, we need something stronger than continuity and which is the Lipschitz condition. First we shall define a Lipschitz map.

Let P and Q are non-empty metric spaces with distance function d. Let k be any positive number. A map f: P → Q is Lipschitz (with constant k) if, for all p,p'ϵ P,
$$d(f(p),f(p')) \leq kd(p,p').$$
Clearly, any Lipschitz map is continuous (in fact, uniformly continuous). A map f is locally Lipschitz if every p ϵ P has a neighborhood on which f is Lipschitz.

Clearly, Lipschitz condition implies continuity and it is satisfied by any C^1 map on U. Thus, Lipschitz condition is about half way between continuity and differentiability.

As before, let p be a point of a Banach space B, X be a vector field on some neighborhood U of p, and V be some other neighborhood of p. Let d' and d are the distance functions of U and V respectively.

172

Theorem 4.4.4 (Picard)   Let f: U → B be Lipschitz with constant k, and let I be the interval [-a,a], where a < (d'-d)/|f|₀ if d'< ∞, and |f|₀ is the Banach norm. Then for each p ε V, there exists a unique integral curve ϕ_p: I → U of f at p. Let ϕ: IxV → U send (t,p) to $\phi_p(t)$. Then for all t ε I, ϕ^t: V → U is uniformly Lipschitz (in t), and C^r if f is C^r.

Theorem 4.4.5   Same assumptions as in the above theorem. If d'< ∞ , then the map ϕ is Lipschitz. And it is locally Lipschitz.

This theorem provide an answer as to the dependence of ϕ on t and x together. It says that ϕ is locally Lipschitz, and ϕ is as smooth as f. Then the main theorem on the smoothness of local integrals follows easily.

Theorem 4.4.6   If the map f of Theorem 4.4.4 is C^r (r ≥ 1), then the local integral ϕ is also C^r. Furthermore, $D_1(D_2)^r\phi$ exists and equals to $(D_2)^r(f\phi)$: I x B → $L_s(B,B)$.

Since any C^1 vector field is locally Lipschitz, we have the following useful corrollary:

Corollary 4.4.7   Any C^r vector field (r ≥ 1) on a manifold has a C^r local integral at each point of the manifold.

So far we have been dealing with local integrability condition and uniqueness. The main tool for extending local integrability condition to global one is the uniqueness of integral curves proved earlier. The next theorem is the global uniqueness theorem.

Theorem 4.4.8   Let X be a locally Lipschitz vector field on a manifold M, and let α: I → M, and β: I → M be integral curves of X. If for some t_o ε I, α(t_o) = β(t_o), then α = β .

Recall that a point p ε M is a singular point of a vector field X if X(p) = 0. One can easily show that all integral curves of a locally Lipschitz vector field X at p are constant functions if p is a singular point of X.

Earlier we commented that if the velocity of a flow is independent of time, we may consider such a flow as a vector field. Conversely, the following theorem states that a local

173

integral of a vector field is locally a flow.

Theorem 4.4.9   Let ϕ: IxU → M be a local integral of a locally Lipschitz vector field on a manifold M. Then for all p ϵ U and all s, t ϵ I such that s+t ϵ I and ϕ(t,p) ϵ U, ϕ(s,ϕ(t,p)) = ϕ(s+t,p).

It follows immediately:

Corollary 4.4.10   If ϕ: IxM → M is a local integral of a locally Lipschitz vector field on a manifold M, then it is a flow on M.

The next theorem shows the uniqueness of an integral flow in a Banach space.

Theorem 4.4.11   Any Lipschitz vector field X on a Banach space B has a unique integral flow ϕ. If X is C^r then ϕ is also C^r.

The next theorem shows that the local integrals are as smooth as the vector field we integrate and this is completely general.

Theorem 4.4.12   Any local integral of a C^r (r ≥ 1) vector field on a manifold M is C^r. Likewise, any integral of a locally Lipschitz vector field is locally Lipschitz.

Theorem 4.4.13   Any locally Lipschitz vector field X on a compact manifold M has a unique integral flow ϕ, which is locally Lipschitz. If X is C^r (r ≥ 1) then ϕ is also C^r.

By virtue of the fact that velocity fields and integral flows are so closely related, the qualitative theories of smooth vector fields and smooth flows can be thought of as one and the same subject. Since it is sometimes easier to describe the qualitative theory in terms of vector fields, and at other times in terms of flow, we shall feel free to use whichever terminology or viewpoints we find more convenient in any given situation.

4.5   Dispersive systems

In Section 4.3 we were interested in the systems of Poisson stable points or non-wandering points, and recursiveness. In this section, we shall discuss dynamical

systems marked by the absence of recursiveness. We shall
briefly describe dispersive dynamical systems and
instability, and we shall also study the theory of
parallelizable dynamical systems.

First we shall introduce the definition of positive
(negative) prolongation of x ϵ X. For any x ϵ X, the
<u>positive (negative) prolongation of x</u>, $D^{\pm}(x)$ = {y ϵ X: there
is a sequence {x_n} in X and a sequence {t_n} in $R^{\pm}$ such that
$x_n \to x$ and $x_n t_n \to y$}. The <u>positive (negative) prolongational
limit set of x</u>, $J^{\pm}(x)$ = {y ϵ X: there is a sequence {x_n} in
X and a sequence {t_n} in $R^{\pm}$ such that $x_n \to x$, $t_n \to \pm \infty$, and
$x_n t_n \to y$}.

It is clear that for any x ϵ X, $\Gamma^{\pm}(x)$ is a subset of
$D^{\pm}(x)$, $\Omega(x)$ is a subset of $J^{+}(x)$, $\alpha(x)$ is a subset of $J^{-}(x)$.
These inclusions can be illustrated by considering the
system, dx/dt = -x, dy/dt = y. This is a dynamical system
with a saddle point at the origin. We will leave to the
reader to graph the orbits and to establish the above
inclusions.

Let x ϵ X, and the partial map ϕ_x is <u>positively Lagrange
unstable</u> if the closure of the positive semi-orbits, $\Gamma^{+}(x)$,
is not compact. It is called <u>negatively Lagrange unstable</u> if
the closure of $\Gamma^{-}(x)$ is not compact. And it is called
<u>Lagrange unstable</u> if it is both positively and negatively
Lagrange unstable. In Section 4.3 we have defined Poisson
stable and wandering points. Now we are ready to define some
of the concepts of dispersiveness of a dynamical system.

A dynamical system is, <u>Lagrange unstable</u> if for each x ϵ
X the partial map ϕ_x is Lagrange unstable, <u>Poisson unstable</u>
if each x is Poisson unstable, <u>completely unstable</u> if every
x is wandering, <u>dispersive</u> if for every pair of points x, y
ϵ X there exist neighborhoods U_x of x and U_y of y such that
U_x is not positively recursive with respect to U_y.

As an example, let us consider a dynamical system in R^2,
whose phase portrait is Fig.4.5.1. The unit circle contains
a fixed point p and an orbit Γ such that for each point q ϵ
Γ we have $\Omega(q) = \alpha(q) = \{p\}$. All orbits in the interior of

175

the unit circle (= {p} U Γ) have the same property as Γ. All
orbits in the exterior of the unit circle spiral to the unit
circle as t → ∞, so that for each point q in the exterior of
the unit circle we have $\Omega(q) = \{p\} \cup \Gamma$, and $\alpha(q) = 0$. Notice
that if we consider the dynamical system obtained from this
one by deleting the fixed point p, now the dynamical system
is defined on $R^2 - \{p\}$, then this new system is Lagrange
unstable and Poisson unstable, but it is not completely
unstable because for each q ε Γ we have $J^+(q) = \Gamma$, i.e., q ε
$J^+(q)$.

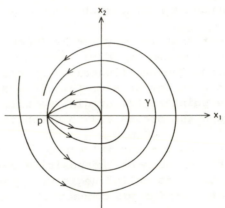

Fig.4.5.1

It should be noted that for dynamical systems defined by
differential equations in R^2, the concepts of Lagrange
instability and wandering are equivalent. This may be easily
proven by using the Poincaré-Bendixson theorem for planar
systems. Note that the above example has the fixed point
removed, i.e., in $R^2 - \{p\}$.

We next consider an example of a dispersive system which
turns out to be not parallelizable when we introduce this
concept shortly. Let the dynamical system in R^2 be,

dx/dt = f(x,y),   dy/dt = 0,

where f(x,y) is continuous, and f(x,y) = 0 whenever the
points (x,y) are of the form (n,1/n), and n is a positive

176

integer. For simplicity, we assume that $f(x,y) > 0$ for all
other points. The phase portrait is shown in Fig.4.5.2. Let
us now consider a new system which is obtained from the
above system by removing the sets
$\quad I_n = \{(x,y) \mid x \leq n, \ y = 1/n\}, \qquad n = 1, 2, 3, \ldots$
from R^2. This system is dispersive.

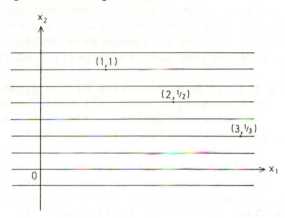

Fig.4.5.2

In the following, we shall state several results
characterize dispersive systems.

Theorem 4.5.1 A dynamical system is dispersive iff for
each $x \in X$, $J^+(x) = 0$.

From this theorem, the above example is clearly
dispersive. A more useful criterion is the following
theorem.

Theorem 4.5.2 A dynamical system is dispersive iff for
each $x \in X$, $D^+(x) = \Gamma^+(x)$ and there are no fixed points or
periodic orbits.

A dynamical system is called <u>parallelizable</u> if there
exists a subset S of X and a homeomorphism h: X → S x R such
that SR = X and h(xt) = (x,t) for every x ∈ S and t ∈ R.
Notice the analogue of this notion of parallelizability for
dynamical systems and the notion for fiber bundles. Indeed,
a notion of cross section is in order. A subset S of X is

177

called a <u>section</u> of the dynamical system if for each $x \in X$ there is a unique $\tau(x)$ such that $x\tau(x) \in S$. Note that not every dynamical system has a section. In fact, <u>any dynamical system has a section iff it has no fixed points or periodic orbits</u>. This is not surprising, because from the fiber bundle view point, the existence of a section implies that for each $\tau(x)$, continuous or not, each section is homeomorphic to each other. Note also, in general $\tau(x)$ is not continuous, but the existence of a section S with continuous $\tau(x)$ implies certain properties of the dynamical system.

<u>Lemma 4.5.3</u>   If S is a section of the dynamical system with $\tau(x)$ continuous on X, then: (i) S is closed in X, (ii) S is connected, arcwise connected, simply connected iff X is connected, arcwise connected, simply connected respectively, (iii) if a subset K of S is closed in S, then Kt is closed in X for every $t \in R$, (iv) if a subset K of S is open in S, then KI is open in X, where I is any open interval in R.

<u>Theorem 4.5.4</u>   A dynamical system is parallelizable iff it has a section S with $\tau(x)$ continuous on X.

This theorem shows that the dynamical system of the second example is not parallelizable. The next theorem is a very important one.

<u>Theorem 4.5.5</u>   A dynamical system on a locally compact separable metric space X is parallelizable iff it is dispersive.

The proof of this theorem depends on certain properties of sections. We shall not go into any detail here. It suffices to say that the concept of tubular neighborhoods is essential to its development. For more details, the reader is directed to Bhatia and Szegö [1970], Ch. 4.

4.6   Linear systems

Naively, one might hope to obtain a complete classification of all dynamical systems on Euclidean space R^n, starting with the simplest types and gradually building up some understanding of more complicated ones. The simplest diffeomorphisms of R^n are translations, and it is easy to show that two such maps are <u>linearly conjugate</u> (topological conjugate by a linear automorphism of R^n) providing the constant vector is non-zero. Furthermore, any two non-zero constant vector fields on R^n are linearly flow equivalent. The next simplest systems are linear discrete systems generated by linear automorphisms of R^n and vector fields on R^n with linear principal part. Even more, the classification problem is far from trivial, nor is it solved.

The difficulties one encounters even with linear problems underline the complexity and richness of the theory of dynamical systems. And indeed this is a major part of its attraction. Meanwhile, these difficulties also indicate the need for a modest approach to the problem. We only attempt to classify a "suitably large" class of dynamical systems instead of the complete solution.

For linear systems on R^n, it is easy to give a precise definition of what a "suitably large" class means. The set $L(R^n)$ of linear endomorphisms of R^n is a Banach space with the norm

$$|L| = \sup\{|L(x)|: x \in R^n, |x|=1\}.$$

A "suitably large" subset of linear vector fields on R^n is the one both open and dense in $L(R^n)$.

Let X be a topological space and $\sim$ is an equivalence relation on X. We say that a point $x \in X$ is <u>stable</u> with respect to $\sim$ if x is an interior point of its $\sim$ class. The <u>stable set</u> Σ of $\sim$ is the set of all stable points in X. Clearly, Σ is an open subset of X. If X has a countable basis, then Σ contains points of only countably many equivalence classes. Nonetheless, Σ may fail to be dense in X. In the present context, $X = L(R^n)$ (or $GL(R^n)$), $\sim$ is topological equivalence (conjugacy), and Σ is called the set of <u>hyperbolic</u> linear vector fields (automorphisms). In such a case, Σ is dense and easily classifiable.

The basic idea of calculus is to approximate a function locally by a linear function. Similarly, when discussing the local properties of a dynamical system we study its "linear approximation", which is a linear system. The important question is whether this linear system is a good approximation in the sense that it is locally qualitatively the same as the original system. And this obviously leads to the stability (in the broader sense) of the linear system. The importance of considering the hyperbolic linear vector fields (automorphisms) is that they are stable not only under small linear perturbations but also under small smooth (but not necessarily linear) perturbations. We shall begin with a review of linear systems on R^n, which provide a background of concrete examples against which the more general results on Banach spaces may be tested and placed in perspective.

Let T be any linear endomorphism of R^n, then we may think of it as a vector field on R^n. The corresponding ordinary differential equation is $x' = T(x)$ where $x \in R^n$. The integral flow can be written down immediately.

Since $L(R^n)$ is a Banach space, then the infinite series
$$\exp(tT) = id + tT + t^2 T^2 /2 + \ldots + t^n T^n /n! + \ldots,$$
$t \in R$, converges for all t and T and the integral flow ϕ of T is $\phi(t,x) = \exp(tT)(x)$. We shall call a flow ϕ _linear_ if ϕ^t is a linear automorphism varying smoothly with t. Thus ϕ is linear iff its velocity field is linear.

As an example, if $T = a(id)$ for some $a \in R$,
$$\exp(tT) = id + at(id) + \ldots + (at)^n id^n /n! + \ldots$$
$$= (1 + at + \ldots + (at)^n /n! + \ldots) id = e^{at} id.$$

A notion of linear equivalence for linear flow can be defined by letting ϕ and μ be linear flows with velocities S and $T \in L(R^n)$ respectively. Then μ is _linearly equivalent_ to ϕ, $\mu \sim_L \phi$, if for some $\alpha \in R$, $\alpha > 0$, and linear automorphism $h \in GL(R^n)$ of R^n, the following diagram commutes:

$$\begin{array}{ccc} R \times R^n & \xrightarrow{\phi} & R^n \\ \alpha \times h \downarrow & & \downarrow h \\ R \times R^n & \xrightarrow{\mu} & R^n \end{array}$$

180

That is, $h\phi(t,x) = \mu(\alpha t, h(x))$ for all $(t,x) \in R \times R^n$, or equivalently, from the standpoint of velocity fields, $\mu \sim_L \phi$ iff $S = \alpha(h^{-1}Th)$. Thus the problems of classifying linear flows up to linear equivalence is the same as classifying $L(R^n)$ up to similarity (linear conjugacy).

The fundamentally important result by Kuiper [1975] states that if ϕ and μ are linear flows given by linear endomorphisms S and T of R^n whose eigenvalues all have zero real part then ϕ is topologically equivalent to μ iff ϕ is linearly equivalent to μ. More generally, two linear flows ϕ and μ on R^n with decompositions $U_+ + U_- + U_0$ and $V_+ + V_- + V_0$, where $U_+(V_+)$, $U_-(V_-)$, $U_0(V_0)$ are subspaces corresponding to positive, negative, zero real part of the eigenvalues respectively, are topologically equivalent iff $\phi|U_0$ is linearly equivalent to $\mu|V_0$, dim U_+ = dim V_+, and dim U_- = dim V_-.

Let us recall again the linear flows in R and R^2. (i) Linear flows in R. Any linear endomorphism of the real line R is of the form $x \rightarrow ax$, $a \in R$. The map is hyperbolic iff $a \neq 0$. The integral flow is $t \cdot x = xe^{at}$ and there are exactly three topological equivalence classes:

$$\longrightarrow\!\!-\ 0\ -\!\!\longleftarrow\quad,\quad \ldots\ 0\ \ldots\quad,\quad \longleftarrow\!\!-\ 0\ -\!\!\longrightarrow.$$

$$a < 0 \qquad\qquad a = 0 \qquad\qquad a > 0$$

(ii) Linear flows on R^2. The real Jordan form for a real 2x2 matrix is one of the following three types:

(a) $\begin{bmatrix} \alpha & 0 \\ 0 & \mu \end{bmatrix}$, (b) $\begin{bmatrix} \alpha & 1 \\ 0 & \alpha \end{bmatrix}$, (c) $\begin{bmatrix} a & -b \\ b & a \end{bmatrix}$,

where α, μ, a, b $\in$ R and b > 0. One can show that a linear vector field on R^2 is hyperbolic iff (a) $\mu\alpha \neq 0$, (b) $\alpha \neq 0$, (c) $a \neq 0$.

181

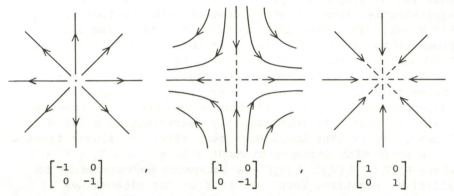

$$\begin{bmatrix} -1 & 0 \\ 0 & -1 \end{bmatrix} \quad , \quad \begin{bmatrix} 1 & 0 \\ 0 & -1 \end{bmatrix} \quad , \quad \begin{bmatrix} 1 & 0 \\ 0 & 1 \end{bmatrix}$$

There are also five topological equivalence classes of non-hyperbolic linear flows [see, e.g., Irwin 1980, p.87].

As we have pointed out earlier, discrete linear dynamical systems on R^n are determined by linear automorphisms of R^n. There are two equivalence relations, topological conjugacy and linear conjugacy, which are similar to the usual sense of linear algebra. Nonetheless, the relation between linear and topological conjugacy is even more difficult to pin down than the relation between linear and topological equivalence of flows. Here we shall restrict ourselves by noting that there is an open dense subset of $GL(R^n)$, (where S & $T \in GL(R^n)$, S is similar to T if for some $P \in GL(R^n)$, $T = PSP^{-1}$, i.e., $GL(R^n)$ is the set of linear similarity mappings), analogous to $GL(R^n)$ in the flow case, $HL(R^n)$. Elements of $HL(R^n)$ are called <u>hyperbolic linear automorphisms of R^n</u>. For any $T \in GL(R^n)$ iff none of its eigenvalues lies on the unit circle S^1 in C. It should be noted that the term "hyperbolic" must carry a dual meaning, this is because T is a hyperbolic vector field iff exp T is a hyperbolic automorphism. Nonetheless, the meaning should be clear from the context.

The simplest example is the automorphisms of R. Since any element T of $GL(R)$ is of the form $T(x) = ax$, where a is a non- zero real number. Thus there are six topological conjugacy classes, (i) a < -1, (ii) a = -1, (iii) -1 < a < 0, (iv) 0 < a < 1, (v) a = 1, (vi) a > 1. Clearly, by definition, except (ii) and (v) are hyperbolic.

The map T: R² → R² by T(x,y) = (x/2,2y) is in HL(R²). Its spectrum is the set {1/2,2} of eigenvalues of T and neither of these points is on S¹. The following figure demonstrates the effect of T. Note that the hyperbola and their asymptotes are invariant submanifolds of T. The x-axis is stable in the sense that positive iterates of T take its points into bounded sequences (they converge to the origin). The y-axis is unstable (meaning stable with respect to T⁻¹). A direct sum decomposition into stable and unstable summands is typical of hyperbolic linear automorphisms.

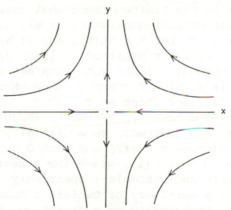

One can show that if T ∈ GL(Rⁿ) and if a ∈ R is sufficiently near, but not equal to, 1, then aT is hyperbolic. From this, one can deduce that HL(Rⁿ) is dense in GL(Rⁿ).

The next is the set of automorphisms in R². The classification is already quite complicated. We shall refer to Irwin [1980, p.88-9].

To embark on studying linear dynamical systems on a Banach space B, we take note of the analogy of the set of eigenvalues of a linear endomorphism T, which is called the spectrum of T, $\sigma(T)$. Recall that the set of linear endomorphisms of B, $L(B)$, is a Banach space with the norm defined by $|T| = \sup\{|T(x)|: x \in B, |x| \leq 1\}$. And the spectrum radius $r(T)$ is $r(T) = \sup\{|\mu|: \mu \in \sigma(T)\}$. It is a measure of the eventual size of T under repeated iteration, i.e., it equals to $\lim_{n\to\infty} |T^n|^{1/n}$. One may also choose a norm of B such that T has norm precisely $r(T)$ with respect to the corresponding norm of $L(B)$. When $r(T) < 1$, we call T a <u>linear contraction</u>, in the sense of metric space theory. If T is an automorphism, and T^{-1} is a contraction, then T is called <u>expansion</u>. Note that T is an automorphism iff $0 \notin \sigma(T)$, and that $r(T^{-1}) = (\inf \{|\mu|: \mu \in \sigma(T)\})^{-1}$. In fact, $\sigma(T^{-1}) = \{\mu^{-1}: \mu \in \sigma(T)\}$.

Analogous to the finite dimensional case, let $GL(B)$ be the open subset of $L(B)$ consisting of all linear automorphisms of B. Let $T \in GL(B)$, and if $\sigma(T) \cap S^1$ is empty, then we call T a <u>hyperbolic linear automorphism</u>. The set of all hyperbolic linear automorphisms of B is denoted by $HL(B)$. It is straightforward to show that if $T \in HL(B)$, $0 \in B$ is the only fixed point of T.

Now let D be a contour, symmetric about the real axis, if B is a real Banach space, such that $\sigma(T) \cap D$ is empty. Furthermore, suppose $0 \notin \sigma(T)$ and that D separates $\sigma(T)$ into two subsets $\sigma_s(T)$ and $\sigma_u(T)$, where the subscripts s and u stand for stable and unstable respectively, as we shall see later. Then by a corollary of Dunford's spectral mapping theorem, B splits uniquely as a direct sum $B_s + B_u$ of T-invariant closed subspaces such that, if $T_s: B_s \to B_s$ and $T_u: B_u \to B_u$ are the restrictions of T, then $\sigma(T_s) = \sigma_s(T)$ and $\sigma(T_u) = \sigma_u(T)$. By the spectral decomposition theorem, one can show that T_s is a contraction and T_u is an expansion. Let c be any number less than one but greater than the larger of the spectral radii of T_s and T_u^{-1}. Then one can prove:

<u>Theorem 4.6.1</u> If $T \in HL(B)$ then there is an equivalent norm $\| \ \|$ such that: (i) for all $x = x_s + x_u \in B$, $\|x\| = \max\{\|x_s\|, \|x_u\|\}$; (ii) $\max\{\|T_s\|, \|T_u^{-1}\|\} = a \leq c$. Thus, with

184

respect to the new norm $\| \ \|$, T_s is a metric contraction, and T_u is a metric expansion.

The stable summand B_s of B with respect to T is characterized as $\{x \in B: T^n(x)$ is bounded as $n \to \infty \}$, and equivalently as $\{x \in B: T^n(x) \to 0$ as $n \to \infty \}$. B_s is also called the stable manifold of O with respect to T, and T_s is called the stable summand of T. Similar characterizations and definitions hold for the unstable summand B, with T^{-n} replacing T^n. We say that $T \in HL(B)$ is isomorphic to $T' \in$ HL(B') if there exists a topological linear isomorphism from B to B' taking $B_s(T)$ onto $B_s'(T')$ and $B_u(T)$ onto $B_u'(T')$. Equivalently, T is isomorphic to T' iff there are linear isomorphisms of $B_s(T)$ onto $B_s'(T')$ and $B_u(T)$ onto $B_u'(T')$. Thus there are exactly n+1 isomorphism classes in $HL(R^n)$.

Theorem 4.6.2 HL(B) is open in GL(B), and hence in L(B).

Theorem 4.6.3 Any $T \in HL(B)$ is stable with respect to isomorphism.

Theorem 4.6.4 Let T and T' belong to the same path component of GL(B) and also have spectral radii < 1. Then T and T' are topologically conjugate.

Corollary 4.6.5 Two hyperbolic linear homeomorphisms of R^n are topologically conjugate iff

(i)   there are isomorphic,

(ii)   their stable components are either both orientation preserving or both orientation reversing, and

(iii) their unstable components are either both orientation preserving or both orientation reversing.

Thus, there are exactly 4n topological conjugacy classes of hyperbolic linear automorphisms of R^n (n $\geq$ 1). The main theorem is:

Theorem 4.6.6 For any Banach space B, any $T \in HL(B)$ is stable in L(B) with respect to topological conjugacy.

Corollary 4.6.7 Let B be finite dimensional, then HL(B) is the stable set of GL(B) with respect to topological conjugacy. Moreover, HL(B) is an open dense subset of GL(B).

Let us now turn to study linear vector fields. We are

185

interested in determining which vector fields are stable in
L(B) with respect to topological equivalence. We call T ∈
L(B) a <u>hyperbolic linear vector field</u> if its spectrum $\sigma(T)$
does not intersect the imaginary axis of **C**, and we denote
the set of all hyperbolic linear vector fields on B by
HLV(B).

Proposition 4.6.8 A linear vector field T is hyperbolic
iff for some non-zero real t, exp(tT) is a hyperbolic
isomorphism. Moreover, T has integral flow ϕ: RxB → B given
by $\phi(t,x)$ = exp(tT)(x).

As before, we shall define stable summands and stable
manifolds at O. Let T ∈ HLV(B), $\sigma_s(T)$ = {$\mu \in \sigma(T)$: Re μ < 0}
and $\sigma_u(T)$ = {$\mu \in \sigma(T)$: Re μ > 0}. Also let D be a contour
consisting of the line segment [-ir, ir] and the semi-circle
{$re^{i\theta}$: $\pi/2 \leq \theta \leq 3\pi/2$} where r is large enough such that D
encloses σ_s. As before, the spectral decomposition theorem
gives a T-invariant direct sum splitting B = $B_s \oplus B_u$ such
that $\sigma(T_s)$ = $\sigma_s(T)$ and $\sigma(T_u)$ = $\sigma_u(T)$, where $T_s \in L(B_s)$ and T_u
$\in L(B_u)$ are the restrictions of T. We call T_s and B_s stable
summand and stable manifold at O respectively. Likewise, T_u
and B_u are unstable summand and unstable manifold at O
respectively. One can easily show that for all t > 0, the
stable summand of the hyperbolic automorphism exp(tT) is
exp(tT_s) and similarly for unstable summands.

One can show that the map exp:L(B) → L(B) is continuous,
and since HL(B) is open in L(B), Proposition 4.5.8 implies
that HLV(B) is also open in L(B). Moreover, we can define
isomorphism for hyperbolic vector fields just as for
hyperbolic automorphisms. Since the stable and unstable
summands of B are the same with respect to T ∈ HLV(B) as
with respect to exp T ∈ HL(B) as noted before, thus T ∈
HLV(B) is stable respect to isomorphism follows immediately
from the corresponding result for HL(B). We then have the
following theorem.

Theorem 4.6.9 Let T ∈ HLV(B), then T = $T_s \oplus T_u$, where
$\sigma(T_s)$ = $\sigma_s(T)$ and $\sigma(T_u)$ = $\sigma_u(T)$. The set HLV(B) is open in
L(B), and its elements are stable with respect to

isomorphism.

As in the case of hyperbolic linear automorphism, we may choose a norm on B such that the stable and unstable summands of exp(tT) are respectively a metric contraction and a metric expansion.

Theorem 4.6.10 Let T ∈ HLV(B) have spectrum $\sigma(T) = \sigma_s(T)$. Then there exists a norm $\|\|$ on B equivalent to the given one such that, for any non-zero x ∈ B, the map from R to (0, ∞) taking t to $\|\exp(tT)(x)\|$ is strictly decreasing and surjective. There also exists b > 0 such that for all t > 0, $\|\exp(tT)\| \le e^{-bt}$.

Corollary 4.6.11 If ϕ: RxB → B is the integral flow of T ∈ HLV(B), then ϕ maps RxS^1 homeomorphically onto B/{o}.

Theorem 4.6.12 Let T, T'∈ HL(B) have spectra in Re z < 0. Then T and T' are flow equivalent.

Corollary 4.6.13 If T and T'∈ HLV(B) are isomorphic, then they are flow equivalent, thus topologically equivalent.

Thus we have the following important result:

Theorem 4.6.14 Any T ∈ HLV(B) is stable in L(B) with respect to flow equivalence, thus with respect to topological equivalence.

The above discussion will provide the basis for comparing the vector field or flow in question with the known linear vector field or flow. The stable and unstable decomposition and summonds are the basis for the discussion of stability.

4.7   Linearization

As we have pointed out earlier we can hardly expect to make much progress in the global theory of dynamical systems on a smooth manifold M, if we ignore completely how systems behave locally. We shall attempt to classify dynamical systems locally, i.e., linearly. But even here, as we have pointed out at the begining of section 5, there are difficulties with linear systems, thus we can only expect partial success.

As we have seen earlier, linear systems exhibit a diverse variety of behavior near the fixed point zero. In order to focus on our classification scheme, we shall only discuss the class of fixed points which often occurs to be the sort that one "usually comes across". For linear systems, we are able to give this vague notion a precise topological meaning in terms of the spaces of all linear systems on a given Banach space B.

Let ϕ be a given dynamical system on a smooth manifold M which has a fixed point p. Since we are only interested in local structure, we can assume M = B and p = 0, after taking a chart at p. Suppose now that by altering ϕ slightly near 0 we end up with another fixed point near 0, with a local phase portrait resembling that of ϕ at 0. Intuitively, the linear approximation of ϕ at 0 (we shall make sense of this term later) must have a phase portrait near 0 resembling that of ϕ near 0 as most all "nearby" linear systems. With the stability theorems of the last section, it is clear that we should consider fixed points whose linear approximations are hyperbolic systems.

A fixed point p of a diffeomorphism f of M is hyperbolic if the tangent map $T_p f: M \to M_p$ is a hyperbolic linear automorphism. Corollary 4.5.5 supplies a classification of hyperbolic linear automorphisms. We shall extend this result to a classification of hyperbolic fixed points using the Hartman theorem [Hartman 1964; Moser 1969] which states that any hyperbolic fixed point p of f is topologically conjugate to the fixed point 0 of $T_p f$. An analogous situation exists with flows. A fixed point p of a local integral $\phi: I \times U \to M$ of a vector field X on M (or equivalently a zero of X) is hyperbolic if for some $t \neq 0$ in I, p is a hyperbolic fixed point of ϕ^t [Grobman 1959, 1962].

Proposition 4.7.1   The point p is a hyperbolic zero of a vector field X iff for any chart μ at p, the differential of the induced vector field $(T\mu) X \mu^{-1}$ at $\mu(p)$ is a hyperbolic vector field.

188

An equivalent but more sophisticated approach is to consider TX: TM → T(TM), which is a section of the vector bundle $T\pi_M$: T(TM) → TM, where π_M: TM → M. But $T\pi_M$ may be identified with the tangent bundle on TM, where π_{TM}: T(TM) → TM by the canonical involution, and TX then becomes a vector field on TM. With such identification, T_pX maps M_p into $T(M_p)$. Thus T_pX is a linear vector field on M_p. Recall that this linear vector field is the Hessian of X at p. Thus:

Proposition 4.7.2   p is a hyperbolic zero of X iff the Hessian of X at p is a hyperbolic linear vector field.

We shall state later that if p is a hyperbolic fixed point of a vector field X then it is flow equivalent to the zero of the Hessian of X at p.

Recall that a point p of M is a regular point of a dynamical system on M if it is not a fixed point of the system. The following two theorems show that regular points are uninteresting from local viewpoint.

Theorem 4.7.3   If p is a regular point of a diffeomorphism f: M → M and g is a translation of the model space B of M by a non- zero vector X_o, then f|p is topologically conjugate to g|0.

Theorem 4.7.4   Let p be a regular point of a C^1 flow ϕ on M. Let x_o be any non-zero vector of the model space B of M, and let μ be the flow on B defined by $\mu(t,x) = x + tx_o$. Then ϕ|p is flow equivalent to μ|0.

It should be noted that the equivalence in the above two theorems are as smooth as the system under consideration.

Let T be a hyperbolic linear automorphism of B with skewness a < 1 with respect to the norm | | on B, i.e., $\max\{|T_s|, |T_n^{-1}|\} = a < 1$. The main piller of the theory of the stability of T with respect to topological conjugacy under smooth small perturbations is the theorem of Hartman [1964].

Lemma 4.7.5   Let η : B → B be Lipschitz with constant k < 1−a. Then T+η has a unique fixed point.

Theorem 4.7.6 (Hartman's linearization theorem)   Let $\eta \in C^0(B)$ be Lipschitz with constant k < min$\{1-a, |T_s^{-1}|^{-1}\}$.

Then T+η is topologically conjugate to T.

A more detailed version can be stated as:

Theorem 4.7.7   Let η, μ ϵ C^o(B) be Lipschitz with constant k < min{1-a, $|T_s^{-1}|^{-1}$}. Then there exists a unique g ϵ C^o(B) such that (T+η)(id+g) = (id+g)(T+μ). Furthermore, id+g is a homeomorphism, and thus is a topological conjugacy from T+μ to T+η.

Note that, the conjugacy id+g in the above theorem cannot always be as smooth as T+η or T+μ. The main application of Hartman's theorem is to compare a diffeomorphism near a hyperbolic fixed point with its linear approximation at the point.

Corollary 4.7.8   A hyperbolic fixed point p of a C^1 diffeomorphism f: M → M is topologically conjugate to the fixed point 0 of $T_p f$.

Corollary 4.7.9   There are 4n topological conjugacy classes of hyperbolic fixed points that occur on n-dimensional manifolds.

In relation to our main interest, we shall say a few words concerning stability. We call a fixed point p of a diffeomorphism f: M → M <u>structurally stable</u> if for each sufficiently small neighborhood U of p in M there is a neighborhood V of f in Diff(M) such that, for all g ϵ V, g has a unique fixed point q in U and q is topologically conjugate to p. In fact, one can prove that <u>fixed points are structurally stable if (and in finite dimensions, only if) they are hyperbolic</u>. The following is an outline of the proof.

Let f: U → f(U) be a diffeomorphism of open subsets of B with a fixed point at 0 is hyperbolic, and that Df(0) has skewness a with respect to the norm | | on B. Choose k, where 0 < k < 1-a, so small that for all T ϵ L(B) with |T - Df(0)| ≤ k, T is hyperbolic and topologically conjugate to Df(0). Let B_b be a closed ball in B with center 0 and radius b (< 1) such that |Df(x) - Df(0)| ≤ k/2 for all x ϵ B_b. Since if η: B_b → B is Lipschitz with constant k and if $|\eta|_o$ ≤ b(1-a) then T+η: B_b → B has a unique fixed point. One can

190

prove that for all C^1 maps g: U → B with $|g - f| \leq kb/2$, g
has a unique fixed point p in B_b. It then easily follows
that g|p is topologically conjugate to f|0, and hence 0 is
structurally stable under C^1-small perturbations of f.
Conversely, one can prove that if B is finite dimensional
and if 0 is structurally stable under C^1-small perturbation
of f then 0 is hyperbolic.

Of the statement we have just outlined, the proof is a
very important one. It forms a basis for our discussion of
structural stability in Chapter 6.

Let us now discuss Hartman's linearization theorem in
the context of flows. It was discovered independently by
Grobman [1959,1962].

Theorem 4.7.10   Let T ∈ HL(B). For all Lipschitz maps ∩
∈ C^0(B) with sufficiently small Lipschitz constant, there is
a flow equivalence from T to T+η.

Let us recall the definition of the Hessian of a map at
a critical point. Let M be $C^∞$, f is $C^∞$ on M. f has a
critical point at p ∈ M if df_p = 0. If p is a critical point
of f, then the Hessian of f at p, H_f, is a bilinear function
on M_p defined as: if u,v ∈ M_p, X ∈ X(M) such that if X(p) =
u, then H_f(u,v) = v(Xf). Since df_p = 0, then H_p(u,v) is
independent of the choice of X and H_f is symmetric. With
Hessian defined, we can state the following corollaries.

Corollary 4.7.11   A hyperbolic zero, p, of a C^1 vector
field on M is flow equivalent to the zero of its Hessian at
p.

Corollary 4.7.12   Flow equivalence and orbit equivalence
coincide for hyperbolic zeros of C^1 vector fields on an
n-dim. manifold M. There are precisely n+1 equivalence
classes of such points with respect to either relation.

Similar to the statement about the relationship between
hyperbolic fixed point and structural stability, one can
prove the following: Let X be a C^1 vector field on B with a
zero at p. If p is hyperbolic then it is stable with respect
to flow equivalence under C^1-small perturbations of X.
Conversely, if B is finite dim. and if p is stable with

191

respect to topological equivalence under C^1 small
perturbation of X, then p is hyperbolic.
    It should be emphasized that hyperbolicity of fixed
points is not an invariant of topological or flow
equivalence. In other words, it is possible for a
non-hyperbolic fixed point to be topological or flow
equivalent to a hyperbolic fixed point. For instance, the
non-hyperbolic zero of the vector field $X(x) = x^3$ on R is
clearly flow equivalent to the hyperbolic zero of the vector
field $W(x) = x$.
    There are three distinct possibilities for a fixed point
p of a flow ϕ to be topologically equivalent to a
hyperbolic fixed point q of a flow μ: (a) $(T_q\mu)_s = T_q\mu$, (b)
$(T_q\mu)_u = T_q\mu$, and (c) niether (a) nor (b). We call p (and q)
(a) a sink, (b) a source, and (c) a saddle point,
respectively.

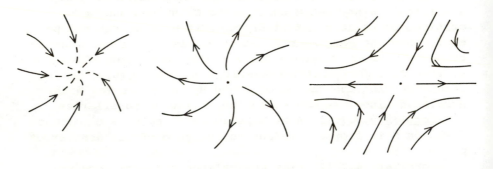

        sink,   source,   saddle.
    It is clear that any sink p has the property that it is
asymptotically stable in the sense of Liapunov (we shall
discuss this in the next chapter). In other words, any
positive half orbit of a sink p, $R^+ \cdot x = \{t \cdot x \; ; \; t \geq 0\}$
starting at a point x near p stays near p and eventurally
ends at p. Recall in the introduction (Chapter 1), in the
language of differential equations, asymptotic stability is
concerned with the way an individual solution of a system

varies as the initial conditions are varied. This is different from structural stability, which deals with the way the set of all solutions varies as we change the system itself.

It is appropriate at this juncture for us to discuss some properties of hyperbolic closed orbits and Poincare maps. Let p be a point on a closed orbit of a C^r flow ϕ ($r \geq$ 1) on a manifold M. Suppose that the orbit $\Gamma = R \cdot p$ of ϕ through p has period τ. Then $\phi^\tau: M \to M$ is a C^r diffeomorphism with p its fixed point. Where the tangent map $T_p \phi^\tau$ is a linear homeomorphism of M_p. It keeps the linear subspace $<X(x)>$ of M_p pointwise fixed, where $X(x)$ is the velocity of ϕ at x, because ϕ^τ keeps the orbit Γ pointwise fixed. Γ is said to be <u>hyperbolic</u> if M_p has a $T_p \phi^\tau$- invariant splitting, i.e., $M_p = <X(p)> \oplus F_p$ where $T_p \phi^\tau | F_p$ is hyperbolic. Thus, hyperbolicity for a closed orbit means that the linear approximation to ϕ^τ is as hyperbolic as possible.

A clear geometrical picture can be obtained in terms of Poincare maps. Let N be some open disk embedded as a submanifold of M through p such that $<X(x)> \oplus N_p = M_p$. Notice that this is exactly the condition of transversality. Indeed, we say that N is <u>transverse to the orbit at p</u>, and call it a <u>cross-section to the flow</u> at p. We assert that for some small open neighborhood U of p in N, there is a unique continuous function $\beta: U \to R$ such that $\beta(p) = \tau$, and $\phi(\beta(y),y) \in N$. Here β is a <u>first return function for N</u>. Any two such functions agree on the intersection of their domains. Intuitively, $\beta(y)$ is the time it takes a point starting at y to move along the orbit of the flow in the positive direction until it hits the section N again. We denote the point at which it hits again by f(y). Thus, f: U $\to$ N is a map defined by $f(y) = \phi(\beta(y),y)$.

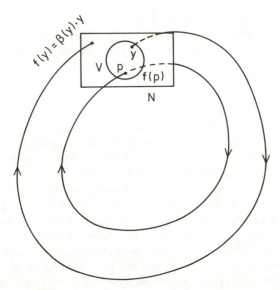

We call f a <u>Poincaré map</u> for N. It is well defined up to the
domain. So far, it is not required that Γ to be hyperbolic.
The next theorem shows that Γ is hyperbolic precisely when
the Poincaré map for some section N at p has a hyperbolic
fixed point at p.

   <u>Theorem 4.7.13</u> For sufficiently small U, the first
return function β is well defined and C^r, and the Poincaré
map f is a well defined C^r diffeomorphism of U onto an open
subset of N. If M_p has a $T_p\phi^\tau$-invariant splitting $<X(p)> \oplus$
F_p, then T_pf is linearly conjugate to $T_p\phi^\tau|F_p$. The orbit Γ is
hyperbolic iff p is a hyperbolic fixed point of f.

   Let f as before, and f': U' → f'(U') be a diffeomorphism
of open subsets of a manifold N'. A <u>topological conjugacy</u>
from f to f' is a homeomorphism h: U <u>U</u> f(U) → U' <u>U</u> f'(U')
such that, for all y ∈ U, hf(y) = f'h(y), here <u>U</u> denotes
union.

   <u>Proposition 4.7.14</u> If two diffeomorphisms f: U → f(U)
and f': U' → f'(U') are topologically conjugate, then their
suspensions are flow equivalent.

194

The above result simplifies the problem of classifying suspension (see Sect.4.1). The main connection between the flow near a closed orbit Γ of a C^r flow ϕ on M and the Poincaré map at a cross-section N of the flow at p of Γ is the following theorem.

Theorem 4.7.15   Let f: U → f(U) be a Poincaré map at p for the cross-section N. Then there is a C^r orbit preserving diffeomorphism h from some neighborhood of the orbit R·p of the suspension $\Sigma(f)$ to some neighborhood of the orbit Γ of ϕ such that h(p) = p.

Since if Γ is hyperbolic, then any Poincaré map f has a hyperbolic fixed point at p. Then by Corollary 4.7.8 and Proposition 4.7.14 we have

Corollary 4.7.16   If Γ is hyperbolic then the flow ϕ is topologically equivalent at Γ to $\Sigma(T_p f)$ at its unique closed orbit.

From Corollary 4.7.9 we have:

Corollary 4.7.17   There are precisely 4n different hyperbolic closed orbits that can occur in a flow on an (n+1)-dim. manifold (n ≥ 1).

Hyperbolic closed orbits are structurally stable, in the sense that, if Γ is such an orbit of a C^1 vector field X on M, and Y is a vector field on M that is C^1-closed to X, then for some neighborhood U of Γ in M, Y has a unique closed orbit in U, and this closed orbit is topologically equivalent to Γ. We shall discuss structural stability in more detail in Chapter 6.

Recall that in Hartman's theorem we altered a hyperbolic linear homeomorphism T by a perturbation η and found a topological conjugacy h = id+g from T to T+η. It was pointed out that h is not necessarily C^1 even when η is C^∞, since differentiating the conjugacy relation would place algebraic restrictions on the first derivatives of T+η. The question arises naturally as to whether further differentiation places further restrictions on higher derivatives, and even if these algebraic restrictions are satisfied, if the smoothness of η has any effect on h. It turns out that, in finite dimensions, further restrictions are the exception rather than the rule. We have some positive results on the smoothness of conjugacy relations. The main theorem is due to Sternberg [1957, 1958], for more details, see for instance, Nelson 1969], and a couple of relevant theorems by Hartman.

Theorem 4.7.18 (Sternberg) Let $T \in L(R^n)$ have eigenvalues $\alpha_1, \ldots, \alpha_n$ (can be complex or degenerated) satisfying $\alpha_i \neq \alpha_1^{m_1} \cdot \ldots \cdot \alpha_n^{m_n}$ for all $1 \leq i \leq n$ and for all non-negative integers $m_1, \ldots, m_n$ with $\Sigma_{i=1} m_i \geq 2$. Let $\eta: U \rightarrow R^n$ be a C^s map ($s \geq 1$) defined on some neighborhood U of O with $\eta(O) = D\eta(O) = 0$. Then $(T+\eta)|O$ is C^r conjugate to $T|O$, where, for a given T, r depends only on s and tends to ∞ as s does.

Notice that the eigenvalue condition implies that $T \in HL(R^n)$. Also, if η is C^∞, the maps T and T+η are C^∞ conjugate at O.

Theorem 4.7.19 Let $T \in L(R^n)$ be a contraction and $\eta: U \rightarrow R^n$ be a C^1 map defined on some neighborhood U of O, with $\eta(O) = D\eta(O) = 0$. Then $(T+\eta)|O$ is C^1 conjugate to $T|O$.

Theorem 4.7.20 (Hartman) Let $T \in HL(R^n)$, where n = 1 or 2, and let η be the same as above. Then $(T+\eta)|O$ is C^1 conjugate to $T|O$.

Nitecki [1979] studied the dynamic behavior of solutions for systems of the form dx/dt = G(F(x)) where F is a real-valued function on R^n near equilibria and periodic orbits. He has found that for n $\geq$ 3, the behavior near periodic orbits is arbitrary, i.e., any diffeomorphism of the (n-1)-disk isotopic to the identity arises as the Poincaré map near a periodic orbit for some choice of F and G. On the other hand, for n $\geq$ 2 the behavior near equilibria is severely restricted. Indeed, (a) if dG/dF $\neq$ 0 at an equilibrium point, the flow in a neighborhood is conjugate to that of a constant vector field multiplied by a function; (b) if an equilibrium point is isolated, it is an extremum of F, and if F satisfies a convexity condition near the equilibrium, then the flow in a neighborhood resembles that

196

described in (a), except that the stable and unstable sets may be cones instead of single orbits; and (c) when n = 2, a stronger condition on F, together with a weaker condition of G, again yields the description in (a).

Finally, let us end this chapter by giving a list of further readings. First of all, the classic review article by Smale [1967] is the one on the "must-read" list. Irwin [1980] is the source of most of the notes in this chapter. A most recent introductory book, which takes the reader a long way, by Ruelle [1989] is highly recommended.

Chapter 5   Stability Theory and Liapunov's Direct Method

5.1   Introduction

Earlier in Chapter 1, we described the general idea of stability in the normal sense. Very early on, the stability concept was specialized in mechanics to describe some type of equilibrium of a particle (or a celestial object) or system (such as our solar system). For instance, consider a particle subject to some forces and posessing an equilibrium point q_o. The equilibrium is <u>stable</u> if after any sufficiently small perturbations of its position and velocity, the particle remains arbitrarily near q_o with arbitrarily small velocity. As we have discussed in Chapter 1, the well known example of a pendulum, and we shall not dwell on it. It suffices to say that the lowest position (equilibrium point), associated with zero velocity, is a stable equilibrium, whereas the hightest one also with zero velocity is an unstable one.

We have also briefly mentioned that when fomulated in precise mathematical terms, this "mechanical" definition of stability was found very useful in many situations, yet inadequate in many others. This is why a host of other concepts have been introduced and each of them relates to the "mechanical" definition and to the common sense meaning of stability. Contrary to the "mechanical" stability, Liapunov's stability has the following features:
1°. it does not pertain to a material particle (or the equation), but to a general differential equation;
2°. it applies to a solution, i.e., not only to an equilibrium or critical point.

In his mémoire, Liapunov [1892] dealt with stability by two distinct methods. His 'first method' presupposes an explicit solution known and is only applicable to some restricted, but important situations. While this 'second method' (also called 'direct method') does not require the prior knowledge of the solutions themselves. Thus, Liapunov's direct method is of great power and advantage. An

198

elementary introduction to Liapunov's direct method can be found in La Salle and Lefschetz [1961].

Suppose for an autonomous system $dx/dt = f(x)$, in which the system is initially in an equilibrium state, then it remains in that state. But this is only a mathematical statement, yet a real system is subject to perturbations and it is impossible to control its initial state exactly. This begs the question of stability, that is, under an arbitrary small perturbation will the system remain near the equilibrium state or not? In the following, we shall make these concepts precise, and we will discuss these questions extensively.

Let $dx/dt = f(x,t)$ where x and f are real n-vectors, $t \in$ R ("time"), f is defined on $R \times R^n$. We assume f is smooth enough to ensure its existence, uniqueness, and continuous dependence of the solutions of the initial value problem associated with the differential equation over $R \times R^n$. For simplicity, we assume that all solutions to be mentioned later exist for every $t \in R$. Let $\|\cdot\|$ denote any norm on R^n.

A solution $x'(t)$ of $dx/dt = f(x,t)$ is <u>stable at t_o</u>, in the sense of Liapunov if, for every $\epsilon > 0$, there is a $\delta > 0$ such that if $x(t)$ is any other solution with $\|x(t_o) - x'(t_o)\| < \delta$, then $\|x(t) - x'(t)\| < \epsilon$ for all $t \geq t_o$. Otherwise, $x'(t)$ is <u>unstable at t_o</u>. Thus the stability at t_o is nothing but continuous dependence of the solution on $x_o = x(t_o)$, uniform with respect to $t \in [t_o, \infty)$.

We can gain some geometric insight into this stability concept by considering the pendulum. As before, the set of first order differential equations have the form: $d\theta/dt = \Omega$, $d\Omega/dt = - g\sin\theta$. The origin of the phase space (θ, Ω) represents the pendulum hanging vertically downward and is at rest. As we have shown before in the phase portrait, all solutions starting near the origin form a family of non-intersecting closed orbits encircling the "origin". Given $\epsilon > 0$, consider an orbit entirely contained in the disk B_ϵ of radius ϵ centered at the "origin". Further choose any other disk B_δ of radius δ contained in this chosen

199

orbit. Due to the non-intersecting nature of the orbits, every solution starting in B_δ at any initial time remains in B_ϵ. This demonstrates stability of the equilibrium for any initial time. On the other hand, any other solution corresponding to one of the closed orbits is unstable. This is because the period of the solution varies with the orbit and two points of (θ, Ω)-plane, very close to each other at t $= t_o$, but belonging to different orbits, will appear in opposition after some time. This happens no matter how small the difference of periods is. Yet, it remains that the orbits are close to each other. This leads to the concept of orbital stability.

We have discussed the dissipative system, such as the damped pendulum, whereby the stable equilibrium becomes asymptotically stable. That is, if all neighboring solutions x(t) of x'(t) tend to x'(t) when t → ∞. We have also said a few words about our Solar system and the notion of Lagrange stability. In the following we shall give various definitions of stability and attractivity for our future use. We shall note that if we replace x by a new variable z $= x - x'(t)$, then $dx/dt = f(t,x)$ becomes $dz/dt = g(t,z) \equiv$ $f(t,z+x'(t)) - f(t,x'(t))$, where $g(t,0) = 0$ for all t ∈ R. The origin is a critical point of the transformed equation and stability of the solution x(t) of the original equation is equivalent to stability of this critical point for the transformed equation. Of course, such a transformation is not always possible, nor is it always rewarding. Nonetheless, for the time being, we shall concentrate on stability of critical points.

Let us consider a continuous function f: I x D → R^n, $(t,x) \to f(t,x)$ where I = (τ, ∞) for some τ ∈ R or $\tau = -\infty$, and D is a domain of R^n, containing the origin. Assume that $f(t,0) = 0$ for all t ∈ I so that for the differential equation $dx/dt = f(t,x)$, the origin is an equilibrium or critical point. Furthermore, assume f to be smooth enough so that through every (t_o, x_o) ∈ IxD, there passes one and only one solution of $dx/dt = f(t,x)$. We represent this solution

200

by $x(t;t_o,x_o)$, thus displaying its dependence on initial conditions. By definition, $x(t_o,t_o,x_o) = x_o$. For the right maximal interval where $x(\cdot;t_o,x_o)$ is defined, we write $J^+(t_o,x_o)$ or simply J^+. We recall that $B_\sigma = \{x \in R^n: \|x\| < \sigma\}$.

The solution $x = 0$ of $dx/dt = f(t,x)$ is <u>stable</u> if for a given $\epsilon > 0$ and $t_o \in I$, there is a $\delta > 0$ such that for all $x_o \in B_\delta$ and $t \in J^+$, one has $\|x(t;t_o,x_o)\| < \epsilon$. The solution is <u>unstable</u> if for some $\epsilon > 0$ and $t_o \in I$ and each $\delta > 0$ there is an $x_o \in B_\delta$ and a $t \in J^+$ such that $\|x(t;t_o,x_o)\| \geq \epsilon$. The solution is <u>uniformly stable</u> if given $\epsilon > 0$, there is a $\delta = \delta(\epsilon)$ such that $\|x(t;t_o,x_o)\| < \epsilon$ for all $t_o \in I$, all $\|x_o\| < \delta$ and all $t \geq t_o$.

Since $J^+ \subset [t_o,\infty)$, thus in principle any solution may cease to exist after a certain finite time. Nonetheless, if $\overline{B_\epsilon} \subset D$, the solutions mentioned in the definitions of stability and uniform stability may continue up to ∞.

The solution $x = 0$ of $dx/dt = f(t,x)$ is <u>attractive</u> if for each $t_o \in I$ there is an $\eta = \eta(t_o)$, and for each $\epsilon > 0$ and each $\|x_o\| < \eta$, there is a $\sigma = \sigma(t_o,\epsilon,x_o) > 0$ such that $t+\sigma \in J^+$ and $\|x(t;t_o,x_o)\| < \epsilon$ for all $t \geq t_o+\sigma$. The solution is <u>equi- attractive</u> if for each $t_o \in I$ there is an $\eta = \eta(t_o)$ and for each $\epsilon > 0$ a $\sigma = \sigma(t_o,\epsilon)$ such that $t_o+\sigma \in J^+$ and $\|x(t;t_o,x_o)\| < \epsilon$ for all $\|x_o\| < \eta$ and all $t \geq t_o+\sigma$. The solution is <u>uniformly attractive</u> if for some $\eta > 0$ and each $\epsilon > 0$ there is a $\sigma = \sigma(\epsilon) > 0$ such that $t_o+\sigma \in J^+$ and $\|x(t;t_o,x_o)\| < \epsilon$ for all $\|x_o\| < \eta$, all $t_o \in I$ and all $t \geq t_o+\sigma$.

As remarked earlier, if $\overline{B_\epsilon} \subset D$, the solutions mentioned immediately above exist over $[t_o,\infty)$. Thus, in the definition of attractivity, all solutions starting from B_η approach the origin as $t \to \infty$. For equi-attractivity, they tend to 0 uniformly with respect to $x_o \in B\eta$ whereas for uniform attractivity they tend to 0 uniformly with respect to $x_o \in B\eta$ and $t_o \in I$.

We can define attraction slightly differently. Let the phase space X be locally compact, and M a non-empty compact subset of X. The <u>region of weak attraction of M</u>, $A_w(M) = \{x$

201

ϵ X: $\Omega(x) \cap M \neq 0$}, the <u>region of attraction of M</u>, A(M) = {x
ϵ X: $\Omega(x) \neq 0$ and $\Omega(x)$ a subset of M}, and the <u>region of
uniform attraction of M</u>, $A_u(M)$ = {x ϵ X: $J^+(x) \neq 0$ and $J^+(x)$
a subset of M}. Furthermore, any point x in $A_w(M)$, A(M),
$A_u(M)$ is said to be <u>weakly attracted, attracted, or
uniformly attracted to M</u> respectively.

    <u>Proposition 5.1.1</u> Given M, a point x is:
(i) weakly attracted to M iff there is a sequence {t_n} in R
with $t_n \to \infty$ and $d(xt_n, M) \to 0$, where d(.,.) is the metric
distance,
(ii) attracted to M iff d(xt,M) $\to$ 0 as t $\to \infty$,
(iii) uniformly attracted to M iff for every neighborhood V
of M there is a neighborhood U of x and a T > 0 with U_t a
subset of V for t $\geq$ T.

    <u>Theorem 5.1.2</u> For any given M, $A_u(M)$ is a subset of
A(M), which is a subset of $A_w(M)$, and they all are
invariant.

    A given set M is said to be: (i) a <u>weak attractor</u> if
$A_w(M)$ is a neighborhood of M, (ii) an <u>attractor</u> if A(M) is a
neighborhood of M, (iii) a <u>uniform attractor</u> if $A_u(M)$ is a
neighborhood of M, (iv) <u>stable</u> if every neighborhood U of M
contains a positively invariant neighborhood V of M, (v)
<u>asymptotically stable</u> if it is stable and is an attractor. A
weak attractor will be called a global weak attractor if
$A_w(M)$ = X. Similarly for global attractor, global uniform
attractor. An attractor is a <u>strange attractor</u> if it
contains a transversal homoclinic orbit. The <u>basin of
attraction</u> of A(M) is the set of initial points p ϵ M such
that $\phi(p)$ approaches A(M) as t $\to \infty$, where ϕ is the flow.

    It is important to point out at this point that
<u>attractivity does not imply stability</u>! For instance, an
autonomous system in R^2 presented in Hahn [1967]. As in
Fig.5.1.1, where Γ is a curve separating bounded and
unbounded orbits. The origin is unstable, in spite of the
fact that every solution tends to it as t $\to \infty$.

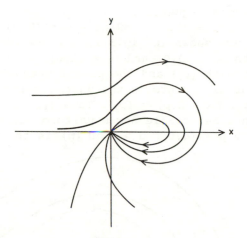

Fig.5.1.1

Here we offer a slightly different definition of attraction. The <u>origin of attraction</u> of the origin at t_o is the set $A(t_o) = \{x \in D: x(t;t_o,x_o) \to 0, t \to \infty\}$. If $A(t_o)$ does not depend on t_o, we say that the region of attraction is <u>uniform</u>. Furthermore, if $D = R^n = A(t_o)$ for every t_o, then the origin is globally attractive. The origin is <u>uniformly globally attractive</u> if for any $\eta > 0$, any $t_o \in I$ and any $x_o \in B_\eta$, $x(t;t_o,x_o) \to 0$ as $t \to \infty$, uniformly with respect to t_o and x_o.

<u>Asymtotic stability</u>: If the solution $x = 0$ of $dx/dt = f(t,x)$ is stable and attractive. If it is stable and equi-attractive, it is called <u>equi-asymptotically stable</u>. If it is uniformly stable and uniformly attractive, it is callec <u>uniformly asymptotically stable</u>. If it is stable and globally attractive, it is called <u>globally asymptotically stable</u>. If it is uniformly stable and uniformly globally attractive, then it is called <u>uniformly globally asymptotically stable</u>.

Example: $dx/dt = -x$, $x \in R$. The origin is uniformly globally asymptotically stable. Check the definitions.

Example. Consider a planar dynamical system defined by

203

the differential equations,

$$dr/dt = r(1 - r), \qquad d\theta/dt = \sin^2 (\theta/2).$$

The orbits are shown in Fig.5.1.2. These consist of two fixed points $p_1 = (0,0)$ and $p_2 = (1,0)$, an orbit Γ on the unit circle with $\{p_2\}$ asd the positive and the negative limit set of all points on the unit circle. All points p, where $p \neq p_2$, have $\Omega(p) = \{p_2\}$. Thus $\{p_2\}$ is an attractor. But it is neither stable nor a uniform attractor. Note also that for any $p = (\alpha,0)$, $\alpha > 0$, $J^+(p)$ is the unit circle.

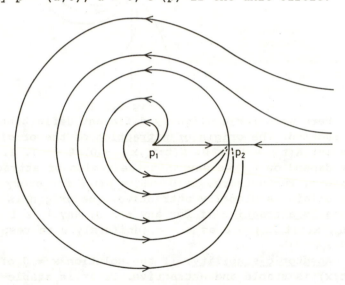

Fig.5.1.2

Now let us state some results about attractors.

<u>Theorem 5.1.3</u> If M is a weak attractor, attractor, or uniform attractor, then the coresponding set $A_w(M)$, $A(M)$, or $A_u(M)$ is open (indeed an open neighborhood of M).

<u>Theorem 5.1.4</u> If M is stable then it is positively invariant. As a consequence, if $M = \{x\}$, then x is a fixed point.

<u>Theorem 5.1.5</u> A set is stable iff $D^+(M) = M$.

In view of the above theorem, we can give the following

204

definition. If a given set M is unstable, the non-empty set
$D^+(M)$ - M will be called the <u>region of instability</u> of M. If
$D^+(M)$ is not compact, then M is said to be <u>globally
unstable</u>.

Theorem 5.1.6   If M is stable and is a weak attractor,
then M is an attractor and consequently asymptotically
stable.

Theorem 5.1.7   If M is positively invariant and a
uniform attractor, then M is stable. Consequently M is
asymptotically stable.

Theorem 5.1.8   If M is asymptotically stable then M is a
uniform attractor.

Theorem 5.1.9   Let $f(t,x)$ in $dx/dt = f(t,x)$ be
independent of t or periodic in t. Then, stability of the
origin implies uniform stability, and asymptotic stability
implies unform asymptotic stability [Yoshizawa 1966].

Note that there is no such theorem for attractivity or
uniform attractivity. The next few theorems determine
whether the components of a stable or asymptotically stable
set inherit the same properties. We would like to point out
(it is very easy to show) that in general the properties of
weak attraction, attraction, and uniform attraction are not
inherited by the components.

Theorem 5.1.10   A set M is stable iff every component of
M is stable.

It should be noted that the above theorem holds even if
X is not locally compact [Bhatia 1970].

Theorem 5.1.11   Let M be asymptotically stable, and let
M^* be a component of M. Then M^* is asymptotically stable iff
it is an isolated component.

The following very important theorem is a corollary to
the above theorem.

Theorem 5.1.12   Let X be locally compact and locally
connected. Let M be asymptotically stable. Then M has a
finite number of components, each of which is asymptotically
stable.

Outline of proof. A(M) is open and invariant. Since X is

205

locally connected, each component of A(M) is also open. The components of A(M) are therefore an open cover for the compact set M. Thus only a finite number of components of A(M) is needed to cover M. Now it is ealy to show that each component of A(M) contains only one component of M.

Theorem 5.1.13   Let M be a compact weak attractor. Then $D^+(M)$ is a compact asymptotically stable set with $A(D^+(M)) \equiv A_w(M)$. Moreover, $D^+(M)$ is the smallest asymptotically stable set containing M.

5.2   Asymptotic stability and Liapunov's theorem

In his Mémoire, Liapunov [1892] gave several theorems dealing directly with the stability problems of dynamical systems. His methods were inspired by Dirichlet's proof of Lagrange's theorem on the stability of equilibrium, is referred to by Russian authors as Liapunov's second method. Liapunov's first method for the study of stability rests on considering some explicit representation of the solutions, particularly by infinite series. What is known as Liapunov's second (or direct) method is making essential use of auxiliary functions, also known as Liapunov functions. The simplest type of such functons are C^1, V: IxD → R,   (t,x) → V(t,x) where I and D as before. Note that the space X is assumed locally compact implicitly.

An interesting introduction to Liapunov's direct method can be found in La Salle and Lefschetz [1961]. A much more complete and detailed treatment of stability as well as attractivity can be found in Rouche, Habets and Laloy [1977]. Many of the material in this chapter are coming from Rouche, Habets and Laloy [1977].

If x(t) is a solution of dx/dt = f(t,x), the derivatives of the time function V'(t) = V(t,x(t)) exists and dV'(t)/dt = $(\partial V/\partial x)(t,x(t))f(t,x(t)) + (\partial V/\partial t)(t,x(t))$.   (5.2-1)
If one introduces the function dV/dt: IxD → R,   (t,x) → dV(t,x)/dt by dV(t,x)/dt = $(\partial V/\partial x)(t,x)f(t,x) + (\partial V/\partial t)(t,x)$, it then follows that dV'(t)/dt =

dV(t,x(t))/dt. Thus, computing dV'(t)/dt at some given time t does not require a knowledge of the solution x(t), but only the value of x at t.

Another useful type of function is the <u>function of class K</u>, i.e., a function a ϵ K, a: $R^+ \to R^+$, continuous, strictly increasing, with a(0) = 0.

We call a function V(t,x) <u>positive definite</u> on D, if V(t,x) is defined as before, and V(t,0) = 0 and for some function a ϵ K, every (t,x) ϵ IxD such that V(t,x) $\geq$ a($\|x\|$). When we say that V(t,x) is positive definite without mentioning D, it means that for some open neighborhood D' $\subset$ D of the origin, V is positive definite on D'.

A well known necessary and sufficient condition for a function V(t,x) to be positive definite on D is that there exists a continuous function V^*: $\overline{D} \to R$ such that $V^*(0) = 0$, $V^*(x) > 0$ for x ϵ $\overline{D}$, x $\neq$ 0, and furthermore V(t,0) = 0 and V(t,x) $\geq$ $V^*(x)$ for all (t,x) ϵ IxD.

The following two theorems are pertinent to the differential equation dx/dt = f(t,x).

<u>Theorem 5.2.1</u> [Liapunov 1892] If there exists a c^1 function V: IxD $\to$ R and for some a ϵ K, and every (t,x) ϵ IxD such that,

(i)   V(t,x) $\geq$ a($\|x\|$) ;   V(t,0) = 0;

(ii) V(t,x) $\leq$ 0; then, the origin is stable.

Proof: Let t_o ϵ I and ϵ > 0 be given. Since V is continuous and V(t_o,0) = 0, there is a δ = $\delta(t_o,\epsilon)$ > 0 such that V(t_o,x_o) < a(ϵ) for every x_o ϵ B_δ. Writing x(t;t_o,x_o) by x(t), and using (ii), one obtains for every x_o ϵ B_δ and every t ϵ J^+ : a($\|x\|$) $\leq$ V(t,x(t)) $\leq$ V(t_o,x_o) < a(ϵ). But since a ϵ K, one obtains that $\|x(t)\|$ < ϵ .

<u>Theorem 5.2.2</u> In addition to the assumptions in Theorem 5.2.1, if for some b ϵ K and every (t,x) ϵ IxD: V(t,x) $\leq$ b($\|x\|$), then the origin is uniformly stable.

Let us use this theorem to derive the well-known stability criteria for linear systems or the linear approximation. Let x be an n-vector in R^n, and the system can be written as:

$dx/dt = Px + q(x,t)$,

where P is a constant nonsingular matrix and the nonlinear term q is quite small with respect to x for all $t \geq 0$. For simplicity, without loss of generality, we assume that the components of q have in some region D and for $t \geq 0$ continuous first partial derivatives in x_k and in t. Thus, in D and for $t \geq 0$, the existence theorem applies. Let us assume that the characteristic roots r_1, r_2, ..., r_n of the matrix P are all distinct and consider first the case where they are all real. Clearly, one can choose coordinates in such a way that P is diagonalized and $P = \text{diag}(r_1,...,r_n)$. Then, (a) if the roots are all negative. Take

$V = x_1^2 + ... + x_n^2$, and

$dV/dt = (r_1 x_1^2 + ... + r_n x_n^2) + s(x,t)$,

where s is small compared with the parenthesis. Thus, in a sufficiently small region D both V and -dV/dt are positive definite functions. Thus they satisfy the conditions of Theorem 5.2.1, thus the origin is asymptotically stable. (b) Some of the roots, say, $r_1,...$, r_p $(p < n)$ are positive, and the remaining are negative. Now take

$V = x_1^2 + ... + x_p^2 - x_{p+1}^2 - ... - x_n^2$.

Then we have,

$dV/dt = (r_1 x_1^2 + ... + r_p x_p^2 - r_{p+1} x_{p+1} - ... - r_n x_n^2) + s(x,t)$

where s is small compared with the parenthesis. At some points arbitrarily near the origin V is positive. Since r_{p+1}, ..., $r_n < 0$, thus dV/dt is positive definite, and the origin is unstable. In other words, a sufficient condition for the origin of the system to be asymptotically stable is that the characteristic roots all have negative real parts. If there is a characteristic root with a positive real part, then the origin is unstable.

As an example, there is an interesting application to the standard closed RLC circuit with nonlinear coupling. The equation of motion of the charge q is:

$L d^2 q/dt^2 + R dq/dt + q/C + g(x,dx/dt)) = 0$,

where dq/dt is the current, g represents nonlinear coupling with terms of at least of second order. We can rewrite this

208

second order differential equation as:

$$dq/dt = i, \quad di/dt = -q/LC - iR/L - g(q,i).$$

Clearly, the origin is a critical point and its characteristic roots are the roots of $R^2 + Rr/L + 1/LC = 0$. Since R, L and C are positive, the roots have negative real parts. If $R^2/L^2 < 4/LC$ or $R^2 < 4L/C$, then the characteristic roots are both complex with negative real part $-R/L$. When this happens, we have spirals as paths and the origin is asymptotically stable. The origin is a stable focus. If $R^2 > 4L/C$, the origin is a stable node and is asymptotically stable.

Next, we want to introduce the concept of partial stability and the corresponding stability theorem. Let two integers m, and n > 0, and two continuous functions f: IxDxR^m → R^n, g: IxDxR^m → R^m, where I = (τ,∞), D is a domain of R^n containing the origin. Assume f(t,0,0) = 0, g(t,0,0) = 0 for all t ϵ I, and f and g are smooth enough so that through every point of IxDxR^m there passes one and only one solution of the differential system

$$dx/dt = f(t,x,y) , \quad dy/dt = g(t,x,y) . \qquad (5.2\text{-}2)$$

To shorten the notation, let z be the vector $(x,y) \epsilon$ R^{n+m} and $z(t;t_o,z_o) = (x(t;t_o,z_o),y(t;t_o,z_o))$ for the solution of the system. Eq.5.2-2 starting from z_o at t_o.

The solution z = 0 of Eq.(5.2-2) is <u>stable with respect to x</u> if given ϵ > 0, and $t_o \epsilon$ I, there exists δ > 0 such that $\|x(t;t_o,z_o)\| < \epsilon$ for all $z_o \epsilon$ B$_\delta$ and all t ϵ J$^+$. <u>Uniform stability with respect to x</u> is defined accordingly.

<u>Theorem 5.2.3</u> If there exists a C^1 function V: IxDxR^m → R such that for some a ϵ K and every (t,x,y) ϵ IxDxR^m,
(i)   V(t,x,y) $\geq$ a($\|x\|$), V(t,0,0) = 0;
(ii) dV(t,x,y)/dt $\leq$ 0; then the origin z = 0 is stable with respect to x. Moreover, if for some b ϵ K and every (t,x,y) ϵ IxDxR^m,
(iii) V(t,x,y) $\leq$ b($\|x\|$ + $\|y\|$); then the origin is uniformly stable in x.

Examples: (i) Consider the linear differential equation $dx/dt = (D(t) + A(t))x$, where D and A are nxn matrices, being continuous functions of t on I ϵ (τ,∞), D diagonal, A skew- symmetric, $x \epsilon R^n$. Choosing $V(t,x) = (x,x)$, (where (x,x) is a scalar product), we get $dV(t,x)/dt = 2(x,D(t)x)$. If the elements of D are ≤ 0 for every $t \epsilon I$, then $V \leq 0$ and from Theorem 5.2.2, the origin is uniformly stable.

(ii) Next, we shall look at the stability of steady rotation of a rigid body. Consider a rigid body with a fixed point O in some inertial frame and no external force applied. Let I, M, N be the moments of inertia with respect to O, and Ω is the angular velocity in the inertial frame. The Euler equations of Ω, in the principal axes of inertia Ω = (p,q,r), are

$$Idp/dt = (M - N)qr,$$
$$Mdq/dt = (N - I)rp, \qquad\qquad (5.2-3)$$
$$Ndr/dt = (I - M)pq.$$

The steady rotations around the first axis correspond to the critical point $p = p_0$, $q = 0$, $r = 0$. Define new variables, $x = p - p_0$, $y = q$, $z = r$, the critical point is shifted to the origin, and Equations (5.2-3) becomes

$$dx/dt = (M - N)yz/I,$$
$$dy/dt = (N - I)(p_0 + x)z/M, \qquad\qquad (5.2-4)$$
$$dz/dt = (I - M)(p_0 + x)y/N.$$

If $I \leq M \leq N$, the steady rotation is around the largest axis of the ellipsoid of inertia. An auxiliary function suitable for Liapunov's theorem is

$$V = M(M - I)y^2 + N(N - I)z^2 + [My^2 + Nz^2 + I(x^2 + 2xp_0)]^2$$
$$(5.2-5)$$

which is a first integral, such that $dV/dt = 0$. It follows that the origin is stable for Eq.(5.2-4). Furthermore, it is even uniformly stable since the system is autonomous. If $I > M \geq N$, one obtains another auxillary function using the first integral,

$$V = M(I - M)y^2 + N(I - N)z^2 + [My^2 + Nz^2 + I(x^2 + 2xp_0)]^2.$$
$$(5.2-6)$$

Therefore, the steady rotations of the body around the longest and shortest axes of its ellipsoid of inertia are stable with respect to p, q, r. Note that the auxiliary

210

functions Eqs.(5.2-5) and (5.2-6) are combinations of the integrals of energy and of moment of mementa. We shall present a general method for constructing such combinations shortly.

(iii) The third example deals with the stability of a glider. This may as well be a hovering bird. First, suppose the plane of symmetry coincides at any moment with a vertical plane in an inertia frame. Let $\mathbf{v}$ be the velocity of its center of mass, and θ the angle between $\mathbf{v}$ and horizontal axis. The longitudinal "axis" of the glider is assumed to make a constant angle α with $\mathbf{v}$. Let m be the mass of the glider and g be the gravitational acceleration. Let $C_D(\alpha)$ and $C_L(\alpha)$ be the coefficients of drag and lift respectively. The equations of motion are:

$mdv/dt = - mg\sin\theta - C_D(\alpha)v^2$, $mvd\theta/dt = - mg\cos\theta + C_L(\alpha)v^2$.

Letting $v_o^2 = mg/C_L$, $\tau = gt/v_o$, $y = v/v_o$, and $a = C_D/C_L$,   we then have transformed the equations into

$$dy/d\tau = - \sin\theta - ay^2$$
$$d\theta/d\tau = (-\cos\theta + y^2)/y. \qquad\qquad (5.2-7)$$

We have introduced a non-vanishing drag for future reference only.

For the moment, let a = 0, then the above equations have the critical points $y_o = 1$, $\theta_o = 2\pi k$ for k an integer, all of them corresponding to one and the same horizontal flight with constant velocity. Let us concentrate on $y_o = 1$, $\theta_o = 0$. One can easily verify that $V(y,\theta) = (y^3/3) - y\cos\theta + 2/3$ is a first integral for Equations (5.2-7) with a = 0. In some neighborhood of (1,0), $V(y,\theta) > 0$ except $V(1,0) = 0$. Therefore, if the critical point (1,0) is translated to the origin, V expressed in the new variables satisfies the hypotheses of Theorem 5.2.2, thus one can prove the uniform stability. For more detail, see e.g., Etkin [1959].

211

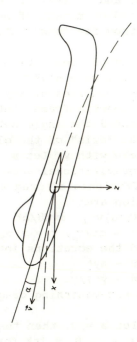

Fig.5.2.1

We now go back to the general setting and state some sufficient conditions for instability.

Theorem 5.2.4 [Chetaev 1934] If there exist $t_o \in I$, $\epsilon >$ 0, an open set $S \subset B_\epsilon$, (with $\overline{B_\epsilon} \subset D$), and a C^1 function V: $[t_o,\infty) \times B_\epsilon \to R$ such that on $[t_o,\infty) \times S$:
(i) $0 < V(t,x) \leq k < \infty$, for some k;
(ii) $dV(t,x)/dt \geq a(V(t,x))$, for some $a \in K$; if further
(iii) the origin of the x-space belong to ∂S;
(iv) $V(t,x) = 0$ on $[t_o,\infty) \times (\partial S \cap B_\epsilon)$;
then the origin is unstable.

The following two corollaries of Theorem 5.2.4 were in fact established long before Theorem 5.2.4 was known.

Corollary 5.2.5 [Liapunov 1892] If there exist $t_o \in I$,

$\epsilon > 0$, an open set $S \subset B_\epsilon$, (with $\overline{B}_\epsilon \subset D$), and a C^1 function
$V: [t_o,\infty) \times B_\epsilon \to R$ such that on $[t_o,\infty) \times S$:
(i) $0 < V(t,x) \leq b(\|x\|)$ for some $b \in K$;
(ii) $dV(t,x)/dt \geq a(\|x\|)$ for some $a \in K$; if further
(iii) the origin of the x-space belongs to ∂S;
(iv) $V(t,x) = 0$ on $\{t_o,\infty) \times (\partial S \cap B_\epsilon)$; then the origin is
unstable.

Corollary 5.2.6 [Liapunov 1892] If in Corollary 5.2.5,
(ii) is replaced by
(ii') $V(t,x) = c \, V(t,x) + W(t,x)$ on $[t_o,\infty) \times S$, where $c > 0$,
and $W: [t_o,\infty) \times S \to R$ is continuous and ≥ 0, then the origin
is unstable.

Note that, if the differential equation is autonomous
and if V depends on x only, then (i) and (ii) in Theorem
5.2.4 can be simplified to: (i) $V(x) > 0$ on S; and (ii)
$dV(x)/dt > 0$ on S.

The main use of Corollary 5.2.6 is to help prove
instability by considering the linear approximation. This is
a very useful way to look at many applications. Suppose,
$dx/dt = f(t,x)$ can be specified as:

$$dx/dt = Ax + g(t,x) \tag{5.2-8}$$

where A is an nxn real matrix and $Ax + g(t,x)$ has all the
properties required from $f(t,x)$. Then we have the following
theorem.

Theorem 5.2.7 [Liapunov 1892] If at least one
eigenvalue of A has strictly positive real parts and if
$\|g(t,x)\|/\|x\| \to 0$ as $x \to 0$ uniformly for $t \in I$, then the
origin is unstable for Eq.(5.2-8).

This theorem has very important applications, in
particular, when the system can be linearized. There are
many practical systems which satisfy the above criteria of
decomposing $f(t,x)$ into $Ax + g(t,x)$. We shall encounter this
theorem later in Chapter 7.

As an example for immediate demonstration, we shall
briefly discuss the Watt's governor. It is a well-known
device and it is sufficient to present it as in Fig.5.2.2.
If we disregard the friction, the equations of motion are:

$$d(Id\theta/dt)/dt = -k(\phi - \phi_0),$$
$$d(I'd\phi/dt)/dt - ((d\theta/dt)^2/2)(\partial I/\partial\phi) = \partial U/\partial\phi,$$
where $I = C + 2ml^2\sin^2\phi$, $I'= 2ml^2$, $U = 2mlcos\phi$ and C, l, m, k and g are positive constants.

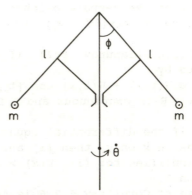

Fig.5.2.2

The steady motion $\phi = \phi_0$, $d\phi/dt = 0$, $(d\theta/dt)^2 = (d\theta_0/dt)^2 = g/lcos\phi_0$ is unstable by Theorem 5.2.7. It is straight forward to show that the eigenvalue equation for the linear part of the equation is:
$$(c/2ml^2 + \sin^2\phi_0)\mu^3 + (d\theta_0/dt)^2\sin^2\phi_0(1 + 3cos^2\phi_0 + c/2ml^2)\mu$$
$$+ (k/2ml^2)\theta_0\sin2\phi_0 = 0.$$
When $d\theta_0/dt < 0$, then $\Delta(0) < 0$, whereas $\Delta(\mu) \to \infty$ when $\mu \to \infty$, thus there is a strictly positive eigenvalue. Likewise, when $d\theta_0/dt > 0$, there is a strictly negative eigenvalue. Thus at least one eigenvalue has a strictly positive real part. Then, by Theorem 5.2.7, the steady motion being considered is unstable. For more details on Watt's governor and the use of friction to stablize its steady motion, see Pontryagin [1962].

The following theorem, due to Liapunov, on uniform asymptotic stability also gives an interesting estimate of the region of attraction. Also, uniform asymptotic stability has been studied long before simple asymptotic stability.

Theorem 5.2.8 [Liapunov 1892]   Suppose there exists a

214

C^1 function V: IxD → R such that for some functions a, b, c
ε K and every (t,x) ε IxD: (i) a(‖x‖) ≤ V(t,x) ≤ b(‖x‖);
(ii) dV(t,x)/dt ≤ - c(‖x‖). Choosing α > 0 such that $\overline{B}_α$ ⊂ D,
let us set for every t ε I, $V^{-1}_{t,x}$ ≡ {x ε D: V(t,x) ≤ a(α)}.
Then (a) for any t_o ε I and any x_o ε $V^{-1}_{t,α}$: x(t;t_o,x_o) → 0
uniformly in t_o, x_o when t → ∞; (b) the origin is uniformly
asymptotically stable.

Note that, assumptions (i) and (ii) are equivalent to
(i) and (ii') dV(t,x)/dt ≤ - c'(V(t,x)) for some c'ε K.
Also, the existence of a C^1 function V(t,x) such that: (i)
V(t,x) ≥ a(‖x‖); V(t,0) = 0; (ii) dV(t,x)/dt ≤ -c(‖x‖),
for some a, c ε K and every (t,x) ε IxD does not imply
uniform asymptotic stability, nor even asymptotic stability!
This is demonstrated by Massera [1949] by a counter-example.

<u>Corollary 5.2.9</u> The origin is uniformly globally
asymptotically stable if the assumptions of Theorem 5.2.8
are satisfied for D = R^n and a(r) → ∞ as r → ∞.

To illustrate Theorem 5.2.8, let us consider the scalar
equation, which represents an RLC circuit with parametric
excitation, (time varying capacitance),

$$d^2x/dt^2 + a\ dx/dt + b(t)x = 0 \qquad\qquad (5.2-9)$$

where a > 0, b = b_o(1 + εf(t)) with b_o ≥ 0, and f(t) is a
bounded function from R → R. Or one can view this equation
representing a mechanical oscillator with viscous friction
and a time-varying spring "constant". Eq.(5.2-9) is
equivalent to

$$dx/dt = y, \qquad dy/dt = -ay - b(t)x.$$

The auxiliary function V(x,y) = $(y + ax/2)^2/2 + (a^2/4 +$
$b_o)x^2/2$, is positive definite. Then according to Theorem
5.2.8, the origin will be uniformly asymptotically stable if
the time derivative of V, dV(t,x)/dt = - $ay^2/2$ - (b - b_o)xy
- $abx^2/2$ is negative definite. Following Sylvester's
criterion, this is so if for some α,

$$ε^2 b_o f(t)^2 - a^2(1 + εf(t)) ≤ - α < 0. \qquad\qquad (5.2-10)$$

This condition is satisfied for any sufficiently small ε.
One can view the differential equation as representing two
opposing forces at work: a parametric excitation

proportional to ϵ, and a load which is the damping force adx/dt. Satisfying Eq.(5.2-10) tantamount to choosing the resistance, a, large enough for the load to absorb all the energy provided by the excitation. In this case, the origin is asymptotically stable. Otherwise, if the load is not large enough, one may expect that the balance of energy of the system increases and the origin becomes unstable. This is, of course, a heuristic view.

Another example is the damped pendulum, which we discussed at the very beginning in Chapter 1. We have shown that the origin is asymptotically stable by choosing the total energy as the auxiliary function. Nonetheless, this is not a good choice because the time derivative of E is not negative definite. It proves stability but not asymptotic stability. Therefore, a "natural" choice may not always fit Theorem 5.2.8. Finding a suitable auxiliary function is often a matter of intuition, experience or an "art".

As we have noticed that it is often difficult to exhibit an auxiliary function whose time derivative is negative definite, so an alternative way of proving asymptotic stability will be to work out some more elaborate theorems and propositions to allow one to use auxiliary functions whose time derivative is non- positive (i.e., ≤ 0), but of course along with some more information.

Theorem 5.2.8 can be used to prove asymptotic stability by considering the linear approximation as illustrated in the RLC- circuit with parametric excitation. Now suppose dx/dt = f(t,x) is of the form dx/dt = Ax + g(t,x) as specified before Theorem 5.2.7, then we have the corresponding theorem.

Theorem 5.2.10 [Liapunov 1892] If all eigenvaleues of A have strictly negative real parts and if $\|g(t,x)\|/\|x\| \to 0$ as $x \to 0$, uniformly for all $t \in I$, then the origin is uniformly asymptotically stable.

As an example, let us go back to the problem of the glider considering non-vanishing drag, i.e., described by Eqs.(5.2-7). These equations admit a critical point,

216

$$y_o = (1 + a^2)^{-1/4}, \quad \theta_o = -\tan^{-1}a,$$

which corresponding to a rectilinear downward motion at constant velocity. Without loss of generality, we assume that $0 > \theta_o > -\pi/2$. Change the variables $y = y_o + y_1$, $\theta = \theta_o + \theta_1$, we transfer the origin to the critical point. By computing the terms of first order in y_1, θ_1, we obtain for the linear variational equation,

$$dy_1/dt = -(2a)y_1/(1+a^2)^{1/4} - \theta_1/(1+a^2)^{1/2},$$
$$d\theta_1/dt = 2y_1 - a\theta_1/(1+a^2)^{1/4}.$$

One can verify that the eigenvalues of this set of equations have strictly negative real part. Therefore, according to Theorem 5.2.10, the critical point is asymptotically stable.

The following couple of theorems are variations of Theorem 5.2.8.

Theorem 5.2.11 [Massera 1949] If we replace (ii) in Theorem 5.2.8 by (ii'), there exist a function $U: IxD \to R$ and a function $c \in K$ such that: $U(t,x) \geq c(\|x\|)$, $U(t,0) = 0$, and for any σ_1, σ_2 with $0 < \sigma_1 < \sigma_2$, $dV(t,x)/dt + U(t,x) \to 0$ as $t \to \infty$, uniformly on $\sigma_1 \leq \|x\| \leq \sigma_2$, then the origin is equi-asymptotically stable.

Theorem 5.2.12 [Antosiewicz 1958] If the origin is uniformly stable, and if there exists a C^1 function $V: IxD \to R$ such that for some functions a, $c \in K$ and every $(t,x) \in IxD$: (i) $V(t,x) \geq a(\|x\|)$; $V(t,0) = 0$; (ii) $dV(t,x)/dt \leq -c(\|x\|)$; then the origin is equi-asymptotically stable.

Next, we shall introduce a theorem which makes use of two auxiliary functions.

Theorem 5.2.13 [Salvadori 1972] Suppose there exist two C^1 functions $V: IxD \to R$ and $W: IxD \to R$ such that for some functions a, b, $c \in K$ and every $(t,x) \in IxD$:
(i) $V(t,x) \geq a(\|x\|)$; $V(t,0) = 0$;
(ii) $W(t,x) \geq b(\|x\|)$; $W(t,0) = 0$;
(iii) $dV(y,x)/dt \leq -c(W(t,x))$;
(iv) $W(t,x)$ is bounded from below or from above. Choosing $\alpha > 0$ such that $\overline{B_\alpha} \subset D$, for any $t \in I$, we put $V^{-1}_{t,\alpha} = \{x \in D: V(t,x) \leq a(\alpha)\}$. Then (a) the region of attraction $A(t_o) \supset V^{-1}_{t,\alpha}$; (b) the origin is asymptotically stable.

Note that, the function W in the above theorem can be of special form. For instance, if we identify $W(t,x)$ with $(x|x)$, or $W(t,x) = V(t,x)$, we obtain Corollary 5.2.14 or 5.2.15 respectively. In fact, these two corollaries were known long before Theorem 5.2.13 was established.

Corollary 5.2.14 [Marachkov 1940]   Suppose there exists a C^1 function $V: IxD \rightarrow R$ such that for some functions a, c ϵ K, and every $(t,x) \epsilon IxD$: (i) $V(t,x) \geq a(\|x\|)$; $V(t,0) = 0$; (ii) $dV(t,x)/dt \leq - c(x|x)$. If, moreover, $f(t,x)$ is bounded on IxD, then (a) for every $\alpha > 0$ such that $B_\alpha \subset D$, the region of attraction $A(t_0) \supset V^{-1}{}_{t,\alpha}$; (b) the origin is asymptotically stable.

Corollary 5.2.15 [Massera 1956]   Suppose there exists a C^1 function $V: IxD \rightarrow R$ such that for some functions a, c ϵ K, and every $(t,x) \epsilon IxD$: (i) $V(t,x) \geq a(\|x\|)$; $V(t,0) = 0$; (ii) $dV(t,x)/dt \leq - c(V(t,x))$; then (a) for every $\alpha > 0$ such that $\overline{B}_\alpha \subset D$, the region of attraction $A(t_0) \supset V^{-1}{}_{t,\alpha}$; (b) the origin is asymptotically stable.

As an example for the application of Theorem 5.2.13, let us generalize the damped pendulum by considering the pendulum with a variable friction $h(t)$, thus the equation of motion becomes: $d^2\theta/dt^2 + h(t)d\theta/dt + \sin\theta = 0$, where h is a C^1 function from $I \rightarrow R$, and we set the gravitational constant $g = 1$ just for simplicity. We are looking for some hypotheses concerning $h(t)$, as mild as possible, and the system will entail asymptotic stability of the origin of the phase plane. Let us try the auxiliary function $V(t,\theta,d\theta/dt) = (d\theta/dt + a\sin\theta)^2/2 + b(t)(1 - \cos\theta)$, which is the sum of a quadratic function. If we put $b(t) = 1 + ah(t) - a^2$, then

$dV(t,\theta,d\theta/dt)/dt = - (h(t) - a\cos\theta)(d\theta/dt)^2 - a \sin^2\theta +$
$ah'(t)(1 - \cos\theta) - a^2(1 - \cos\theta)\sin\theta \; d\theta/dt$
$= - (h(t) - a)(d\theta/dt)^2 - a(2 - h'(t))(1 - \cos\theta) + O_3,$

where O_3 contains terms of at least third order in θ and $d\theta/dt$, all independent of t. Now, if there exist two constants $\alpha > a$, and $\beta < 2$ such that (i) $h(t) \geq \alpha > a > 0$; (ii) $h'(t) \leq \beta < 2$; then V and $- dV/dt$ are positive definite. Furthermore, let us define $W(\theta,d\theta/dt) = (d\theta/dt)^2/2$

218

+ $(1- \cos\theta)$ and $dW(t,\theta,d\theta/dt)/dt = - h(t)(d\theta/dt)^2 \leq 0$ is bounded from above. Then from Theorem 5.2.13, the origin is asymptotically stable.

Note that, because of the requirement of $f(t,x)$ be bounded, therefore a bound should be assumed for $h(t)$. Thus, Corollary 5.2.14 could not be used. Corollary 5.2.15 could not be used either, at least not the $V(t,\theta,d\theta/dt)$ chosen above, because satisfying (ii) of Corollary 5.2.15 would also require further hypotheses on $h(t)$.

One can also show equi-asymptotical stability of the origin by using an "variation" of Theorem 5.2.13.

Theorem 5.2.16 [Salvadori 1972]   Suppose there exist two C^1 functions $V: IxD \rightarrow R$, and $W: IxD \rightarrow R$ such that for some a ϵ R, some function b, $c \epsilon$ K, and every $(t,x) \epsilon$ IxD:

(i) $V(t,x) \geq a$;
(ii) $W(t,x) \geq b(\|x\|)$; $W(t,0) = 0$;
(iii) $dV(t,x)/dt \leq - c(W(t,x))$;
(iv) $dW(t,x)/dt \leq 0$;

then the origin is equi-asymptotically stable.

Return to partial stability by getting back to Eq.(5.2-2) along with the accompanying assumptions as presented there. We suppose further that all solutions of Eq.(5.2-2) exist on $[t_o,\infty)$. The solution $z = 0$ of Eq.(5.2-2) is said to be <u>uniformly asymptotically stable with respect to x</u>, if it is uniformly stable with respect to x, and if for given $\epsilon > 0$, $t_o \epsilon$ I, there exist $\sigma > 0$ and $\eta > 0$ such that $\|x(t,t_o,z_o)\| < \epsilon$ for all $\|z_o\| < \eta$, and all $t \geq t_o + \sigma$.

Theorem 5.2.17   Suppose there exists a C^1 function $V: IxDxR^m \rightarrow R$ such that for some functions a, b, $c \epsilon$ K and every $(t,z) \epsilon$ IxDxR^m:

(i) $a(\|x\|) \leq V(t,z) \leq b(\|x\|)$;
(ii) $dV(t,z)/dt \leq - c(\|x\|)$.   Then
(a) for any $\alpha > 0$ and any $(t_o,z_o) \epsilon$ Ix$(B_\alpha \cap D)$xR^m, $x(t;t_o,z_o) \rightarrow 0$ uniformly in t_o, z_o when $t \rightarrow \infty$;
(b) the origin is uniformly asymptotically stable with repsect to x.

Theorem 5.2.18   Assume $u \epsilon$ R^{n+k}, $0 \leq k \leq m$, as a vector

219

containing all components of x and k components of y. Suppose there exists a C^1 function V: IxDxR^m → R such that for some functions a, b, c ∈ K and every (t,z) ∈ IxDxR^m:

(i) a($\|x\|$) ≤ V(t,z) ≤ b($\|x\|$);

(ii) dV(t,z)/dt ≤ - c($\|x\|$);

then the origin is uniformly asymptotically stable with respect to x.

When the whole space is the region of asymptotic stability, we say that we have <u>complete stability</u>. Then we have the following:

<u>Theorem 5.2.19</u> For an autonomous system, let there exists a C^1 function V: IxD → R such that, V(x) > 0 for all x ≠ 0 and dV(x)/dt ≤ 0. Let E be the locus dV/dt = 0 and let S be the largest invariant set contained in E. Then all solutions bounded for t > 0 tend to S as t → ∞.

In addition, if we know that V(x) → ∞ as $\|x\|$ → ∞, then each solution is bounded for t ≥ 0, and one can conclude that all solutions approach S as t → ∞. Furthermore, if S is the origin, we then have complete stability. Thus, in order to establish complete stability we need to show that every solution is bounded for t ≥ 0 and that S is the origin.

Suppose that V(x) → ∞ as $\|x\|$ → ∞ and that dV(x)/dt < 0 for x ≠ 0. Then S is certainly the origin, and it is easy to show that every solution is bounded for t ≥ 0. Let x(t) be the solution through x_0, then for some sufficiently large r, V(x) > V(x_0) for all $\|x\|$ ≥ r. But since V(x(t)) decreases with t, we note that $\|x(t)\|$ < r for all t ≥ 0. Hence every soulution is bounded. Thus,

<u>Theorem 5.2.20</u> Let V be C^1. Suppose (i) V(x) > 0 for x ≠ 0; (ii) dV/dt < 0 for x ≠ 0; and (iii) V → ∞ as $\|x\|$ → ∞. Then the autonomous system is completely stable.

In many applications, it occurs that one can construct a Liapunov function V satisfying Theorem 5.2.20. Examples will be given in Section 7.7.

The question arises whether it is true that stability, asymptotic stability, etc., imply the existence of Liapunov functions such as described in various theorems in this

220

section. We shall answer this in the next section. It should be pointed out that even though this is a good mathematical problem, it is not of great practical importance as we shall see in the next section.

5.3   Converse theorems

So far, in most theorems given, the existance of a Liapunov- like function is assumed. The question arises naturally whether such a function actually exists, i.e., given some stability or attractivity properties of the origin, can one build up an auxiliary function suitable for the corresponding theorem? It is answered in the so-called converse theorems.

Before we get to the converse theorems, there are several comments are in order. First of all, most converse theorems are proved by actually constructing the suitable auxiliary function. Such construction almost always assumes the knowledge of the solutions of the differential equation. This is why converse theorems give no clue to the practical search for Liapunov's functions. Second, when the existence of such and such function is necessary and sufficient for certain stability property, then any other sufficient condition will imply the originial one. Nonetheless, this observation should not prevent anyone from looking for other sufficient conditions which might be more practical, easier to apply. Third, sometimes a stability property of a system can be studied by considering first a simplified system. Let us suppose that the stability of a simplified system can be established easily. Then we can deduce from a converse theorem the existence of a suitable auxiliary function. Then under appropriate conditions, it might also be a good auxiliary function for the original system and prove the stability property of the original system.

Hence we shall give the converses of three important theorem given earlier. The general setting is the same as before, i.e., considering a continuous function f: IxD → R^n,

221

I = (τ,∞), D ⊂ R^n containing the origin, and the differential equation dx/dt = f(t,x), where the origin is an equilibrium or critical point.

Theorem 5.3.1 (converse of Theorem 5.2.1) [Persidski 1933] If the function f is C^1 and if the origin is stable, then there exist a neighborhood U ⊂ D of the origin and a C^1 function V: IxU → R such that for some a ∈ K and every (t,x) ∈ IxU: (i) V(t,x) ≥ a($\|x\|$); V(t,0) = 0; (ii) dV(t,x)/dt ≤ 0.

Theorem 5.3.2 (converse of Theorem 5.2.2) [Kurzweil 1955] If the function f is C^1 and if the origin is uniformly stable, then there exist a neighborhood U ⊂ D of the origin and a C^1 function V: IxU → R such that for some a, b ∈ K, and every (t,x) ∈ IxU:
(i) b($\|x\|$) ≥ V(t,x) ≥ a($\|x\|$);   V(t,0) = 0;
(ii) V(t,x) ≤ 0.

Theorem 5.3.3 (converse of Theorem 5.2.8) [Massera 1949, 1956] If the function f on IxD is locally Lipschitzian in x uniformly with respect to t, and if the origin is uniformly asymptotically stable, then there exist a neighborhood U ⊂ D of the origin and a function V: IxU → R possessing partial derivatives in t and x of arbitrary order, such that for some functions a, b, c ∈ K and every (t,x) ∈ IxU:
(i) a($\|x\|$) ≤ V(t,x) ≤ b($\|x\|$);
(ii) dV(t,x)/dt ≤ - c($\|x\|$).

Corollary 5.3.4 If f on IxD is Lipschitzian in x uniformly with respect to x, V can be chosen such that all partial derivatives of any order of V are bounded on IxU, with the same bound for all of them. If on IxD f is independent of t, or periodic in t, V can be chosen independent of t or periodic in t respectively.

5.4   Comparison methods
In calculus or even in algebra, we have learned that there are many methods of testing the convergence of a series. Among them is the comparison test. Similarly, in determining the stability of a equation at a point, one can

222

find the relationship between the said equation and a given
equation with stability properties known at the origin. More
explicitly, consider the equation

$$dx/dt = f(t,x),\hspace{4cm}(5.4-1)$$

and suppose that there exists a scalar differential equation

$$du/dt = g(t,u)\hspace{4cm}(5.4-2)$$

with a critical point at the origin $u = 0$ and some known
stability properties. The comparison method studies the
relationship which should exist between Eqs.(5.4-1) and
(5.4-2) in order that the stability properties of Eq.(5.4-2)
entail the corresponding properties for Eq.(5.4-1).

In order to avoid needless intricacies in computation,
we shall define $g(t,u)$ of Eq.(5.4-2) on IxR^+, where $R^+ =$
$[0,\infty)$, in this section. Of course, the uniqueness of the
solutions of the comparison equation is assumed here for
convenience. Furthermore, when it is supposed that
Eq.(5.4-2) admits of $u = 0$ as a critical point and that
point is stable, or asymptotically stable, only positive
perturbations will have to be considered.

Lemma 5.4.1 [Wazewski 1950] Let $g: IxR^+ \to R^+$, $(t,u) \to$
$g(t,u)$ be continuous and such that Eq.(5.4-2) has a unique
solution $u(t;t_o,u_o)$ through any $(t_o,u_o) \in IxR^+$. Let $[t_o,b)$ be
the maximal future interval where $u(t;t_o,u_o)$ is defined. Let
$v: [t,b) \to R$ be (i) $v(t_o) \le u_o$; (ii) $dv(t)/dt \le g(t,v(t))$ on
$[t_o,b)$; then $v(t) \le u(t)$ on $[t_o,b)$.

Theorem 5.4.2 [Corduneanu 1960] Suppose that there
exists a function g as in Lemma 5.4.1, with $g(t,0) = 0$, and
a C^1 function $V: IxD \to R$, such that for some function $a \in K$
and every $(t,x) \in IxD$:
(i) $V(t,x) \ge a(\|x\|)$, $V(t,0) = 0$;
(ii) $dV(t,x)/dt \le g(t,V(t,x))$;   then
(a) stability of $u = 0$ implies stability of $x = 0$;
(b) asymptotic stability of $u = 0$ implies equi-asymptotic
stability of $x = 0$.
Furthermore, if for some function $b \in K$ and every $(t,x) \in$
IxD: (iii) $V(t,x) \le b(\|x\|)$;   then
(c) uniform stability of $u = 0$ implies uniform stability of

223

x = 0;

(d) uniform asymptotic stability of u = 0 implies uniform asymptotic stability of x = 0.

Let us apply this theorem to some particular cases. For $g(t,u) = 0$, then (a) is reduced to Liapunov's theorem 5.2.1, (c) yields Persidski's theorem 5.2.2 on uniform stability. By choosing $g(t,u) = - c(u)$ for some $c \in K$, we notice that for the equation $du/dt = - c(u)$, the origin $u = 0$ is uniformly asymptotically stable, i.e., (b) reduces to Massera's theorem [1956], and (d) is reduced to Liapunov's theorem 5.2.8. This is because $dV(t,x)/dt \le - c(a\|x\|)$ and that $c \cdot a$ is a function of class K.

Various ways of choosing the function $g(t,u)$ can be illustrated in the following:

(1) If $c \in K$ and if $\phi \colon I \to R^+$ is continuous, then the origin $u = 0$ is uniformly stable for the equation $du/dt = -\phi(t)c(u)$. If, moreover, $\int_0^\infty \phi(t)dt = \infty$, the origin is equi-asymptotically stable.

(2) If $\beta \colon I \to R$ is continuous, the equation $du/dt = \beta(t)u$ has a critical point at the origin, which is stable, uniformly stable, or equi-asymptotically stable according to:

(a) for all $t_0 \in I$ there exist $A > 0$ and $t \ge t_0$ such that $\int_{t_0}^t \beta(s)ds \le A$, or

(b) therer exist $A > 0$, for all $t_0 \in I$ and $t \ge t_0$ such that $\int_{t_0}^t \beta(s)ds \le A$, or

(c) for all $t_0 \in I$ such that $\int_{t_0}^t \beta(s)ds \to -\infty$, as $t \to \infty$, respectively.

(3) Suppose there exist two C^1 functions $V \colon IxD \to R$, $k \colon I \to R$ and a continuous function $g \colon IxR^+ \to R$ such that $g(t,0) = 0$, that the solutions of $du/dt = g(t,u)$ are unique, and for some function $a \in K$ and every $(t,x) \in IxD$:

(i) $V(t,x) \ge a(\|x\|)$, $V(t,0) = 0$;

(ii) $k(t) \, dV(t,x)/dt + dk(t)/dt \, V(t,x) \le g(t,k(t)V(t,x))$;

(iii) $k(t) > 0$; and moreover,

(iv) $k(t) \to \infty$ as $t \to \infty$;

then stability of u = 0 implies equi-asymptotic stability of

224

x = 0 [Bhatia and Lakshmikantham 1965].
(4) The following result pertains to the system of equations
Eq. (5.2-2) concerning partial stability. Suppose there
exist a function g as in Lemma 5.4.1, with g(t,0) = 0, and a
C^1 function V: IxDxR^m → R, such that for some function a ε
K and every (t,x,y) ε IxDxR^m:
(i)   V(t,x,y) ≥ a(‖x‖),   V(t,0,0) = 0;
(ii) dV(t,x,y)/dt ≤ g(t,V(t,x,y));   then
(a) stability of u = 0 implies stability with respect to x
of x = y = 0;
(b) asymptotic stability of u = 0 implies equi-asymptotic
stability with respect to x of x = y = 0, provided that the
solutions of Eq.(5.2-2) do not approach ∞ in a finite time.
If moreover, for some function b ε K and every (t,x,y) ε
IxDx R^m: (iii) V(t,x,y) ≤ b(‖x‖ + ‖y‖);   then
(c) uniform stability of u = 0 implies uniform stability
with respect to x of x = y = 0;
(d) uniform asymptotic stability of u = 0, along with the
existence of solutions of Eq.(5.2-2) which do not approach ∞
in a finite time, implies uniform asymptotic stability with
respect to x of x = y = 0.
      For more details on the comparison method, see, for
instance, Rouche, Habets and Laloy [1977], Ch. 9.

5.5   Total stability

      Up to now, all the considerations on stability pertain
to variations of the initial conditions. Here we shall
consider another type of stability which takes into account
the variations of the second member of the equation. For
most practical problems, significant perturbations do occur
not only at the initial time, but also during the evolution.
      We shall still assume that the differential equation
      dx/dt = f(t,x)   (5.5-1)
is such that f(t,0) = 0 for all t ε I. Also, we shall
consider
      dy/dt = f(t,y) + g(t,y)   (5.5-2)

225

along with Eq.(5.5-1), where g: IxD → R^n satisfies the same regularity conditions as f. This ensures global existence and uniqueness for all solutions of Eq.(5.5-2). This function g will play the role of a perturbation term added to the second member of Eq.(5.5-1). Particularly, it will not be assumed that g(t,0) = 0, and therefore the origin will not in general be a solution of Eq.(5.5-2). The solution x = 0 of Eq.(5.5-1) is called <u>totally stable</u> whenever for any ϵ > 0 there exist δ_1, δ_2 > 0 for all $t_o \epsilon$ I and any $y_o \epsilon B_\delta$ and for any g such that for all t ≥ t_o and for any x ϵB_ϵ where $\|g(t,x)\|$ < δ_2, then for any t ≥ t_o, we have y(t;t_o,y_o) ϵB_ϵ.

  <u>Theorem 5.5.1</u> [Malkin 1944] If there exist a C^1 function V: IxD → R, three functions a, b, c ϵ K and a constant M such that, for every (t,x) ϵ IxD:
(i)   a($\|x\|$) ≤ V(t,x) ≤ b($\|x\|$);
(ii)   dV(t,x)/dt ≤ - c($\|x\|$), dV/dt computed along the solution of Eq.(5.5-1);
(iii) $\|$ V(t,x)/ x$\|$ ≤ M;
then the origin is totally stable for Eq.(5.5-1).

  Indeed, Malkin [1952] showed that asymptotic stability implies total stability.

  <u>Theorem 5.5.2</u> [Malkin 1944] If f is Lipschitzian in x uniformly with respect to t on IxD and if the origin is uniformly asymptotically stable, then the origin is totally stable.

  Clearly, the hypotheses of Theorem 5.5.1 do not imply that any solution y(t;t_o,y_o) tends to 0 as t → ∞; in fact, g(t,y) does not vanish, nor does it fade down as t → ∞. Nonetheless, some kind of asymptotic property can be found.

  <u>Theorem 5.5.3</u> [Malkin 1952] In the hypotheses of Theorem 5.5.1, for any ϵ > 0 there exist δ_1 > 0 and for any η > 0, there exists δ_2' > 0 and for all $t_o \epsilon$ I, if $y_o \epsilon B_\delta$, and if for any t ≥ t_o and x ϵB_ϵ such that $\|g(t,y)\|$ < δ_2', then there is a T > 0 such that for any t ≥ T, we have y(t;t_o,y_o) ϵB_η.

  Note that in the definition of total stability, one can

226

replace δ_1 and δ_2 by a single $\delta_1 = \delta_2 = \delta$. Also, in Theorem 5.5.1 the condition $V(t,x) \leq b(\|x\|)$ can be replaced by $V(t,0) = 0$ [Hahn 1967]. Massera [1956] has shown that if the origin is totally stable for a linear differential equation $dx/dt = A(t)x$, where A is a continuous nxn matrix, then it is uniformly asymptotically stable. Nonetheless, in the same paper, it is demonstrated that for an equation $dx/dt = f(x)$, $f \in C^1$, $f(0) = 0$, total stability does not imply uniform asymptotic stability.

Auslander and Seibert [1963] relates hitherto unrelated points of view of stability. The first is a generalized version of Liapunov's second method. The second is the concept of prolongation, which is a method of continuing orbits beyond their Ω-limit sets. And the third is the concept of total stability. We shall briefly relate them via some of the theorems.

A generalized Liapunov function for a compact invariant set C is a nonnegative function V, defined in a positively invariant neighborhood W of M, and satisfying the following: (a) if $\epsilon > 0$, then there exists $\mu > 0$ such that $V(x) > \mu$, for x not in $S_\epsilon(C) \equiv \{y \in X|\ d(y,C) < \epsilon\}$; (b) if $\mu > 0$, there exists $\eta > 0$ such that $V(x) < \mu$, for $x \in S_\eta(C)$; (c) if $x \in W$, and $t \geq 0$, $V(xt) \leq V(x)$. By a gereralized Liapunov function at infinity we mean a nonnegative function V defined on all of X satisfying the following: (a) V is bounded on every compact set; (b) the set $\{x|\ V(x) \leq \mu\}$ is compact, for all $\mu \geq 0$; (c) if $x \in X$, and $t \geq 0$, then $V(xt) \leq V(x)$.

Theorem 5.5.4 [Lefschetz 1958] The compact set M is stable iff there exists a generalized Liapunov function for M.

Theorem 5.5.5 Let M be compact. Then the following statements are equivalent:
(a) M is absolutely stable.
(b) M possesses a fundamental sequence of absolutely stable compact neighborhoods.
(c) There exists a continuous generalized Liapunov function

for M.

We have not formally defined the term absolutely stable. The definition has to be based on some mathematical machinary which we will not use later on. Thus, one can consider the above theorem also as a definition. One can consider the following two theorems in the same light.

Theorem 5.5.6 A dynamical system is 1-bounded iff there exists a generalized Liapunov function at infinity.

Theorem 5.5.7 The following statements are equivalent:
(a) A dynamical system is absolutely bounded.
(b) Every compact set is contained in an absolutely stable compact set.
(c) There exists a continuous Liapunov function at infinity.

A dynamical system is said to be ultimately bounded if there exists a compact set A such that $\Omega(X)$ is a subset of A. The following two theorems relate asymptotic stability, ultimate boundedness, absolute stability, and absolute boundedness.

Theorem 5.5.8 If the compact set M is asymptotically stable, it is absolutely stable. In fact, M is asymptotically stable iff there exists a continuous Liapunov function V for M such that, if x does not belong to M, and t > 0, then V(xt) < V(x).

Theorem 5.5.9 If a dynamical system is ultimately bounded, it is absolutely bounded. Furthermore, there exists a compact set M which is globally asymptotically stable.

The following theorem is another definition of total boundedness and also relates it to prolongation.

Theorem 5.5.10 The following statements are equivalent:
(a) The dynamical system is totally bounded.
(b) If A is a compact subset of X, then the prolongation of A is also compact.

The result of Milkin [1952] together with the following theorem gives an interesting relationship between asymptotic stability and absolute stability.

Theorem 5.5.11 (a) Total stability implies absolute stability. (b) Boundedness under perturbations implies absolute boundedness.

Let M be a positively invariant set of the flow ϕ. M is
strongly stable under perturbations if (a) it is weakly
stable under perturbations and (b) there exists a constant α
with the following property: given any ϵ, there exist τ and
δ such that the flow is ultimately bounded whenever $\phi*$
satisfies $\phi(\tau) = \delta$.

Theorem 5.5.12   Uniform asymptotic stability and strong
stability under perturbations are equivalent.

For a detailed discussion of the connections between
asymptotic stability and stability under perturbations, see
Seibert [1963]. For higher prolongations and absolute
stability, see Ch. 7 of Bhatia and Szegö [1970].

As an example, let us consider the system
$d^2x/dt^2 + f(x^2 + (dx/dt)^2)dx/dt + x = 0$. To every zero of
the function $f(r^2) \equiv f(x^2 + (dx/dt)^2)$ corresponds a limit
cycle $x^2 + (dx/dt)^2 = r^2$. The orbits between two neighboring
limit cycles are spirals with decreasing or increasing
distance from the origin, depending on the sign of f. (a) If
$f(r^2) = \sin(\pi/r^2)$, the origin is totally stable, therefore
absolutely stable, but not asymptotically stable. (b) If
$f(r^2) = \sin(\pi r^2)$, the system is totally bounded, therefore
absolutely bounded, but not ultimately bounded.

Examples for the non-autonomous system under persistent
perturbations will be discussed and illustrated in Ch.7, in
particular, in Section 7.2.

5.6   Popov's frequency method to construct a Liapunov function

In this section we shall use a nuclear reactor as an
example to illustrate one of the very practical methods,
namely the frequency method, to construct a Liapunov
function. This method is due to Popov [1962]. For more
details, see for instance, Lefschetz [1965] and Popov
[1973].

Basic variables to describe the state of a nuclear
reactor are the fast neutron's density D, D > 0, and the

reactor are the fast neutron's density D, D > 0, and the temperatures $T \in R^n$ of its various constituents. The neutron's density satisfies

$$dD/dt = kD \qquad (5.6-1)$$

where the reactivity k is a linear function of the state $k = k_o - r^t T - \eta D$, where $r \in R^n$, r^t the transpose of r, $\eta \in R$. By Newton's law of heat transfer, one gets the temperature equations

$$dT/dt = AT - bD \qquad (5.6-2)$$

where A is a non-singular real nxn matrix and $b \in R^n$. The system of Equations (5.6-1) and (5.6-2) has the equilibrium values

$$T_o = k_o A^{-1} b / (\eta + r^t A^{-1} b),$$
$$D_o = k_o / (\eta + r^t A^{-1} b).$$

Recall that D > 0, and changing of the variables

$$x = T - T_o, \qquad \mu = - D_o (\ln(D/D_o) + r^t A^{-1} x) / k_o,$$

after some tedious computations, one obtains

$$dx/dt = Ax - b\phi(\sigma), \qquad d\mu/dt = \phi(\sigma),$$
$$\sigma = - r^t A^{-1} x - k_o \mu / D_o, \qquad (5.6-3)$$

where $\phi(\sigma) = D_o(e^\sigma - 1)$. Note that $\sigma \phi(\sigma) > 0$ when $\sigma \neq 0$. A system like Eq.(5.6-3) is also known as an indirect control system [see Lefschetz 1965]. We shall not get into the technical details about the reactor, but rather prove some sufficient conditions for the global asymptotic stability of the critical point at the origin for Eq.(5.6-3).

<u>Lemma 5.6.1</u>   Let A be a non-singular nxn matrix, whose eigenvalues have strictly negative real parts, D a symmetric, positive definite non-singular matrix of order n. Let $b \in R^n$, $b \neq 0$, $k \in R^n$, and let τ and ϵ be real, where $\tau \geq 0$, $\epsilon > 0$. Then a necessary and sufficient condition for the existence of a symmetric positively definite, non-singular matrix B of order n and a $g \in R^n$ such that (a) $A^t B + BA = - qq^t - \epsilon D$, (b) $Bb - k = q\tau$, is that ϵ be small enough and that the inequility

$$\tau + 2 \, Re(k^t (i\Omega I - A)^{-1} b) > 0$$

be satisfied for all real Ω [Yacuborich 1962; Kalman 1963].

<u>Theorem 5.6.2</u>   Pertain to Eq.(5.6-3), suppose that all

and $c^tb + \Gamma > 0$. If furthermore, there exist real constants α and β such that (i) $\alpha \geq 0$, $\beta \geq 0$, $\alpha + \beta > 0$; (ii) for all real Ω,

$$\text{Re}[(2\alpha\Gamma + i\Omega\beta)(c^t(i\Omega I - A)^{-1}b + \Gamma/i\Omega)] > 0;$$

then the origin $x = 0$, $\sigma = 0$ is globally asymptotically stable for the system Eq.(5.6-3).

Remark: The results obtained by this frequency method are stronger and more effective than those considered earlier. And in fact they apply to a whole class of systems corresponding to any function ϕ with $\phi(\sigma)\sigma > 0$ when $\sigma \neq 0$, and they yield the auxiliary function explicitly. Moreover, the frequency criterion is the best possible result in the sense that the proposed Liapunov function proves asymptotic stability iff this criterion is satisfied. Nonetheless, the scope of the method is undoubtedly narrower than the Liapunov's direct method.

Before ending this chapter, we would like to point out that all auxiliary functions introduced up to now are C^1 functions. Nonetheless, it may happen, such as the example of a transistorized circuit may show, that the "natural" Liapunov function is not that regular. Thus, it is nature to generalize the theorems of Liapunov's second method to emcompass the case of a less smooth function V. Using some results of Dini derivatives, one can prove most theorems in this chapter, while imposing on the auxiliary functions only a local Lipschitz condition with respect to x and continuity in V(t,x). For Dini derivatives see, e.g., McShane [1944], Rouche, Habets and Laloy [1977]. For non- continuous Liapunov functions, see for instance, Bhatia and Szegö [1970].

5.7   Some topological properties of regions of attraction

In this section, we will present some additional properties of weak attractorsand their regions of attraction. This section can be skipped over for the first reading, or if the reader is not particularly interested in the topological properties of attraction.

As before, the space X is locally compact. From the definition of stability, it is clear that if a singleton {x} is stable, then x is a critical or fixed point. In particular, if {x} is asymptotically stable, then x is a fixed point. The next theorem concerns an important topological property of the region of attraction of a fixed point in R^n.

<u>Theorem 5.7.1</u> If a fixed point $p \epsilon R^n$ is asymptotically stable, then A(p) is homeomorphic to R^n.

<u>Corollary 5.7.2</u> If p is an asymptotically stable fixed point in R^n, then A(p) - {p} is homeomorphic to $R^n - \{0\}$, where O is the origin in R^n.

<u>Theorem 5.7.3</u> Let a subset M of R^n be a compact invariant globally asymptotically stable set in R^n. Then R^n - M = C(M) is homeomorphic to $R^n - \{0\}$.

<u>Theorem 5.7.4</u> Let a subset M of R^n be a compact positively invariant set, which is homeomorphic to the closed unit ball in R^n. Then M contains a fixed point.

<u>Theorem 5.7.5</u> Let a subset M of R^n be a compact set which is a weak attractor. Let the region of weak attraction $A_w(M)$ of M be homeomorphic to R^n. Then M contains a fixed point. In particular, when $A_w(M) = R^n$ (i.e., M is a global weak attractor), then M contains fixed point.

<u>Corollary 5.7.6</u> If the dynamical system defined in R^n is admitting a compact globally asymptotically stable set (equivalent to the system being ultimately bounded), then it contains a fixed point.

<u>Corollary 5.7.7</u> The region of attraction of a compact minimal weak attractor M cannot be homeomorphic to R^n, unless M is a fixed point.

<u>Corollary 5.7.8</u> If M is compact minimal and a global weak attractor, then M is a singleton. Consequently, M consists of a fixed point.

It should be interesting to note that if M is a compact invariant asymptotically stable set in X, then the restriction of the dynamical system to the set A(M) - M is parallelizable, thus dispersive. Now we shall discuss asymptotic stability of closed sets and its relation with the Liapunov functions.

Let $S(M,\delta) = \{y: d(y,M) < \delta$, where M is a subset of X, and $\delta > 0\}$. A closed subset M of X will be called: (i) a <u>semi-weak attractor</u>, if for each $x \in M$, there is a $\delta_x > 0$, and for each $y \in S(x,\delta_x)$ there is a sequence $\{t_n\}$ in R, $t_n \to \infty$, such that $d(yt_n,M) \to 0$, (ii) a <u>semi-attractor</u>, if for each $x \in M$, there is a $\delta_x > 0$, such that for each $y \in S(x,\delta_x)$, $d(yt,M) \to 0$ as $t \to \infty$, (iii) a <u>weak attractor</u>, if there is a $\delta > 0$ and for each $y \in S(M,\delta)$, there is a sequence $\{t_n\}$ in R, $t_n \to \infty$, such that $d(yt_n,M) \to 0$, (iv) an <u>attractor</u>, if there is a $\delta > 0$ such that for each $y \in S(M,\delta)$, $d(yt,M) \to 0$ as $t \to \infty$, (v) a <u>uniform attractor</u>, if there is an $\alpha > 0$, and for each $\epsilon > 0$ there is a $T = T(\epsilon) > 0$, such that $x[T,\infty)$ is a subset of $S(M,\epsilon)$ for each $x \in S[M,\alpha]$, (vi) an <u>equi-attractor</u>, if it is an attractor, and if there is an $\mu > 0$ such that for each ϵ, $0 < \epsilon < \mu$. and $T > 0$, there exists a $\delta > 0$ with the property that $x[0,T] \cap S(M,\delta) = 0$ whenever $\epsilon \leq d(x,M) \leq \mu$, (vii) <u>semi-asymptotically stable</u>, if it is stable and a semi-attractor, (viii) <u>asymptotically stable</u>, if it is uniformly stable and is an attractor, (ix) <u>uniformly asymptotically stable</u>, if it is uniformly stable and a uniform attractor. The following figures are orbits of certain dynamical systems in R^2. See the following figures Fig.5.7.1.

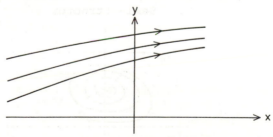

Stable but not equistable

233

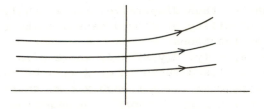

Equistable but not stable

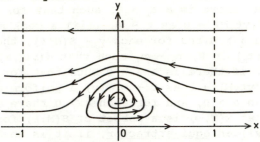

Semi-weak attractor

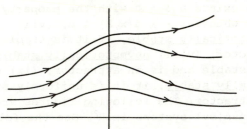

Semi-attractor

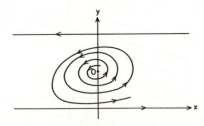

Weak attractor

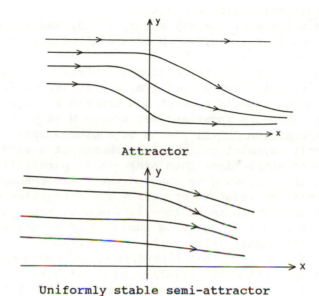

Attractor

Uniformly stable semi-attractor

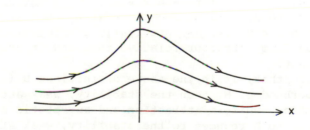

Stable attractor

Theorem 5.7.9   Let M be a closed set. Then M is
asymptotically stable iff there is a function $\phi(x)$ defined
in X with the following properties:
(i) $\phi(x)$ is continuous in some neighborhood of M which
contains the set $S(M,\delta)$ for some $\delta > 0$,
(ii) $\phi(x) = 0$ for $x \in M$, $\phi(x') > 0$ for x' not belong to M,
(iii) there exist strictly increasing functions $\alpha(\mu)$, $\beta(\mu)$,
where $\alpha(0) = \beta(0) = 0$, defined for $\mu \geq 0$, such that

235

$\alpha(d(x,M)) \leq \phi(x) \leq \beta(d(x,M))$,
(iv) $\phi(xt) \leq \phi(x)$ for all $x \in X$, $t > 0$, and there is a $\delta > 0$ such that if $x \in S(M,\delta) - M$ then $\phi(xt) < \phi(x)$ for $t > 0$, and $\phi(xt) \to 0$ as $t \to \infty$.

Theorem 5.7.10   Let a closed invariant set M be asymptotically stable. Let A(M) - M (or in particular the space X) be locally compact and contain a countable dense subset. Then the invariant set A(M) - M is parallelizable.

Proposition 5.7.11   Let M be a closed invariant uniformly asymptotically stable subset of X with A(M) as its region of attraction. Then A(M) - M is parallelizable.

In the following we shall discuss the concepts and properties of relative stability and attraction of a compact set. X is assumed to be locally compact. For a given compact set M, a subset of X, and a subset U of X, the set M is said to be: (i) stable relative to U, if given an $\epsilon > 0$ there exists $\delta > 0$, such that $\Gamma^+(S(M,\delta) \cap U)$ is a subset of $S(M,\epsilon)$, (ii) a weak attractor relative to U, if $\Omega(x) \cap M \neq 0$, for each $x \in U$, (iii) an attractor relative to U, if $\Omega(x) \neq 0$, $\Omega(x)$ is a subset of M, for each $x \in U$, (iv) a uniform attractor relative to U, if $J^+(x,U) \neq 0$, $J^+(x,U)$ is a subset of M, for each $x \in U$, (v) asymptotically stable relative to U if M is a uniform attractor relative to U and it is positively invariant.

Note that, if in the above definitions U is a neighborhood of M, then the stability, weak attraction, attraction, uniform attraction and asymptotic stability of M relative to U reduces to the stability, weak attraction, attraction, uniform attraction and asymptotic stability of M respectively.

Theorem 5.7.12   A compact subset M of X is stable relative to a subset U of X iff M contains $D^+(M,U)$.

Theorem 5.7.13   Let a subset M of X be compact and such that $A_w(M) - M \neq 0$. Let U be a subset of $A_w(M)$ be a set with the following properties: (i) U is closed and positively invariant; (ii) $U \cap M \neq 0$. Then the set $D^+(M,U)$ is compact and asymptotically stable relative to U.

Theorem 5.7.14 Let a subset M of X be compact and positively invariant and let M' be the largest invariant subset contained in M. Then M' is a stable attractor, relative to M.

For example, consider a limit cycle Γ in R^2 with the property that all orbits outside the unit disk bounded by the limit cycle Γ, has Γ as their sole positive limit set, and all orbits in the interior of the disk tend to the equilibrium point O. See Fig.5.7.2. We shall meet this auto-oscillatory system in Section 7.2. Suffice to note that if U is the complement of the disk bounded by Γ, then Γ is relatively stable with respect to U. Note also that if Γ is an asymptotically stable limit cycle, then Γ is stable with respect to every component of $R^2 - \Gamma$.

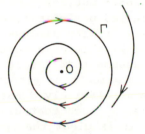

Fig.5.7.2

These considerations lead to the following definition and theorem. Let M be a compact subset of X. M is said to be component-wise stable if M is relative stable with respect to every component of X - M.

Theorem 5.7.15 Let a compact subset M of X be positively stable. Then M is component-wise stable.

The converse is not true in general. The following example illustrates this conclusion. Let X be a subset of R^2 given by X = {(x,y) ϵ R^2: y = 1/n, n is any integer, or y = 0}. Clearly the space is a metric space with the distance between any two points being the Euclidean distance between the points in R^2. We can define a dynamical system on X by dx/dt = $|y|$, dy/dt = 0. Then the set {(0,0)} in X is

237

component-wise stable, but is not stable.

It is then nature to ask under what conditions the converse of Theorem 5.7.15 is true. We first introduce the following definition, then followed by a couple of theorems.

Let M be a compact subset of X. We say the pair (M,X) is stability-additive if the converse of Theorem 5.7.15 holds for every dynamical system defined on X which admits M as an invariant set.

Theorem 5.7.16 The pair (M,X) is stability-additive if X - M has a finite number of components.

Theorem 5.7.17 The pair (M,X) is stability-additive if X - M is locally connected.

5.8 Almost periodic motions

We have discussed periodicity and recurrence. In the following we shall briefly discuss the concept intermediate between them, namely that of almost periodicity. For convenience, we assume that the metric space X is complete.

A motion ϕ_x is said to be almost periodic if for every ϵ > 0 there exists a relatively dense subset of numbers $\{\tau_n\}$ called displacements, such that $d(xt, x(t+\tau_n)) < \epsilon$ for all t $\in$ R and each τ_n. It is clear that periodic motion and fixed points are special cases of almost periodic motions. And it is also easy to show that every almost periodic motion is recurrent. We shall show that not every recurrent motion is almost periodic, and an almost periodic motion need not be periodic. The following theorems show how stability is deeply connected with almost periodic motions.

Theorem 5.8.1 Let the motion ϕ_x be almost periodic and let the closure of $\Gamma(x)$ be compact. Then:
(i) every motion ϕ_y with y $\in$ closure of $\Gamma(x)$, is almost periodic with the same set of displacements $\{\tau_n\}$ for any given ϵ > 0, but with the strict inequality < replaced by $\leq$;
(ii) the motion ϕ_x is stable in both directions in the closure of $\Gamma(x)$.

Corollary 5.8.2 If M is a compact minimal set, and if

one motin in M is almost periodic, then every motion in M is almost periodic.

Corollary 5.8.3   If M is a compact minimal set of almost periodic motions, then the motions through M are uniformly stable in both directions in M.

Theorem 5.8.4   If a motion ϕ_x is recurrent and stable in both directions in $\Gamma(x)$, then it is almost periodic.

Theorem 5.8.5   If a motion ϕ_x is recurrent and positively stable in $\Gamma(x)$, then it is almost periodic.

Theorem 5.8.6   If the motions in $\Gamma(x)$ are uniformly positively stable in $\Gamma(x)$ and are negatively Lagrange stable, then they are almost periodic.

A semi-orbit $\Gamma^+(x)$ is said to <u>uniformly approximate</u> its limit set, $\Omega(x)$, if given any $\epsilon > 0$, there is a $T = T(\epsilon) > 0$ such that $\Omega(x)$ is a subset of $S(x[t,t+T],\epsilon)$ for each $t \in R^+$. We want to find out under what conditions a limit set is compact and minimal. We have the following:

Theorem 5.8.7   Let the motion ϕ_x be positively Lagrange stable. Then the limit set $\Omega(x)$ is minimal iff the semi-orbit $\Gamma^+(x)$ uniformly approximates $\Omega(x)$.

Theorem 5.8.8   Let the motion ϕ_x be positively Lagrange stable, and let the motions in $\Gamma^+(x)$ be uniformly positively stable in $\Gamma^+(x)$. If furthermore $\Gamma^+(x)$ uniformly approximates $\Omega(x)$, then $\Omega(x)$ is a minimal set of almost periodic motions.

It should be pointed out that no necessary and sufficient conditions are known yet. In closing, let us discuss an example of an almost periodic motion which is neither a fixed point nor a periodic motion.

Consider a dynamical system defined on a torus by the following set of simple differential equations:

$$d\phi/dt = 1, \qquad d\theta/dt = \alpha,$$

where α is irrational. For any point p on the torus, the closure of $\Gamma(p) =$ the torus, and since α is irrational, no orbit is periodic. The torus is thus a minimal set of recurrent motions. To show that the motions are almost periodic, we note that if $p_1 = (\phi_1,\theta_1)$, and $p_2 = (\phi_2,\theta_2)$, then

$$d(p_1t,_2t) = (\phi_1 - \phi_2)^2 + (\theta_1 - \theta_2)^2 = d(p_1,p_2),$$
where the values of $\phi_1 - \phi_2$ and $\theta_1 - \theta_2$ ae taken as the
smallest in absolute value of the differences, and also note
that any motion on the torus is given by $\phi = \phi_0 + t$, $\theta = \theta_0$
+ αt. Thus the motions are uniformly stable in both
directions in the torus. Thus, from Theorem 5.8.4, the torus
is a minimal set of almost periodic motions.

In Section 4.4 we have derived the van der Pol's
equation of the nonlinear oscillator. Cartwright [1948] and
her later series of papers dealt with a generalized van der
Pol's equation for forced oscillations. The periodic and
almost periodic orbits are obtained. For detail, see also
Guckenheimer and Holmes [1983]. Krasnosel'skii, Burd and
Kolesov [1973] discusses broader classes of nonlinear almost
periodic oscillations. The book by Nayfeh and Mook [1979]
gives many detailed discussions on nonlinear oscillations.
The Annual of Mathematics series on nonlinear oscillations
are highly recommended for further reading [Lefschetz 1950,
1956, 1958, 1960]. Indeed, there are many current problems,
such as in phase locked laser arrays, where some of the
results are very applicable to the problems. We shall
briefly point this out in the next chapter.

Most of this chapter is based on several chapters of
Rouche, Habets and Laloy [1977]. This is still one of the
best sources for Liapunov's direct method.

Chapter 6   Introduction to the General Theory
of Structural Stability

6.1   Introduction

However complex it may seem, our universe is not random, otherwise it would be futile to study it. Instead, it is an endless creation, evolution, and annihilation of forms and patterns in space which last for certain periods of time. One of the central goals of science is to explain, and if possible, to predict such changes of form. Since the formation of such structures or patterns and their evolutional behavior are "geometric" phenomena, uncovering their common bases is a topological problem. But the existence of topological principles may be inferred from various analogies found in the critical behavior of systems. It should be emphasized that recognizing analogies is an important source of knowledge as well as an important methodology of acquiring knowledge.

There is a striking similarity among the instabilities of convection patterns in fluids, cellular solidification fronts in crystal growth, vortex formation in superconductors, phase transitions in condensed matter, particle physics, laser physics, nonlinear optics, geophysical formations, biological and chemical patterns and diffusion fronts, economical and sociological rhythms, and so forth. Their common characteristic is that one or more behavior variables or order parameters undergo spontaneous and discontinuous changes or cascades, if competing, but slowly and continuously control parameters or forces cross a bifurcation set and enter conflicting regimes. Consequently, an initially quiescent system becomes unstable at critical values of some control variables and then restabilize into a more complex space or time-dependent configuration. If other control parameters cause the disjoint bifurcation branches to interact, then multiple degenerate bifurcation points produce higher order instabilities. Then, the system undergoes additional transitions into more complex states, giving rise to hysteresis, resonance and entramment effects. These ultimately lead to states which are intrinsically chaotic.

In the vicinity of those degenerate bifurcation points, a system is extremely sensitive to small changes, such as imperfection, or external fluctuations which lead to symmetry breaking. Consequently, the system enhances its ability to perceive and to adapt its external environment by forming preferred patterns or modes of behavior. Prigoqine's concept of dissipative structures [1984], Haken's synergetics [1983], and Thom's catastrophe theory [1973] are most prominent among the theoretical study of these general principles.

As we have discussed earlier, the guiding idea of a stable system is to find a family of dynamical systems which contains "almost all" of them, yet can be classified in some qualitative fashion. It was conjectured that structurally stable systems would fit the bill. Although it turns out not to be true except for low dimensional cases, structural stability is such a natural property, both mathematically and physically, that it still holds a central place in the theory of dynamical systems.

As we have pointed out in Section 1.2, there is the doctrine of stability in which structurally unstable systems are regarded as suspicious. This doctrine states that, due to measurement uncertainties, etc., a model of a physical system is valuable only if its qualitative properties do not change under small perturbations. Thus, structural stability is imposed as a prior restriction on "good" models of physical phenomena. Nonetheless, strict adherence to such doctrines is arguable to say the least. It is very true that some model dynamical systems, such as an undamped harmonic oscillator, the Lotka-Volterra equations of the predator-prey model, etc., are not good models for the phenomena they are supposed to represent because perturbations give rise to different qualitative features. Nonetheless, these systems are indeed realistic models for

242

the chaotic behavior of the corresponding deterministic systems since the presumed strange attractors of these systems are not structurally stable.

If we turn to the other side of the coin, as we have also pointed out in Section 1.2, since the systems are not structurally stable, details of their dynamical evolutions which do not persist under perturbations may not correspond to any verifiable physical properties of the systems. Consequently, one may want to reformulate the stability doctrine as <u>the only properties of a (or a family of) dynamical system(s) which are physically (or quantitatively) relevant are those preserved under perturbations of the system(s)</u>. Clearly, the definition of (physical) relevance depends on the specific problem under study. Therefore, we will take the spirit that the discussions of structural stability requires that one specify the allowable perturbations to a given system.

The two main ingredients of structural stability are the topology given to the set of all dynamical systems and the equivalence relation placed on the resulting topological space. The former is the C^r topology ($1 \leq r \leq \infty$). This topology has been discussed in Chapter 3, and the idea in our context is clear. For instance, two diffeomorphisms are C^r-close when their values and values of corresponding derivatives up to order r are close at every point. Once we have defined the C^r topology, we may be able to be more specific about what we mean by "almost all" of the systems. The latter attribute is topological equivalence for flows and topological conjugacy for diffeomorphisms.

Before we get into a more general discussion of structural stability for manifolds, diffeormorphisms, function spaces of maps, and so forth, let us discuss the concept for R^n.

Let $F \in C^r(R^n)$, we want to specify what we mean by a perturbation G of F. Let F be as above, r, k are positive integers, $k \leq r$, and $\epsilon > 0$, then G is a <u>C^k perturbation of size ϵ</u> if there is a compact subset K of R^n such that $F = G$

243

on the set $R^n - K$ and for all $(i_1, \ldots, i_n)$ with $i_1 + \ldots + i_n = i \le k$ we have $|(\partial^i/\partial x_1^{i_1} \ldots \partial x_n^{i_n})(F - G)| < \epsilon$. F and G can also be vector fields. Note also that this definition can be stated in terms of k-jets as introduced in Chapter 3.

Let us recall that two C^r maps F and G are C^k $(k \le r)$ equivalent or C^k conjugate if there exists a C^k homeomorphism h such that $h \cdot F = G \cdot h$. C^0 equivalence is called topological equivalence. Two C^r vector fields f and g are C^k $(k \le r)$ equivalent if there exists a C^k diffeormorphism h which takes orbits of f to orbits of g, preserving senses but not necessarily parameterization by time. If h does preserve parameterization by time, then it is called a conjugacy. Intuitively, it is not difficult to see that parameterization by time cannot be preserved in general. This is because the periods of closed orbits in flows can be different.

We say a map $F \in C^r(R^n)$ (or a C^r vector field f respectively) is structurally stable if there is an $\epsilon > 0$ such that all C^1 ϵ- perturbations of F (resp. of f) are topologically equivalent to F (resp. f).

Let us consider a small perturbation of a linear system $dx/dt = Ax$, $x \in R^2$: $dx/dt = Ax + \epsilon f(x)$, where f has support in some compact set. Since A is invertible, then by the implicit function theorem, the equation $Ax + \epsilon f(x) = 0$ continues to have a unique solution $\underline{x} = 0 + O(\epsilon)$ near $x = 0$, for sufficiently small ϵ. Since the matrix of the linearized system $d\Phi/dt = (A + \epsilon Df(\underline{x}))\Phi$ has eigenvalues which depend continuously on ϵ, no eigenvalues can cross the imaginary axis if ϵ remains small with respect to the magnitude of the real parts of the eigenvalues of A. Thus the perturbed system has a unique fixed point and invariant manifolds of the same dimensions as those of the unperturbed system, and which are ϵ-close locally in position and slope to the unperturbed manifolds. The problem is finding a homeomorphism which takes orbits of the linear system to those of the perturbed nonlinear system.

Structurally stable systems have "nice" properties,

namely, any sufficiently close system has the same qualitative behavior. Nonetheless, structurally stable behavior can be very complex for flows of dimension ≥ 3 or diffeomorphisms of dimension ≥ 2. It should be stressed that the definition of structural stability is relative to the class of systems we will be dealing with. In fact, structural stability is not even a generic property, we shall come to this in Section 6.3, because we can find structurally unstable systems which remain unstable under small perturbations, and some, in fact, continually change their topological equivalence class as we perturb them. We shall meet such systems shortly.

The idea is the following. Let $\text{Diff}^r(M)$ $(1 \leq r \leq \infty)$ be the set of all C^r-diffeomorphisms of the C^∞-manifold M, provided with the C^r-topology. We say that $f \in \text{diff}^r(M)$ is <u>structurally stable</u> if it is in the interior of its topological conjugacy class. In other words, f is structurally stable iff any C^r- "small" perturbation takes it into a diffeomorphism that is topologically conjugate to f. Similarly, let $X^r(M)$ be the set of all C^r vector fields on M topologized with the C^r topology. Then the vector field $X \in X^r(M)$ (or the corresponding integral flow ϕ) is structurally stable if it is in the interior of its topological equivalence class. This type of definition of structural stability is due to Andronov and Pontryagin [1937]. This concept of structural stability has played a dominant and extremely important role in the development of dynamical systems during the past thirty years. It not only has helped to emphasize the global point of view, but also has been very fruitful in stimulating conjectures culminating in the powerful theorems we shall discuss shortly.

Before we get into any details of the notion of structural stability, we would like to discuss the stable manifolds of the dynamical system at a fixed point.

6.2   Stable manifolds of diffeomorphisms and flows

If we look at a phase diagram of a saddle point
(Fig.6.2.1) and try to analyze which qualitative features
give the saddle point its characteristic appearance, we are
bound to single out the four special orbits that begin or
end at the fixed point. These four orbits together with the
fixed point form the stable manifolds and unstable manifolds
of the dynamical system at the fixed point. In Chapter 4 we
have noted the importance of hyperbolicity in the theory of
dynamical systems, and a hyperbolic structure always implies
the presence of such manifolds. Thus, if we know where the
"singular elements" of a system are (including periodic
points for diffeomorphisms, fixed points and closed orbits
of flows), and if we also know the way in which their stable
and unstable manifolds fit together, we then have a very
good idea on the orbit structure of the system.

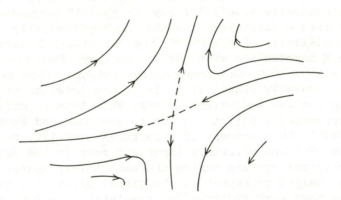

Fig. 6.2.1

Let f: M → M be a diffeomorphism, p be a fixed point of
f. The stable set of f at p is the set {x ∈ M : $f^n(x)$ → p as
n → ∞ }. Note that this set is always non-empty because it
contains p. The unstable set of f at p is the stable set of
f^{-1} at p. For any open neighborhood U of p, the local stable
set of f|U at p is the set of all x ∈ U such that {$f^n(x)$: n

≥ 0} is a sequence in U converging to p. One can show that
if p is hyperbolic then the global stable set is an immersed
submanifold of M and is at least as smooth as the
diffeomorphism f. This submanifold is called the <u>stable</u>
<u>manifold of f at p</u>, and denoted by $W_s(p)$. Clearly, it need
not be an embedded submanifold, for instance, as in Fig.
6.2.2.

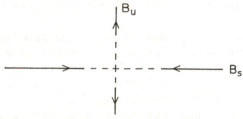

Now we shall deduce the global version of the stable
manifold theorem from the local one. Let T be a hyperbolic
linear automorphism of a Banach space B. From Chapter 4, B
splits into stable and unstable summands $B = B_s(T) \oplus B_u(T) =$
$B_s(T) \times B_u(T)$. The norm we use is the max $\{|\cdot|_s, |\cdot|_u\}$ on B.
As in Chapter 4, suppose T has skewness a and let k be any
number where $0 < k < 1 - a$. Let $B_b = (B_b)_s \times (B_b)_u$ be the
closed ball with center at O and radius b (possibly $b = \infty$)
in B. Let the graph for T be

After a Lipschitz perturbation η with constant k satisfying
$|\eta|_o \le b(1-a)$, the local stable set of $(T + \eta)|B$ is, as
shown in the picture below, the graph of a map, h: $(B_b)_s \to$
$(B_b)_u$. Furthermore, h is as smooth as η.

247

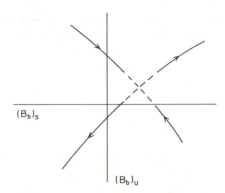

$(B_b)_s$

$(B_b)_u$

Let T subject to a Lipschitz perturbation η with constant k such that $|\eta|_o \leq b(1-a)$. One can show that $T+\eta$ has a unique fixed point in B_b, and by transferring the origin to this point one may assume, for simplicity, that $\eta(O) = 0$. Then we have:

Theorem 6.2.1 (local stable manifold theorem)   Let η: B_b → B be Lipschitz with constant k, where k < 1 - a. Suppose $\eta(O) = 0$. Then there is a unique map h: $(B_b)_s$ → $(B_b)_u$ such that graph h is the stable set of $(T + \eta)|B_b$ at 0, and h is Lipschitz, and is C^r when η is C^r.

Note that the stable manifold theorem works under much weaker assumptions than Hartman's theorem (Theorem 4.6.6), i.e., it is not necessary to assume η to be bounded as b = ∞, nor a condition to ensure T + η to be homeomorphism is imposed.

There are two additional features of the local stable manifold we would like to establish. First, the tangent space at O to the local stable manifold is the stable summand of the tangent map $T_o(T + \eta)$. Second, iterations under $(T + \eta)$ of points of the local stable manifold do not drift in towards O; they approach it exponentially.

Theorem 6.2.2   (i) If η is C^1 in Theorem 6.2.1, then the tangent to graph h at O is parallel to the stable manifold of the hyperbolic linear map T + $D\eta(O)$. (ii) The map h and f|(graph h) are Lipschitz with constant μ where f = T + η, μ

248

= a + k < 1. For any norm $\| \ \|$ on B equivalent to norm $| \ |$ on
B, there exists A > 0 such that for all n ≥ 0 and for all
x,y ϵ graph h,
$$\|f^n(x) - f^n(y)\| \le A\mu^n\|x - y\|.$$

   Theorem 6.2.3   Let p be a hyperbolic fixed point of a C^r
diffeomorphism (r ≥ 1) f of M. Then, for some open
neighborhood U of p, the local stable set of f|U at p is a
C^r imbedded submanifold of M, tangent at p to the stable
summand of T_pf.

   Theorem 6.2.4 (global stable manifold theorem)   Let p be
a hyperbolic fixed point of C^r diffeomorphism f (r ≥ 1) of
M. Then the global stable set S of f at p is a C^r immersed
submanifold of M, tangent at p to the stable summand of T_pf.

   If one looks at the proof of the above theorem, one
notices that the nature of the charts ϕ_i strongly suggests
that the global stable manifold of f at p is an immersed
copy of the stable manifold of the linear approximation.
Indeed, by extending the map η to the whole of B, we can
construct a locally Lipschitz bijection of the Banach space
onto the stable manifold of f. With the help of Theorem
6.2.3, one can show that the bijection gives a C^r immersion
of $(B_b)_s$ in X.

   In fact, one may extend the above theorem from fixed
points to periodic points of a diffeomorphism with little
extra effort. Indeed, if p is a periodic point of a
diffeomorphism f: M → M, then p is a fixed point of the
diffeomorphism f^k, where k is the period of p. Thus:

   Theorem 6.2.5   Let p be a hyperbolic periodic point of
period k of a C^r diffeomorphism f of M (r ≥ 1). Then the
stabe set of f at p is a C^r immersed submanifold of M
tangent at p to the stable summand of T_pf^k.

   Now we turn to the stable manifold theory for flows.
Note that the stable manifold theorem for a hyperbolic fixed
point of a flow is a simple corollary of the corresponding
theorem for diffeomorphisms. Let ϕ be a flow on M, and let d
be an admissible distance function on M. The <u>stable set of ϕ
at p ϵ M</u> is {x ϵ M : d(ϕ(t,x),ϕ(t,p)) → 0 as t → ∞} and the

unstable set of ϕ at p is $\{x \in M : d(\phi(t,x),\phi(t,p)) \to 0 \text{ as } t \to -\infty\}$. Recall that if p is a fixed point of ϕ then $\phi(t,x) \to$ p as $t \to \infty$ iff $\phi(n,x) \to p$ as $n \to \infty$, where $t \in R$ and $n \in Z$. Thus the stable set of the flow ϕ at p is precisely the stable set of the diffeomorphism ϕ^1 at p, or of ϕ^t for any other $t > 0$. Thus we obtain the following theorem.

Theorem 6.2.6 (global stable manifold theorem)   Let p be a hyperbolic fixed point of a C^r flow $(r \geq 1)$ ϕ on M. Then the stable set of ϕ at p is a C^r immersed submanifold of M, tangent at p to the stable summand of $T_p\phi^1$.

Theorem 6.2.7 (stable manifold theorem for closed orbits)   Let Γ be a hyperbolic closed orbit of a C^r $(r \geq 1)$ flow ϕ on a manifold M with distance function d. The stable set $W_s(p)$ of ϕ at p is a C^r immersed submanifold tangent at p to the stable summand of M_p with respect to $T_p\phi^\tau$, where τ is the period of Γ. If $q = \phi^t(p)$, then $W_s(q) = \phi^t(W_s(p))$. The stable set $W_s(\Gamma)$ of ϕ at Γ is a C^r immersed submanifold which is C^r foliated by $\{W_s(q): q \in \Gamma\}$.

Here the submanifold $W_s(p)$ is called the stable manifold of ϕ at p. Since $W_s(p)$ is also the stable manifold of the diffeomorphism ϕ^τ at p, it is independent of the distance function d, as also is $W_s(\Gamma) = U_{q \in \Gamma} W_s(q)$.

We have seen that when a point $p \in X$ is periodic and X has a hyperbolic structure with respect to f, its stable set is a manifold. We want to know whether we can extend this notion of hyperbolicity to a non-periodic point p and get a stable manifold theorem for such p. The problem is that $T_p f^n$ does not map $T_p X$ to itself for $n > 0$, and thus we cannot define hyperbolicity in terms of eigenvalues of this map.

Let A be any invariant subset of a Riemannian manifold X, and let $T_A X$ be the tangent bundle of X over A. We say that A has a hyperbolic structure with respect to f if there is a continuous splitting of $T_A X$ into the direct sum of Tf-invariant sub-bundles E_s and E_u such that, for some constants K and α and for all $v \in E_s$, $w \in E_u$ and $n \geq 0$, $|Tf^n(v)| \leq K\alpha^n|v|$, $|Tf^{-n}(w)| \leq K\alpha^n|w|$, where $0 < \alpha < 1$. That is, Tf will eventually contract on E_s and expand on E_u. A

<u>hyperbolic subset of X</u> (with respect to f) is a closed
invariant subset of X that has a hyperbolic structure. For a
general manifold, the existence of a hyperbolic structure on
the manifold implies there are some topological restrictions
(or requirements) on the manifold. For instance, the
frequently used hyperbolic structure on a manifold is the
Lorentz structure, which requires the existence of a nowhere
zero vector field on the manifold. For more complex
hyperbolic structures (locally, indefinite metric), the
topological requirements are given by a series of theorems
in Steenrod [1951]. We shall not consider these here, only
that such a hyperbolic structure exists.

Before we state the generalized stable manifold theorem,
let us make several definitions so that the theorem can be
stated. Let $B(x,a)$ be an open ball in X with center x and
radius a, with respect to the Riemannian distance function
d, and let $D(x,b)$ be the set $\{y \in X: d(f^n(x),f^n(y)) < b$ for
all $n \geq 0\}$. For $b \geq a \geq 0$ we define the <u>stable set of size</u>
<u>(b,a) of f at x</u> to be $B(x,a) \cap D(x,b)$. A map of an open subset
of the total space of a vector bundle into a manifold is
denoted by F^r (r times continuously fiber differentiable)
if, with respect to admissible atlases, all partial
derivatives in the fiber direction up to order r exist and
are continuous as functions on the total space.

<u>Theorem 6.2.8</u>   (Generalized stable manifold theorem)
Let f be a C^r diffeomorphism of X, and let A be a compact
hyperbolic subset of X, with associated decomposition $T_A X = E_s \oplus E_u$. Then there exists an open neighborhood W of the
zero section in E_s and an F^r map h: $W \to X$ such that for some
$b \geq a \geq 0$ and for all $x \in A$, g restricted to the fiber W_x
over x is a C^r imbedding with image $W_s^{loc}(x)$, the stable set
of size (b,a) at x. The tangent space to $W_s^{loc}(x)$ at x is
$(E_s)_x$.

The above theorem is a corollary of an even more general
theorem formulated by Smale [1967], proved by Hirsch and
Pugh [1970], with addendum by Hirsch, Pugh and Shub [1970].

6.3 Low dimensional stable systems

For the simplest example of a low dimensional stable system, consider a vector field $\mathbf{v}$ on S^1 with no zeros. Then there is precisely one orbit and it is periodic. In fact, we may identify $T(S^1) = S^1 \times R$. Let $v: S^1 \to R$ be the principal part of $\mathbf{v}$. Since v is continuous and S^1 is compact, $|v(x)|$ is bounded below by some constant $a > 0$. Then any perturbation of v less than a does not introduce any zero. Thus the perturbed vector field still has only one orbit, and is topologically equivalent to $\mathbf{v}$. So $\mathbf{v}$ is structurally stable (in fact, C^0-structurally stable).

Now suppose that $\mathbf{v}$ has finitely many zeros, all of which are hyperbolic. We shall call such a vector field on S^1 a Morse-Smale vector field on S^1. (We shall give a definition of a Morse-Smale vector field on a higher dimensional space later. Here we merely want to demonstrate the idea for the simplest case first). Then there must be an even number (2n) of zeros, with sources and sinks alternating around S^1. See, for instance:

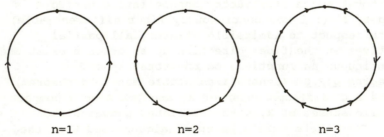

| n=1 | n=2 | n=3 |

Note that the hyperbolic zeros are each individually structurally stable, so that a sufficiently small C^1-perturbation of $\mathbf{v}$ leaves the orbit configuration unchanged on some neighborhood of the zeros. Furthermore, a sufficiently small C^0-perturbation introduces no further zeros outside this neighborhood. Thus the perturbed vector field is topologically equivalent to the original one, and $\mathbf{v}$ is C^1 structurally stable.

It should be noted that there are vector fields which

are not Morse-Smale but nonetheless have the above orbit structures. One can show, through the construction of examples, that structurally stable vector fields on S^1 with finitely many zeros are Morse- Smale. In fact the following stronger result is true.

   Theorem 6.3.1 A vector field on S^1 is C^1 structurally stable iff it is Morse-Smale. Moreover, Morse-Smale vector fields form an open, dense subset of $\Gamma^r S^1$ ($1 \leq r \leq \infty$).

   In attempting to prove that certain properties are open and dense (i.e., it holds for systems in an open, dense subset of Diff(M) or $\Gamma^r(M)$) or generic, one uses transversality theory (Chapter 2). Recall that this theory investigates the way submanifolds of a manifold cross each other, and how a map of one manifold to another throws the first across a submanifold of the second. Sard's theorem (Theorem 2.4.14) usually provides the opening shots. Also note that if a property is dense, then clearly a stable system is equivalent to a system has the property. For instance, the property of having only hyperbolic zeros is open and dense for flows on a compact manifold of any dimension, and any structurally stable flow possesses this property. The proof is essentially the one for S^1. Similarly,

   Theorem 6.3.2 [Kupka 1963, Smale 1963] For diffeomorphisms of a compact manifold M the following properties are generic, and satisfied by structurally stable diffeomorphisms: (i) all periodic points are hyperbolic, (ii) for any two periodic points x and y, the stable manifold of x and the unstable manifold of y intersect transversally. (That is, the tangent spaces of the two submanifolds at any point of intersection p generate (span) the whole tangent space of the manifold M at p. In other words, $(W_s(x))_p + (W_u(y))_p = M_p$ and it is denoted by $W_s(x) \pitchfork W_u(y)$.

   We say that a dynamical system is Morse-Smale if (i) its non- wandering set is the union of a finite set of fixed points and periodic orbits, all of which are hyperbolic, and

(ii) if G·x and G·y are any two orbits of the non-wandering set then $W_s(G·x) \pitchfork W_u(G·y)$. Thus the non-wandering set of a 2-dim. flow consists of at most five types of orbits, nmely, hyperbolic source, sinks, saddle points, and hyperbolic expanding and contracting closed orbits.

Peixoto shows that Theorem 6.3.1 also holds for vector fields on compact orientable 2-manifolds and for diffeomorphisms of S^1. That is:

Theorem 6.3.3 [Peixoto 1962] A vector field on a compact orientable 2-dim. manifold M (resp. a diffeomorphism of S^1) is C^1 structurally stable iff it is Morse-Smale. Morse-Smale systems are open and dense in $\Gamma^r(M)$ (resp. Diff$^r S^1$) for $1 \leq r \leq \infty$.

The theorem for non-orientable 2-manifolds is still an open question.

Although the results above failed to generalize to higher dimensions, it is fortunate to find that for all dimensions:

Theorem 6.3.4 A Morse-Smale system on a compact manifold is structurally stable.

In fact, it gets even better. Since every compact manifold admits a Morse-Smale system [Smale 1961], thus every compact manifold admits a structurally stable system [Palis and Smale 1970].

Theorem 6.3.4 and the above statement are corollaries of the following theorem due to Palis and Smale [1970]:

Theorem 6.3.5 Let M be a compact C^∞ manifold without boundary, and for $r \geq 1$ let Diff(M) be the set of C^r diffeomorphisms of M with the uniform C^r topology. For f ϵ Diff(M) we denote by $\Omega(f)$ the set of nonwandering points of f. If f ϵ Diff(M) satisfies: (a) $\Omega(f)$ is finite, (b) $\Omega(f)$ is hyperbolic, and (c) transversality condition, then f is structurally stable.

One can extend the result to flows or vector fields. Using the above theorem and a converse known for some time, we have:

Theorem 6.3.6 Let f ϵ Diff(M), with $\Omega(f)$ being finite.

254

Then f is structurally stable iff f satisfies (a)-(c) in
Theorem 6.3.5.

This result gives a good characterization of structural
stability in the Ω-finite case and leads to the question of
finding a similar characterization for the general case.
Palis and Smale also proposed the following conjecture. f ϵ
Diff(M) is structurally stable iff f satisfies: (a) axiom A,
i.e., $\Omega(f)$ is hyperbolic and the set of periodic points of f
is dense in $\Omega(f)$; (b) strong transversality condition, i.e.,
for all x, y ϵ $\Omega(f)$, W_s and W_u intersect transversally.

Armed with these results, Afraimovich and Sil'nikov
[1974] considered the system:

$$dx/dt = f(x), \qquad\qquad\qquad (6.3.1)$$

with f ϵ C^r, r $\geq$ 1, in some region D $\subset$ R^n. Let G be a
bounded region which is homeomorphic to a ball and has a
smooth boundary. Then they considered the system,

$$dx/dt = f(x), \qquad d\theta/dt = 1, \qquad\qquad (6.3.2)$$

defined in G x S^1. They have the following theorem.

Theorem 6.3.7 If system (6.3.1) is a Morse-Smale system
in G, then for sufficiently small δ in C^1, the
δ-neighborhoods of system (6.3.2) will be everywhere dense
Morse-Smale systems.

From Palis and Smale [1970] and Robbin [1971] one
obtains the following useful result.

Corollary 6.3.8 If system (6.3.1) is structurally stable
in G and does not have periodic motions, then system (6.3.2)
is also structurally stable.

In the following, we shall briefly discuss perturbation
problems for the two dimensional system dx/dt = f(x). By a
perturbation of the system, we mean a system

$$dx/dt = f(x) + \mu g(x,t), \qquad\qquad\qquad (6.3.3)$$

where x = (u,v) ϵ R^2, μ << 1. Equivalently, we have the
suspended system:

$$dx/dt = f(x) + \mu g(x,\theta),$$
$$d\theta/dt = 1, \quad (x,\theta) \epsilon R^2 \text{ x } S^1. \qquad\qquad (6.3.4)$$

Here f(x) is a Hamiltonian vector field and $\mu g(x,t)$ is a
small perturbation which need not even be Hamiltonian

255

itself. We want to discuss the bahavior of the orbits of
(6.3.3) for (x,μ) in a neighborhood of $\Gamma x\{0\}$ where Γ is the
periodic orbit of the unperturbed system.

The basic idea is due to Melnikov [1963]. It is to make
use of the globally computable solutions of the unperturbed
integrable system in the computation of perturbed solutions.
In order to do this, we must first ensure that the
perturbation calculations are uniformly applicable on
arbitrarily long time intervals. Let us make the assumptions
precise. Consider the system (6.3.4) which is sufficiently
smooth, C^r, $r \geq 2$, and bounded on the bounded region G, and
g is periodic in t with period T, and $f = (f_1(x), f_2(x))^t$, and
$g = (g_1(x,t), g_2(x,t))^t$. For simplicity, we assume that the
unperturbed system is Hamiltonian with $f_1 = \partial H/\partial v$, $f_2 =$
$-\partial H/\partial u$. Furthermore, we assume the unperturbed flow is: (a)
For $\mu = 0$, the system (6.3.4) possesses a homoclinic orbit
$q^0(t)$, to a hyperbolic saddle point p_0; (b) Let $\Gamma^0 = \{q^0(t) |$
$t \in R\}U\{p_0\}$, the interior of Γ^0 is filled with a continuous
family of periodic orbits $q^\alpha(t)$, $\alpha \in (-1,0)$. Letting $d(x,\Gamma^0)$
$= \inf_{q \in \Gamma^0}|x - q|$ we have $\lim_{\alpha \to 0} \sup_{t \in R} d(q^\alpha(t),\Gamma^0) = 0$. (c) Let
$h_\alpha = H(q^\alpha(t))$ and T_α be the period of $q^\alpha(t)$. Then T_α is a
differentiable function of h_α and $dT_\alpha/dh_\alpha > 0$ inside Γ^0.

Before we state any results, we would like to remark
that many of them can be proved under less restrictive
assumptions. Note that assumption (a) implies that the
unperturbed Poincaré map P_0 has a hyperbolic saddle point p_0
and that the closed curve $\Gamma^0 = W_u(p_0) \cap W_s(p_0)$ is filled with
nontransverse homoclinic points for P_0. Finally, the
<u>Melnikov function</u> is defined as:
$$M(t_0) = \int_{-\infty}^{\infty} f(q^0(t - t_0)) \wedge g(q^0(t - t_0), t) \, dt \qquad (6.3.4)$$
Then we have the following important theorem which allows us
to test for the existence of transverse homoclinic orbits
for specific systems:

<u>Theorem 6.3.9</u> If $M(t_0)$ has simple zeros and is
independent of μ, then for sufficiently small $\mu > 0$, $W_u(p_\mu)$
and $W_s(p_\mu)$ intersect transversely. If $M(t_0)$ remains away from
zero then $W_u(p_\mu) \cap W_s(p_\mu) = 0$.

Remarks: (i) $M(t_o)$ is periodic with period T.
(ii) If the perturbation g is derived from a time-dependent Hamiltonian function G(u,v) such that $g_1 = \partial G/\partial v$, $g_2 = -\partial G/\partial u$; then:
$M(t_o) = \int_{-\infty}^{\infty} \{H(q^o(t - t_o)), G(q^o(t - t_o),t)\}\, dt,$ (6.3.5)
where $\{H,G\}$ denotes the Poisson bracket (Section 2.8).
(iii) If $g = g(x)$, not explicitly time dependent, then using Green's theorem we obtain: $M(t_o) = \int_{int\ \Gamma^o}$ trace Dg(x) dx.
(iv) By changing variables $t \to t + t_o$, we have a more convenient Melnikov integral:
$M(t_o) = \int_{-\infty}^{\infty} f(q^o(t))\ \Lambda\ g(q^o(t),t+t_o)dt.$
Now, let us return to the more general case, where $g = g(x,t;\theta)$.

<u>Theorem 6.3.10</u> Let the system dx/dt = f(x) + μg(x,t;θ), $\theta \in$ R, satisfies the assumptions (a)-(c) before we defined the Melnikov function. Suppose that the Melnikov function $M(t_o,\theta)$ has a quadratic zero $M(\tau,\theta_b) = (\partial M/\partial t_o)(\tau,\theta_b) = 0$ but $(\partial^2 M/\partial t_o^2)(\tau,\theta_b) \neq 0$ and $(\partial M/\partial \theta)(\tau,\theta_b) \neq 0$. Then $\theta_B = \theta_b + 0(\mu)$ is a bifurcation value for which quadratic homoclinic tangencies occur in the family of systems.
If we rewrite system (6.3.3) as: dx/dt = f(x,μ), where x = (x_1,x_2), and for μ = 0, the system becomes dx/dt = f(x,0) has a periodic orbit Γ with period T. Let us define
$\sigma_o = $ tr[∂f(0,0)/∂x} $\neq$ 0.

<u>Theorem 6.3.11</u> The homoclinic orbit Γ is asymptotically stable if $\sigma_o < 0$ and unstable if $\sigma_o > 0$. There can be at most one periodic orbit bifurcating from Γ and it is asymptotically orbitally stable if $\sigma_o < 0$ and unstable if $\sigma_o > 0$.

Thus, if $\sigma_o = 0$, one can have either periodic orbits in any neighborhood of Γ, or Γ can be asymptotically stable or unstable. To illustrate this, consider: dx/dt = 2y, dy/dt = 12x $-3x^2$. This system has a potential V(x,y) = $x^3 - 6x^2 + y^2$, and the solution curves are given as

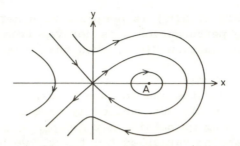

Clearly, the origin O is a hyperbolic saddle and point A =
(4,0) is a center. There is a periodic orbit in any
neighborhood of the homoclinic orbit Γ and σ_0 = 0. Take note
that $\Gamma \subset \{(x,y): V(x,y) = 0\}$. Consider the perturbed system
to be:

$$dx/dt = 2y - \mu V(x,y)(12x - 3x^2) \equiv f_1(x,y,\mu),$$
$$dy/dt = 12x - 3x^2 + \mu V(x,y)2y \equiv f_2(x,y,\mu).$$

It is easy to show that
$\sigma_0(\mu) = \operatorname{tr} \partial(f_1(0,0,\mu),f_2(0,0,\mu))/\partial(x,y) = 0$ for all μ. The
perturbed system has only the equilibrium points O and A
with O a hyperbolic saddle and A a hyperbolic focus for $\mu \neq$
0 sufficiently small. A is stable for $\mu > 0$ and unstable for
$\mu < 0$. The solution curves in (x,y)-space defined by V(x,y)
= 0 is invariant under the perturbation. Thus, Γ is again a
homoclinic orbit for the perturbed system for every μ.
Noting that the perturbed system is obtained from the
original system by a rotation through an angle
arctan($\mu V(x,y)$), it follows that no curve V(x,y) = constant
inside the curve Γ can be tangent to the vector field of the
perturbed system. Thus, there can be no periodic orbits of
the perturbed system inside Γ for $\mu \neq 0$. Since $\mu > 0$ ($\mu < 0$)
implies the focus A is stable (unstable), it follows that Γ
is unstable (asymptotically stable) for $\mu > 0$ ($\mu < 0$).

Chan [1987] considered the equation
$$d^2x/dt^2 + g(x) = -\alpha dx/dt + \mu f(t),$$
where x is real, g is smooth, f is periodic, i.e., f(t+1) =
f(t), and α and μ are small parameters. By using periodic
Melnikov functions and the method of Liapunov, all the

258

Floquet multipliers of the bifurcating subhamonic solutions are determined. Recently, Ling and Bao [1987] have deviced a numerical implementation of Melnikov's method. The procedure is based on the convergence of the integral and the uniqueness of the boundary of the horseshoe region in the parameter space under the conditions that it be inverse symmetric, i.e., $f(x) = -f(-x)$, and there exists more than one homo(hetero)clinic orbits. Meanwhile, Salam [1987] presented explicit calculations which extended the applicability of the Melnikov's method to include a general class of highly dissipative systems. The only required condition is that each system of this class possesses a homo(hetero)clinic orbit. Furthermore, it was shown that sufficiently small time-sinusoidal perturbation of these systems resulted in transversal intersection of stable and unstable manifolds for all but at most discretely many frequencies.

For further details for bifurcations of autonomous or periodic planar equations, and their applications, see, e.g., Chapter 4 of Guckenheimer and Holmes [1983]; Chapter 9-11 of Chow and Hale [1982]; Chapter 7 of Lichtenberg and Lieberman [1983]; and for applications, see, e.g., Smoller [1983].

Recently, Wiggins [1988] has developed a global perturbation technique similar to that of Melnikov [1963] for detecting the presence of orbits homoclinic to hyperbolic periodic orbits and normally hyperbolic invariant tori in a class of ordinary differential equations. This technique is more general then Melnikov's because it applies to systems undergoing large amplitude excitation at low frequencies and to systems undergoing quasiperiodic excitation.

When the paper by Palis and Smale appeared, it was known that there were other structurally stable systems beside Morse-Smale systems. One of them was the toral automorphisms (see Section 4.1). These are structurally stable, but their non-wandering sets are the whole of the tori, so they certainly fail condition (i) of Morse-Smale. Next, we shall very briefly discuss the Anosov systems.

6.4 Anosov systems

A diffeomorphism f: M → M of a manifold M is <u>Anosov</u> if M has a hyperbolic structure with respect to f. Recall that this means that the tangent bundle TM splits continuously into a Tf- invariant direct-sum decomposition $TM = E_s + E_u$ such that Tf contracts E_s and expands E_u with respect to some Riemannian matric on M. Trivially, hyperbolic linear maps f possess this property, since one has the identification $TR^n = R^n \times R^n$ and $Tf(x,v) = (f(x),f(v))$. In the case of toral automorphisms, this splitting is carried over to the tours when the identification was made. So, toral automorphisms are Anosov.

Similarly, a vector field on M is <u>Anosov</u> if M has a hyperbolic structure with respect to it. As examples of such, we have all suspensions of Anosov diffeomorphisms.

<u>Theorem 6.4.1</u> [Anosov 1962]   Anosov systems on compact manifolds are C^1 structurally stable.

There are some unsolved problems about Anosov diffeomorphisms. For instance, is their non-wandering set always the whole manifold? Do they always have a fixed point? Since not all manifolds admit Anosov diffeomorphism, do all n-dim. manifolds which do admit them have R^n as universal covering space?

6.5 Characterizing structural stability

Realizing the diagonal differences of systems such as Morse-Smale and Anosov systems, it is a challenging problem to characterize structural stability. The essential link comes about when Smale recognizes the fact that by replacing the term "closed orbits" in Morse-Smale definition by "basic sets" (to be defined shortly), then Anosov and other systems are also encompassed.

260

A dynamical system has an Ω-decomposition if its non-wandering set Ω is the disjoint union of closed invariant sets Ω_1, Ω_2, ..., Ω_k. If the system is topologically transitive on Ω_i (i.e., Ω_i is the closure of the orbit of one of its points) for all i, we say that $\Omega_i \cup ... \cup \Omega_k$ is a spectral decomposition, and that the Ω_i are basic sets. One can also define a basic set individually by saying that a closed invariant set, $\Lambda \subset \Omega$ is basic if the system is topologically transitive on Λ but Λ does not meet the closure of the orbit of $\Omega | \Lambda$. Note that a basic set is indecomposable since it is not the disjoint union of two non-empty closed invariant sets.

A dynamical system satisfies Axiom A if its non-wandering set (a) has a hyperbolic structure, and (b) is the closure of the set of closed orbits of the system. It was conjectured that (a) implies (b). Newhouse and Palis [1970] have shown that it is true for diffeomorphisms of 2-dim manifolds, but false for higher dimension manifolds [Danberer 1977].

Theorem 6.5.1 (Spectral Decomposition Theorem) The non-wandering set of an Axiom A dynamical system on a compact manifold is the union of finitely many basic sets.

In order to visualize an Axiom A system, one thinks of a system of 2-dim manifold with finitely many fixed points and periodic orbits, all hyperbolic, such as the gradient of the height function on the torus, or a Morse-Smale system for S^2 (see Fig. 6.5.1). But in higher dimensions, one replace the fixed points and periodic orbits by more general basic sets.

It should be pointed out that there is no need for a basic set to be a submanifold. Such basic sets are termed strange or exotic.

Following Smale, we say that a system is AS if it satisfies both Axiom A and the strong transversality condition, which is, for all x and y in the non-wandering set of the system, the stable manifold of the orbit of x and the stable manifold of the orbit of y intersect transversally. The strong transversality c dition is the

261

general version of the second condition in the definition of
a Morse-Smale system. For C^2 diffeomorphisms, the best set
of criteria of structural stability is due to Robbin [1971]
and for C^1 diffeomorphisms and flows Robinson [1975a; 1975b;
1976; 1977; 1980].

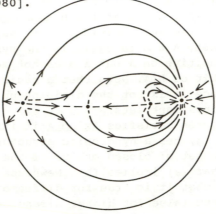

Fig.6.5.1

<u>Theorem 6.5.2</u> Any AS system is C^1 structurally stable.
    It is known that structural stability is equivalent to
AS system when $\Omega(f)$ is finite [Palis and Smale 1970], and
that structural stability and Axiom A imply strong
transversality [Smale 1967]. It seems that the converse of
Theorem 6.5.2 is also true, which would be very
satisfactory. Nonetheless, the closest to this is the result
due to Franks [1973]. A diffeomorphism f: M → M is
<u>absolutely C^1-structurally stable</u> if for some C^1-
neighborhood N ⊂ Diff(M) of f, there is a map σ associating
with g ∈ N a homeomorphism $\sigma(g)$ of Diff(M) such that (i)
$\alpha(f) = id_M$, (ii) for all g ∈ N, g = hfh^{-1} where h = $\sigma(g)$,
(iii) σ is Lipschitz at f with respect to the C^0 matric d
(i.e., for some k > 0 and all g ∈ N, $d(\sigma(g),id_M) \leq kd(g,f)$.
    <u>Theorem 6.5.3</u> Any diffeomorphism is absolutely C^1
structurally stable iff it is AS.
    As an example, in order to classify the transitions
between interacting time-periodic and steady-state solutions

of nonlinear evolution equations by using the normal forms
of the imperfect bifurcation theory (Section 6.6), two types
of bifurcation of solutions of general evolution equations
are fundamental: the bifurcation of a steady state with
amplitude x and the Hopf bifurcation of a time-periodic
solution with amplitude y from a stationary one.
Interactions between them lead to secondary bifurcations of
periodic solutions and to tertiary bifurcations of
double-periodic motions lying on tori and eventually to
chaotic motions. Such interactions occur if a control
parameter μ in the evolution equation crosses some critical
values. By Liapunov-Schmidt reduction we can show that x and
y satisfy two algebraic normal form equations $a(x,\mu,y^2) = 0$
and $yb(x,\mu,y^2) = 0$. The solution of the first is a two
dimensional multisheeted surface $y = \phi(x,\mu)$ in
(x,y,μ)-space, and that of the second equation is another
surface $y = \beta(x,\mu)$. The lines along which both surfaces
intersect are the bifurcation diagrams of the evolution
equations from which the behavior of the system can be
inferred as μ varies. The intersection of the two surfaces
may be transversal so that any perturbation of a and b,
i.e., a slight deformation or shifting of the surfaces,
causes no new type of intersections. In this case structural
stability of the bifurcation diagram is ensured at the
outset. Nonetheless, if the two surfaces intersect with
tangential contact, or just touching, then a slight
deformation or shifting of them produces new intersections
and gives rise to new bifurcation diagrams which then are
stable against any further perturbations. These
perturbations can be thought to be induced by variations of
system-imminent imperfection parameters (such as impurity of
material parameters etc.) in the original evolution
equations. Since the forms of the perturbed polynomials a
and b can be classified into a finite set by imperfect
bifurcation theory, the problem of interacting spatial and
temporal patterns has thus been reduced to linking together
the possible basic bifurcation diagrams. Then a variety of

263

new phenomena, such as gaps in Hopf branches, periodic
motions not stably connected to steady states, and the
discovery of formation of islands, which one can expect to
find in general systems of evolution equations. Many of the
new phenomena, predicted on topological grounds alone, still
await experimental confirmation [Dangelmayr and Armbruster
1983; Armbruster 1983].

As we have mentioned in Sections 1.2 and 6.1,
structurally stable systems may not be dense. Indeed,
Peixoto and Pugh [1968] have shown that structurally stable
systems are not dense on any noncompact manifold of
dimension ≥ 2. Finally, Williams [1970] showed that
structurally stable diffeomorphisms are not dense on the
two-dimensional torus. Thus, we are left with two courses of
approach. We can either alter the equivalence relation on
the space of all dynamical systems hoping that stability
with respect to the new equivalence relation may be dense,
or we can ask for some structures which are less than dense
in the given topology. One of the new equivalence relation
which aroused most interest is Ω-stability. This is based on
Ω-equivalence discussed in Chapter 4. Ω-stability is
stability with respect to Ω-quivalence. Unfortunately
Ω-stability is not any more successful than structural
stability as far as the dense of the structure is concerned.
For examples, see Abraham and Smale [1970] and Newhouse
[1970a,b].

For the second approach, it is natural to ask the
following question: Given an arbitrary dynamical system, can
we deform it into a structurally stable system? If we can,
how small a deformation is necessary? Clearly, we cannot
make it arbitrarily C^1-small, otherwise it would imply
C^1-density of structural stability. Thus we may be able to
deform it by an arbitrarily C^0- small deformation. Here we
only talk about the size of the deformation needed to
produce structural stability, and we leave the smoothness of
the maps and the definition of structural stability as
before. The following theorem by Smale [1973] and Shub

[1972] answers these questions.

   Theorem 6.5.4   Any C^r diffeomorphism ($1 \leq r \leq \infty$) of a compact manifold is C^r isotopic to a C^1 structurally stable system by an arbitrarily C^0-small isotopy.

   Thus the structural stability is dense in $\text{Diff}^r X$ with respect to the C^0-topology. It should be noted that the structural stability is no longer open in this topology. De Oliviera [1976] showed an analogous theorem for flows.

   Recall that the motivation of studying the structural stability of a dynamical system is because one is required to make measurements, but since the measurements are limited by their measurement uncertainties (the measured systems are only approximations of the true systems. It is important to know whether the qualitative behavior of the approximate system and the true system are the same. Structural stability of a system guarantees this if the approximation is sufficiently good. To make things more complicated (also more interesting), in most of the situations the measured quantities would not be completely time independent, but only be approximately constant during the measuring interval. In other words, the true dynamical system is not really autonomous but to a certain extent, time dependent. Thus, we are asking under what conditions an autonomous system is structurally stable when it is perturbed to a time dependent system. Franks [1974] gave a solution for C^2 diffeomorphisms on compact manifolds:

   Theorem 6.5.5   If f: M → M is a C^2 diffeomorphism of a compact manifold, then f is time dependent stable iff f satisfies Axiom A and the strong transversality condition.

   In closing, we would like to point out that several other notions of stability have been proposed in hope that they might be generic, nonetheless, none as yet has been completely sucessful. After all, maybe it is too optimistic to expect to find a single natural equivalence relation with respect to which stability is dense. More recently, attention has been focused on the interesting and important question of bifurcation of systems due mainly to R. Thom

265

[1975]. Which we shall discuss very briefly in the next
section.

In a series of papers, Hirsch [1982, 1985, 1988, 1989a,
1989b, 1989c] has studied a vector field in n-space
determines a competitive (or cooperative) system of
differential equations provided all the off-diagonal terms
of its Jacobian matrix are nonpositive (or nonnegative). He
has found that orthogonal projection along any positive
direction maps a limit set homeomorphically and
equivariantly onto an invariant set of a Lipschitz vector
field in a hyperplane. And limit sets are nowhere dense,
unknotted and unlinked. In other words, most trajectories
are stable and approach stationary points, and limit sets
are invariant sets of systems in one dimension lower. In
dimension 2 every trajectory is eventually monotone, and in
dimension 3 a compact limit set which does not contain an
equilibrium is a closed orbit or a cylinder of closed
orbits. Furthermore, Hirsch [1985] has found that a
cooperative system cannot have nonconstant attracting
periodic solutions. The persistent trajectories of the
n-dimensional system are studied under the assumptions that
the system is competitive and dissipative with irreducible
Jacobian matrices. Then it is shown that there is a
canonically defined countable family of disjoint invariant
open (n-1)-cells which attract all nonconvergent persistent
trajectories. These cells are Lipschitz submanifolds and are
transverse to positive rays. Furthermore, if the Jacobian
matrices are strictly negative then there is a closed
invariant (n-1)-cell which attracts every persistent
trajectory. In 3 dimensional system, the existence of a
persistent trajectory implies the existence of a positive
equilibrium. It is then shown that among 3-dimensional
systems which are competitive or cooperative, those
satisfying the generic conditions of Kupka- Smale also
satisfy the conditions of Morse-Smale and are therefore
structurally stable. This provides a new and easily
recognizable class of systems which can be approximated by

structurally stable systems. For three-dimensional systems, a certain type of positive feedback loop is shown to be structurally stable.

6.6   Bifurcation

As we have discussed earlier at the beginning of Chapter 5, the most important systems are the ones which can be used to model the dynamics of real life situations. But rarely can real life situations ever be exactly described, and we should expect to lead to slight variations in the model system. Consequently, a theory making use of qualitative features of a dynamical system is not convincing nor has its utility unless the features are shared by "nearby" systems. That is to say that good models should possess some form of qualitative stability. Hence our contempt for extremely unstable systems. Furthermore, in a given physical situation, there may be factors present which rule out certain dynamical systems as models. For instance, conservation laws or symmetry have this effect. In this case, the subset of these dynamical systems that are admissible as models may be nowhere dense in the space of all systems, and thus the stable systems that we are considering are really irrelevant. Thus one has to consider afresh which properties are generic in the space of admissible systems! On the other hand, even if the usual space of systems is the relevant one, the way in which a system loses its stability due to perturbation may be of importance, since the model for an event consists of a whole family of systems. In his theory of morphogenesis, Thom envisions a situation where the development of a biological organism, say, is governed by a collection of dynamical systems, one for each point of space time.

Bifurcation is a term which has been used in several areas of mathematics. In general, it refers to a qualitative change of the object under study due to change of parameters on which the object depends. For the kinds of applications

267

we have in mind, the following more precise definition
suffices. Let X and Y be Banach spaces, U ⊂ X, and F: U → Y.
Suppose there is a one-to-one curve Γ = {x(t): t ∈ (0,1)} ⊂
U such that for z ∈ Γ, F(z) = 0. A point p ∈ Γ is a
<u>bifurcation point for F with respect to Γ</u> (more simply a
<u>bifurcation point</u>) if every neighborhood of p contains zeros
of F not in Γ. In most applications, possibly after making a
change of variables, one usually has X = RxB where B is a
real Banach space, F = F(α,u), and Γ = {(α,0): α ∈ (a,b) ⊂
R}. Here, the members of Γ will be called trivial solutions
of F(α,u) = 0. Thus, we are interested in nontrivial zeros
of F.

    We would like to mention several models of phenomena to
illustrate the motivation for studying bifurcation.

    First, an infinite horizontal layer of a viscous
incompressible fluid lies between a pair of perfectly
conducting plates. A temperature gradient T is maintained
between the plates, the lower plate being warmer. If T is
appropriately small, the fluid remains at rest, the heat is
transported through the fluid solely by conduction, and the
temperature is a linear function of the vertical height.
When T exceeds a certain value, the fluid undergoes
time-independent motions called <u>convection current</u> and heat
is transpoted through the fluid by convection and
conduction. In actual experiments, the fluid breaks up into
cells whose shape depends in part on the shape of the
container. This is called Bénard instability.
Mathematically, the equilibrium configuration of the fluid
is described by a system of nonlinear partial differential
equations. Formulated in the general Banach space framework,
the pure conduction solutions correspond to the trivial
solutions, while the value of T at which convection begins
corresponds to a bifurcation point. For reference, see, for
instance the classic, Chandrasekar [1961]. See also
Kirchgässner and Kielhöfer [1973].

    Another interesting problem in fluid motion is the
Taylor problem of rotating fluid. A viscous incompressible

fluid lies between a pair of concentric cylinders whose axis
of rotation is the symmetry axis which is vertical. The
inner cylinder rotates at a constant angular velocity Ω
while the outer one remains at rest. If Ω is sufficiently
small, the fluid particles move in circular orbit with
velocity depending on their distance from the axis of
rotation. Equilibrium states of the fluid are solutions of
the time-independent Navier-Stokes equations, and they are
called <u>Couette flow</u>. When Ω exceeds a critical value, the
fluid breaks up into horizontal bands called <u>Taylor vortices</u>
and a new periodic motion in the vertical direction is
superimposed on the Couette flow. Here Couette flow
corresponds to the trivial solutions in the general
framework, and the values Ω at which the onset of Taylor
vortices taking place corresponds to a bifurcation point.

Buckling phenomena of a flat plate is another example of
bifurcation. A thin, planar, clamped elastic plate is
subjected to a compressive force along its edges. If the
magnitude of this compressive force f is small enough, the
plate remains motionless and in equilibrium. But if f
exceeds a certain value, the plate deflects out of the plane
and assumes a nonplanar equilibrium position called a
<u>buckled state</u>. Equilibrium configurations of the plate
satisfy a system of nonlinear partial differential equation
called the von Kärmän equations. The unbuckled states are
trivial solutions of these equations, while the value of f
at which buckling taking place corresponds to a bifurcation
point. See, for example, Friedrichs and Stoker [1941];
Berger and Fife [1968]; Keller and Autman [1969]; Berger
[1977]. Thompson [1979] has shown that elastic structure
under dead and rigid loadings can assess the stable regions
through a succession of folds, and the examples of buckling
of elastic arches, shallow domes and the incipient
gravitational collapse of a massive cold star are
demonstrated. A closely related mechanical phenomenon has
been described by Duffing's equation. The book by
Guckenheimer and Holmes [1983] gives a very detailed study

of the four systems, namely Van der Pol's equation for
nonlinear electronic oscillator, Duffing's equation for
stiffed spring with cubic stiffness, Lorenz equation for
two-dimensional fluid layer heated from below, and a
bouncing ball on a vibrating table. This book is highly
recommended. The applications of these systems are far
beyond their original problems. We shall see this in Chapter
7. For an elementary yet detailed discussion of low
dimensional bifurcation, the book by Iooss and Joseph [1980]
is recommended. A very recent book by Ruelle [1989] is also
highly recommended. A much more advanced treatment of
bifurcation theory, Chow and Hale [1982] is indispensible.

There are also many interesting applications of
bifurcation theory in other physical and nonphysical
sciences, such as chemical reactions, geophysics,
atmospherical science, biology, and social science.

Let us return to the general theory of bifurcation,
where there are three main questions of interest: (a) What
are the necessary and sufficient conditions for $(\sigma, 0) \in \Gamma$ to
be a bifurcation point? (b) What is the structure of the set
of zeros of $F(\alpha, u)$ near $(\sigma, 0)$? (c) In problems such as
described above, where there is an underlying evolution
equation of which the solutions described are equilibrium
solutions, determine which solutions are stable or unstable.
For the detailed "mechanism" of the bifurcation, see Holmes
and Rand [1978]. For the forced van der Pol-Duffing
oscillator applies to the trubulence flow, the routes to
turbulence are discussed in Coullet, Tresser and Arneodo
[1980].

A <u>catastrophe</u> is a point where the form of the organism
changes discontinuously, this corresponds to topological
change in the orbit structure of the dynamical systems. We
say that the family of dynamical systems <u>bifurcates</u> at the
point where the change is discontinuous.

Let us give some simple examples of bifurcations of
flows. There are local changes which can happen on any
manifold. For convenience and simplicity, let us take a

suitable chart in R^n. First, a vector field v on R by v(x) = $\alpha + x^2$ and $\alpha \epsilon$ R. We are interested in how the orbit structure varies with α. We find that: (i) for $\alpha > 0$, there is no zero, and the whole of R is an orbit oriented positively; (ii) for $\alpha = 0$, we have a zero at x = 0 which is one way zero in the positive direction; (iii) for $\alpha < 0$, we have two zeros, a sink at $-\sqrt{-\alpha}$ and a source at $\sqrt{-\alpha}$. Thus the bifurcation occurs at $\alpha = 0$.

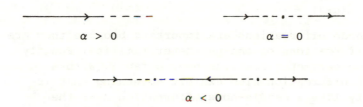

$\alpha > 0$ $\alpha = 0$

$\alpha < 0$

If we take the product of v_α with a fixed (i.e., independent of α) vector field on R^{n-1} having a hyperbolic fixed point at 0, we obtain a bifurcation of the resulting vector field on R^n. All such bifurcations are known as saddle-node bifurcations such as the following pictures for n = 2 depicted. A saddle point and node come together, amalgamate, and cancel each other out!

($\alpha < 0$)

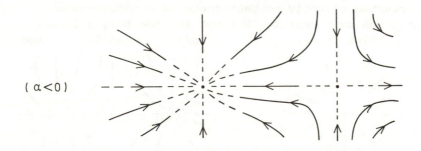

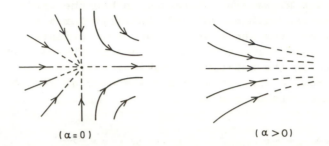

$(\alpha = 0)$ $(\alpha > 0)$

Saddle-node bifurcations are important because they are stable as bifurcations of one-parameter families. Roughly speaking, one-parameter families near a family with a saddle- node bifurcation also exhibit something that is topologically like a saddle-node bifurcation near the original one. One speaks of them as codimension one bifurcations; one can visualize the set of systems exhibiting zeros of the $\alpha = 0$ type in the above example (in some sense) as a submanifold of codimension one in $\Gamma^r(M)$, and the one-parameter families are being given by an arc in $\Gamma^r(M)$ crossing the submanifold transversally.

Note that, the bifurcation illustrated below (a node, i.e., $\alpha \leq 0$, bifurcating into two nodes and a saddle point, $\alpha > 0$) is not stable for the one-parameter families. It can be perturbed slightly so that there is a saddle-node bifurcation pair near to, but not at, the original node.

sink (node)   node   saddle   node

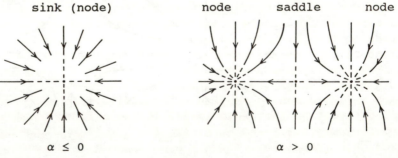

$\alpha \leq 0$ $\alpha > 0$

272

The saddle-node bifurcation is the typical bifurcation resulting when the sign of a real eigenvalue of the differential at a zero is changed by varying a single parameter governing the system. There is a typical codimension one bifurcation which comes about when the sign of the real part of a complex conjugate pair of eigenvalues is changed by varying a single parameter. This is known as the <u>Hopf bifurcation</u> [Hopf 1943].

For instance, consider the vector field v_α on R^2 given by $v_\alpha(x,y) = (-y - x(\alpha + x^2 + y^2), x - y(\alpha + x^2 + y^2))$. Here α is a single real parameter. For all $\alpha \in R$, v_α has a zero at the origin, and the linear terms make this a spiral source for $\alpha < 0$ and a spiral sink for $\alpha > 0$. For $\alpha = 0$, the linear terms would give a center, but the cubic terms make the orbits spiral weakly inwards. The interesting feature is the unique closed orbit one obtains at $x^2 + y^2 = -\alpha$ for each $\alpha < 0$. The bifurcation is illustrated in the following:

$$\alpha < 0 \qquad\qquad \alpha \leq 0$$

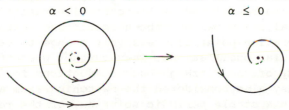

That is, the periodic attractor (or orbit) decreases in size until it amalgamates with the spiral source to form a spiral sink. It is intriguing to note that the reverse bifurcation can happen where a spiral sink splits into a spiral source and a periodic attractor. This is because when one has something from an inert (or dead) source, one creates a pulsating (and alive) periodic orbit.

Even for a simple recurrence equation,
$$x(t+1) = Ax(t)[1 - x(t-1)],$$
which is analogous to the logistic model, is found to show Hopf bifurcation [Morimoto 1988]. It has a fixed point at zero for $A \in [0,1)$, and at $1 - 1/A$ for $A \in [1,2)$, and the fixed point is destabilized at $A = 2$. For $A > 2$ the

oscillatory behavior appears, which is Hopf bifurcation.

Recently, Hale and Scheurle [1985] investigated the smoothness of bounded solutions of nonlinear evolution equations and they have found that in many cases globally defined bounded soutions of evolution equations are as smooth in time as the corresponding operator, even if a general solution of the initial value problem is much less smooth. In other words, initial values for bounded solutions are selected in such a way that optimal smoothness is attained. In particular, solutions which bifurcate from certain steady states, such as periodic orbits, almost periodic orbits, homo- and heteroclinic orbits, have this property.

Recently, Baer and Erneux [1986] have studied the singular Hopf bifurcation from a basic steady state to relaxation oscillation characterized by two quite different time scales of the form $dx/dt = f(x,y,\alpha,\epsilon)$ and $dy/dt = \epsilon g(x,y,\alpha,\epsilon)$ where $\epsilon << 1$ and is the control parameter. Their bifurcation analysis shows how the harmonic oscillations near the bifurcation point progressively change to become pulsed, triangular oscillations. They further presented a numerical study of the FitzHugh-Nagumo equations for nerve conduction. They also considered the switching from a stable steady state to a stable periodic solution, or the reverse transition. Baer et al [1989] further expanded their study of the FitzHugh-Nagumo model of nerve membrane excitability as a delay or memory effect. It can occur when a parameter passes slowly through a Hopf bifurcation point and the system's response changes from a slowly varying steady state to slowly varying oscillations.

Next, let us briefly discuss and state the center manifold theorem, which provides a mean for systematically reducing the dimension of the state spaces needed to be considered when analyzing bifurcations. Later in this chapter, we shall use the Lorenz system and its bifurcations as an example to illustrate the role of center manifold theorem in bifurcation calculations.

Suppose we have an autonomous dynamical system $dx/dt = f(x)$ such that $f(0) = 0$. If the linearization of f at the origin has no pure imaginary eigenvalues, then Hartman's linearization theorem (Theorem 4.6.6) states that the numbers of eigenvalues with positive and negative real parts determine the topological equivalence of the flow near the origin. If there are eigenvalues with zero real parts, then the flow can be quite complicated near the origin. We have seen such situations before. Let us consider the following system: $dx/dt = xy + x^3$, $\quad dy/dt = -y - x^2 y$. We will not go into any detail to analyze the above system. It suffices to say that one of the eigenvalues is -1 and hence a one-dimensional stable manifold exists (indeed, the y-axis). Direct calculation shows that the x-axis is a second invariant set tangent to the center eigenspace E^c. This is an example of a center manifold, an invariant manifold tangent to the center eigenspace. Let us give a simple example before stating the main theorem. The following example is due to Kelley [1967] which also gave the first full proof of the main theorem we shall state shortly.

Let us consider the very simple system:
$$dx/dt = x^2, \qquad dy/dt = -y.$$
The parametric solutions to this system have the following form:
$$x(t) = a/(1 - at), \qquad y(t) = be^{-t}.$$
By eliminating t, we have the solution curves which are graphs of the functions $y(x) = (be^{-1/a})e^{1/x}$. Clearly, for $x < 0$, all of these solution curves approach the origin in such a way that all of their derivatives vanish at $x = 0$. While for $x \geq 0$, the only solution curve approaches the origin is the x-axis. Thus, the center manifold is not unique. Indeed, we can obtain a C^∞ center manifold by piecing together any solution curve in the left half plane with the positive half of the x-axis. The center manifolds (heavy curves) are shown in the following figure. Nonetheless, the only analytic center manifold is the x-axis itself.

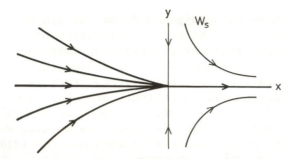

Theorem 6.6.1 (Center manifold theorem) Let f be a C^r vector field on R^n vanishing at the origin and let A = Df(0). Let us divide the spectrum of A into three parts, namely, σ_s, σ_c, σ_u with Re $\mu < 0$ if $\mu \in \sigma_s$, Re $\mu = 0$ if $\mu \in \sigma_c$, and Re $\mu > 0$ if $\mu \in \sigma_u$. Let us denote the eigenspaces of σ_s, σ_c, and σ_u by E^s, E^c, and E^u respectively. Then there exist C^r stable and unstable invariant manifolds W^s and W^u tangent to E^s and E^u at 0 and a C^{r-1} center manifold W^c tangent to E^c at 0. The manifolds Ws are all invariant to the flow of f, and both the stable and unstable manifolds are unique, but the center manifold need not be.

In general, the center manifold method isolates the complicated asymptotic behavior by locating an invariant manifold tangent to the subspace spanned by the eigenspace of eigenvalues on the imaginary axis. As we have noted in the example and in the theorem, there are technical difficulties involving the nonuniqueness and the loss of smoothness of the invariant center manifold which are not present in the invariant stable manifold.

For further examples and detailed discussion including the existence, uniqueness, and smoothness of center manifolds and the proof of the above theorem, see, e.g., Marsden and McCracken [1976], Carr [1981], Chow and Hale [1982], Guckenheimer and Holmes [1983].

After Guckenheimer and Holmes [1983], Guckenheimer published a long paper about multiple bifurcations with

multiple degeneracy in some features of the system and a
multi-parameter in its definition. Multiple bifurcations
occur in the mathematical descriptions of many natural
phenomena, and more importantly provide a means of
organizing the understanding of simple bifurcations and also
provide a powerful analytic tool for locating complicated
dynamical behavior in some models. This paper is highly
recommended, and one may consider this paper as an appendix
to the book mentioned.

Although the phenomena of Hopf bifurcations depending on
some autonomous external parameters are well understood,
nonetheless, the parametrically perturbed Hopf bifurcations
have not received enough attention. The effects of periodic
perturbation of a bifurcating system have been considered by
Rosenblat and Cohen [1980, 1981] and Kath [1981],
nonetheless, they neglected to examine any possible
secondary bifurcations which may exist in these systems. Sri
Namachchivaya and Ariaratnam [1987] studied small periodic
perturbations on two-dimensional systems exhibiting Hopf
bifurcations in detail, and obtained explicit results for
various primary and secondary bifurcations, and their
stabilities. Here the center manifold theorem and other
techniques are utilized.

In addition to the above suggested reading list, Iooss
and Joseph [1980] and Ruelle [1989] are recommended.

Before we end this section, let us briefly discuss the
unfolding of singularities and describe the elementary
catastrophes. Intuitively, unfoldings means that we embed a
singularity of a map in a higher dimensional domain, so that
the "bigger" map offers some insights and advantages.

Let f be a finite sum of products of two elements, each
of which from those germs $f: R^n \to R$ for which $f(O) = 0$. So,
f has a singularity at O. An unfolding of f is a germ f':
$R^{n+r} \to R$, with $f'(O) = 0$, such that if $x \in R^n$, $f'(x,O) =$
$f(x)$. Of course, here $O = (0,\ldots,0)$ with r entries. The
unfolding f' have r-parameters. Note that the constant
unfolding f', defined by $f(x,y) = f(x)$. We say that f' (with

277

r parameters) is <u>versal</u> if any other unfolding of the germ f
is induced from f'. A versal unfolding of a germ f is
<u>universal</u> if the number of parameters is minimal.

    <u>Theorem 6.6.2</u> (a) A germ f of sum of products of two
elements has a versal (thus a universal) unfolding iff f has
finite codimension; (b) If f' of r-parameter is a universal
unfolding of f, then r = codim f. All universal unfoldings
are isomerphic; (c) The universal unfolding of a germ is
stable (even if the germ is not).

    The elementary theorem of Thom classifies singular germs
of codim $\leq$ 4, and these are the elementary catastrophes.
This result is stated in germs, that is, as a local theorem.

    <u>Theorem 6.6.3</u> (Thom's elementary catastrophes) Let f be
a smooth germ (f(**O**) = 0 with **O** a singularity). Let $1 \leq c \leq 4$
be the codim of f. Then f is 6-degree-determined. Up to sign
change, and the addition of a non-degenerate quadratic form,
f is (right) equivalent to one of the germs in the table.

Germ	Codim	Universal manifold	Popular name
x^3	1	$x^3 + ux$	fold
x^4	2	$x^4 + ux^2 + vx$	cusp
x^5	3	$x^5 + ux^3 + vx^2 + wx$	swallow-tail
x^3+y^3	3	$x^3 + y^3 + wxy - ux - vy$	hyperbolic umbilic
x^3-xy^2	3	$x^3 - xy^2 + w(x^2+y^2) - ux - vy$	
			elliptic umbilic
x^6	4	$x^6 + tx^4 + ux^3 + vx^2 + wx$	butterfly
x^2y+y^4	4	$x^2y + y^4 + wx^2 + ty^2 - ux - vy$	
			parabolic umbilic

    There are several places where the details of this
classification theorem are carried out. Brocker's lecture
notes [1975] are excellent. For more details including
comments on higher dimensions, see for instance: Wasserman
[1974]; Zeeman [1976]. Zeeman's article points out that
while germs such as x^3 are not locally stable, their
universal unfoldings are. Furthermore, the global point of
view, as well as genericity for maps $R^n \to R$ are also

discussed.

For those readers who may want to pursue the details and specifics of this theory, one needs at least the following background:

(i) Some ring theory to get to the basic results of Mather et al on finite determination and codimension.

(ii) The Malgrange preparation theorem which generalizes to the smooth case a famous theorem of Weierstrass from several complex variables. See for instance: Brocker [1975], Malgrange [1964]; a chapter in Golubitsky and Guillemin [1973].

(iii) Some basic algebraic geometry.

Catastrophe theory has brought with it a wealth of applications, however the risk of oversimplification in applications are enormous. Prudent caution is required! There are several concrete examples of applications of catastrophe theory, such as: in optics [Berry and Upstill 1980], relativity [Barrow 1981, 1982a,b], geophysics [Gilmore 1981], particle scattering from surface [Berry 1975], rainbow effect in ion channeling in very thin crystal [Neskovic and Perovic 1987], elastic structure under dead and rigid loadings assess the stable regions of an equilibrium path which exhibits a succession of folds [Thompson 1979], just to name a few. Gilmore [1981] also provides some other applications of catastrophe theory.

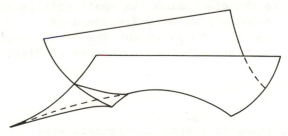

swallow tail

Elliptic umbilic

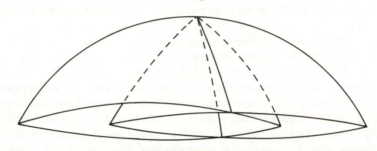

Hyperbolic umbilic

As we have given some references for further reading on the subject of bifurcation theory, we will not discuss the more well- known applications which can be found in most texts or references. Instead, in the next chapter we shall discuss some nonlinear dynamical systems in various disciplines. In the discussion, we shall utilize the concepts and techniques we have discussed here and earlier, and we would also like to point out the common mathematical structures which transcend the boundaries of diverse disciplines.

6.7   Chaos

Recently, chaos is a very fashionable word in the natural sciences. Roughly speaking, chaos is defined as an irregular motion stemming from deterministic equations. Nonetheless, there are somewhat different definitions of the term chaos in the literature. The difference is mainly due to different ways of defining "irregular motion". Irregular motion of stochastic processes, such as Brownian motion, occurs due to random, i.e., unpredictable, causes or sources. Thus these random irregular motions are uncorrelated. We do not consider this type of irregular motion as chaotic. Instead, one can discuss the atypical behavior of the correlation function of chaotic processes.

There are a number of introductory articles on chaos. The following are a partial list. Chernikov, Sagdeev and Zaslavsky [1988], Chua and Madan [1988], Bak [1986] Crutchfield et al [1986], May [1976] and Gleick [1987]. The following review articles on chaos and routes to chaos are highly recommended: Eckmann [1981], Ott [1981], and Tomita [1982] for nonlinear oscillators. There are also many introductory or popular books on chaos. Holden [1986] is highly recommended. Prigogine and Stengers [1984] gives a nontechnical account of order and chaos and raises some philosophical issues. It stimulates and challenges numerous questions and thoughts, nonetheless, due to its lack of specifics, it does not help the "advanced beginners" to find ways or approaches for the solutions.

The idea of chaos has been applied to almost any subject. In addition to the classical applications in hydrodynamics (for instance, even in the periodical laminar flows through curved pipes, under certain conditions the flows exhibit period- tripling, which is reminiscent of one of the routes to chaos [Hamakiotes and Berger 1989]), plasma physics, classical mechanics, nonlinear feedback control, there are chaotic phenomena observed or predicted in optics, chemical and biochemical reactions, semiconductor physics, interacting population and delayed feedback, competitive economy, just to name a few. We shall discuss these applications in next chapter. Here we just want to point out a few examples which we will not go into any further detail, but give references for interested readers to pursue the subjects. For instance, Papantonopoulos, Uematsu and

Yanagida [1987] presented a chaotic inflationary model of the universe, in which nonlinear interaction of dilaton and axion fields, in the context of the super-conformal theory, can dynamically give rise to the initial conditions for the inflation of the universe. Buchler and Eichhorn [1987] discussed various chaotic phenomena in astrophysics. Also, Fesser, Bishop and Kumar [1983] have shown numerically that there are parameter ranges of radio frequency (rf) superconducting quantum interference device (SQUID) for chaotic behavior. They have shown that the strange attractor characterizing the chaotic regimes can be described by a one-dimensional return map. Recently, Herath and Fesser [1987], motivated by the rf-SQUID device and using different mode expansions, investigated nonlinear single well oscillators driven by a periodic force with damping. Very recently, Miles [1988] investigated the symmetric oscillations of an inverted, lightly damped pendulum under direct sinusoidal force, and analytically predicted symmetry-breaking bifurcations and numerically confirmed. Similar results were also obtained for Josephson junction circuit by Yeh and Kao [1982], Kautz and Macfarlane [1986], Yao [1986] and Hadley and Beasley [1987].

Damped, driven pendulum systems have been used to model complicated behavior in nonlinear systems successfully. Varghese and Thorp [1988] chose to study the transiently forced pendulums and they shifted their emphasis to the composition of the boundaries separating the domains of attraction of the various asymptotically stable fixed points. A simple proof of the existence of diffeomorphisms from connected basins to striated basins is also presented.

Since we are on the subject of effects of superconducting material, it is interesting to note that a series of papers on the nonlinear hysteretic forces due to superconducting materials have been published by the Cornell group [Moon, Yanoviak and Ware 1989, Moon, Weng and Chang 1989]. Moon [1988] has demonstrated the period doubling and chaos for the forced vibration due to the nonlinear

282

hysteretic force of a small permanent magnet near the surface of a high temperature superconducting disk. The forces are believed to be related to flux pinning and flux dragging effects in the superconductor of type II state. Based on the displacement of the magnet, a return map under iteration exhibits a bifurcation structure similar to the experimental results obtained.

In fact, much earlier, Huberman and Crutchfield [1979], Huberman, Crutchfield and Packard [1980] have shown that the nonlinear dynamics of anharmonically interacting particles under periodic fields resulted in a set of cascading bifurcations and into chaos. Turschner [1982] has presented an analytic calculation of the Poincaré section of the driven anharmonic oscillator based on proper canonical transformations. Linsay [1981] has demonstrated periodic doubling and chaotic behavior of a driven anharmonic oscillator and the experimental results are in quantitative agreement with the theory by Feigenbaum [1978, 1979]. Testa, Perez and Jeffries [1982] have also experimentally observed successive subharmonic bifurcations, onset of chaos, and noise band merging from a driven nonlinear semiconductor oscillator. See also, [Wiesenfeld, Knobloch, Miracky and Clarke 1984]. Rollins and Hunt [1982] have shown that both a finite forward bias and a finite reverse recovery time are required if the diode resonator is to exhibit chaos. Furthermore, this anharmonic oscillator also exhibits period tripling and quintupling. Nozaki and Bekki [1983] have shown that a nonlinear Schrodinger soliton behaves stochastically with random phases in both time and space in the presence of small external oscillating fields and emits small-amplitude plane waves with random phases. They also have found that the statistical properties of random phases give the energy spectra of the soliton and plane waves. Bryant and Jeffries [1984] studied a forced symmetric oscillator containing a saturable inductor with magnetic hysteresis, approximated by a noninvertible map of the plane. The system displays a Hopf bifurcation to quasiperiodicity, entrainment horns, and chaos. Within an entrainment horn, they observed symmetry breaking, period doubling, and complementary band merging. Very recently, Gunaratne et al [1989] have studied the chaotic dynamics from a nonlinear electronic cirsuit which has shown to exhibit the universal topological structure of maps on an annulus. They further suggested that low-dimensional strange attractors fall into a few classes, each characterized by distinct universal topological features.

Recently, Holmes [1986] (using the averaging method) and Melnikov [1963] (using the perturbation technique) have shown that an N-degree of freedom model of weakly nonlinear surface waves due to Miles [1976] has transverse homoclinic orbits. And this implies that sets of chaotic orbits exist in the phase space. Relevance of their results to experimental work on parametrically excited surface waves [Ciliberto and Gollub 1985] are also briefly discussed. Funakoshi and Inoue [1987] have experimentally demonstrated the chaotic behavior of the surface water wave, modeled by Miles [1976] with the assumption of weak nonlinearity and linear damping, when a container is oscillated resonantly in a horizontal direction with appropriate amplitude and frequency. The experimental work of Funakoshi and Inoue [1987] are directly relevant to Holmes' [1986] results, yet they are not aware of Holmes' work.

Melnikov's method detects transverse homoclinic points in differential equations which are small perturbations of integrable systems. This together with the Smale-Birkhoff homoclinic theorem [Smale 1967] implies the existence of chaotic motions among the solutions of the equation with qualitative information. Recently, Brunsden and Holmes [1987] proposed a method which provides quantitative statistical measures of solutions. They compute power spectra of chaotic motions which are perturbations of homoclinic orbits and their approach relies on the existence of global homoclinic structures, verified by Melnikov theory, and derived from the notion of coherent structures in turbulence theory.

Another example of the route to chaos via period-doubling is a linear, viscously damped oscillator which rebounds elastically whenever the displacement drops to zero. It exhibits a family of subharmonic resonant peaks between which there are cascades of period-doubling bifurcations leading to chaotic regimes [Thompson and Ghaffari 1982].

Yet another example of routes to chaos is the chaotic scattering. Bleher et al [1989] have shown that the onset of chaotic behavior in a class of classical scattering problems could occur in two possible ways. One is abrupt and is related to a change in the topology of the energy surface, while the other is a result of a complex sequence of saddle-node and period- doubling bifurcations. The former, the abrupt bifurcation represents a new generic route to chaos and yields a characteristic scaling of the fractal dimension associated with the scattering function as $[\ln(E_c - E)^{-1}]^{-1}$, for particle energies E near the critical value E_c at which the scattering becomes chaotic.

It is known that steady planar propagation of a combustion front is unstable to disturbances corresponding to pulsating and spinning waves. Recently, Margolis and Matckowsky [1988] considered the nonlinear evolution equations for the amplitudes of the pulsating and spinning waves in a neighborhood of a double eigenvalue of the system, in particular, near a degenerate Hopf bifurcation point, and new quasi-periodic modes of combustion were also described.

$1/f$ noise is found ubiquitously in various scientific and engineering disciplines. It has been speculated whether $1/f$ noise can be explained as a chaotic phenomenon. Recently, Geisel, Zacherl and Radons [1987] proposed a new mechanism for $1/f$ noise as a generic phenomenon in the velocity fluctuations of a particle in a 2-dim periodic potential, and is closely related to the generic structure of phase space of nonintegrable Hamiltonian systems. On the

285

other hand, Bak, Tang and Wiesenfeld [1987] have shown that dynamical systems with spatial degrees of freedom naturally evolve into a self-organized critical state and 1/f (or flicker) noise can be identified with the dynamics of the critical state. They also related this to fractal objects.

Phase transitions of trapped particles and ions, initially observed in 1959 by collisional cooling and more recently by laser cooling, can also be explained as order-chaos transitions [Hoffnagle et al 1988].

In the following we shall briefly discuss one of the well-known examples of deterministic chaos, the Lorenz model of turbulence. For details, see Guckenheimer and Holmes [1983]. C. Sparrow [1982] gives an extensive treatment of Lorenz equations. There are several other well-known examples of deterministic chaos, such as Van der Pol's equations of damped nonlinear oscillator [see, e.g., Holmes 1979, Holmes and Rand 1978], Duffing's equations of nonlinear mechanical oscillator with a cubic stiffness [Novak and Frehlich 1982, Liu and Young 1986], we either have touched upon earlier or we shall meet in next chapter. For a nice review, see, e.g., Holmes and Moon [1983]. We will not be able to get into any detail to discuss their routes to chaos. For details, once again the reader is referred to Guckenheimer and Holmes [1983] and Moon [1987]. Recently, Byatt-Smith [1987] studied the 2π period solution of the forced-damp Duffing's equation with negative stiffness, with linear damping proportional to the velocity. It has very rich structures.

Many problems in physical systems involve the nonlinear interaction of two oscillators with different frequencies. When these frequencies are incommensurate, the interaction involves the amplitude rather than the phases of each oscillator. On the other hand, when the frequencies are in a ratio closely corresponding to a rational fraction of a small denominator, phase locking occurs and the dynamics is much richer, see, e.g., Perez and Glass [1982] and Coppersmith [1987]. Wiesenfeld and Satija [1987] studied the

286

effect of noise on systems having two competing frequencies.
When the system is mode-locked, random perturbations are
effectively suppressed, while outside the locking interval
relatively high levels of broadband noise are presented.
Such effects have also been observed in mode-locked lasers.
We shall discuss this in the next chapter. Van Buskirk and
Jeffries [1985] have observed chaotic dynamics from the
nonlinear charge storage of driven Si p-n junction passive
resonators. And the behavior is in good agreement with
theoretical models. Yazaki, Takashima and Mizutani [1987]
investigated the Taconis oscillations, which are spontaneous
oscillations of gas columns thermally induced in a tube with
steep temperature gradients. Near the overlapping region,
the intersection of the stability curves for two different
modes with incommensurate frequencies has been found that
both modes can be excited simultaneously and competition
between them can lead to complex quasiperiodic and chaotic
states. These problems are of particular interest for some
technologies, such as phase locking of several independent
lasers of the same kind, called the phased array. We shall
come to this in the next chapter. Knobloch and Proctor
[1988] have investigated fully the special case where the
frequencies are in the ratio of 2:1. On the subject of
resonance, Parlitz and Lauterborn [1985] have shown
numerically a periodic recurrence of a specific fine
structure in the bifurcation set of the Duffing equation
which is closely related with the nonlinear resonance of the
system. Gray and Roberts [1988a,b,c,d], in a series of
papers, re-examined chemical kinetic models described by two
coupled ordinary differential equations containing at most
three control parameters, originally studied by Sal'nikov
[1949]. They have found some interesting effects and details
which have missed earlier. Recently, Wiesenfeld and Hadley
[1989] have described a novel feature of certain arrays of N
coupled nonlinear oscillators. They have found that the
number of stable limit cycles scales as (N-1)!. In order to
accommodate this very large multiplicity of attractors, the

287

basins of attraction crowd more tightly in phase space with increasing N. Their simulations have shown that for large enough N, even minute levels of noise can cause the system to hop freely among the many coexisting stable attractors.

Lorenz [1963] presented an analysis of a coupled set of three quadratic ordinary differential equations, one in fluid velocity and two in temperature, for fluid convection in a two-dimensional layer heated from below to modeling atmospherical dynamics. The Lorenz equations are:

$$dx/dt = \sigma(y - x),$$
$$dy/dt = \alpha x - y - xz, \qquad\qquad (6.7\text{-}1)$$
$$dz/dt = -\beta z + xy,$$

where $(x,y,z) \in R^3$, σ (the Prandtl number), α (the Rayleigh number), and β (an aspect ratio) are real positive parameters. For any further detail of Lorenz equations, see, e.g., Sparrow [1982], Guckenheimer and Holmes [1983], Moon [1987]. Amazingly, equations completely equivalent to the set of Lorenz equations, Eq. (6.7-1), occur in laser physics explaining the phenomenon of irregularly spiking of lasers. For the discussion of Lorenz equations in laser physics, see for instance, Haken [1983], Haken [1975a], Sparrow [1986]. For a more comprehensive review of cooperative phenomena in systems far from thermal equilibrium including lasers, nonlinear wave interactions, tunnel diodes, chemical reactions, and fluid dynamics, see Haken [1975b] as well as his book on synergetics [1983].

It suffices to say that for $\alpha < 1$, the origin $(0,0,0)$ is globally attracting, i.e., the fluid is at rest and with linear temperature gradient (corresponding to no laser action in lasre physics). This is because the trace of the Jacobian (the divergence of the vector field) is equal to $-(\sigma + 1 + \beta) < 0$. In fact, for $\alpha < 1$, the origin is a hyperbolic sink and is the only attractor. For $\alpha = 1$, one of the eigenvalues of the linearized system is zero, and the others, $\mu_1 = -\beta$, $\mu_2 = -(\sigma + 1)$, thus a pitchfork bifurcation occurs, that is, the pure conductive solution becomes unstable and the convective motion starts (corresponding to

laser at threshold and lasing starts). As for $\alpha > 1$, the origin is a saddle point with a one-dimensional unstable manifold, and there is a pair of nontrivial steady solutions (or fixed points S_+ and S_-) at $(x,y,z) = (\pm\beta(\alpha-1), \pm\beta(\alpha-1), \alpha-1)$, and these are sinks for $1 < \alpha < \alpha_h = \sigma(\sigma + \beta + 3)/(\sigma - \beta - 1)$. At $\alpha = \alpha_h$ a Hopf bifurcation occurs at the fixed points, since the eigenvalues of the matrix of the linear system are:
$\mu_1 = -(\sigma + \beta + 1)$, and $\mu_2 = \pm i2\sigma(\sigma+1)/(\sigma-\beta-1)$.
To allow imaginary roots, here we assume that $\sigma > 1 + \beta$. Otherwise, μ_2 are real. For $\alpha > \alpha_h$, the nontrivial fixed points are saddle points with two-dimensional unstable manifolds. Thus, for $\alpha > \alpha_h$, all three fixed points are saddles, i.e., unstable. Nonetheless, an attracting set does exist and may contain complicated bounded solutions. One may think that the Hopf bifurcation occurring as α passes through α_h will give rise to stable periodic orbits, but subsequently it has been found that the bifurcation is subcritical [Marsden and McCracken 1976], so that unstable periodic orbits shrink down to the sinks as α increases towards α_h and no closed orbits exist near these fixed points for $\alpha > \alpha_h$.

Thus, qualitatively we can anticipate that all the solutions will have the following behavior in the phase space. For all positive values of the parameters, all solutions of the equations eventually lie in some bounded region and they all tend towards some set in three-dimensional phase space with zero volume. This follows from the dissipative nature of the flow, and implies that solutions do not wander about the whole three-dimensional space but eventually come close to point-like, line-like, or sheet-like objects in the phase space.

Note also that the fixed points S_+ and S_- will be stable for $\sigma < \beta + 1$, but lose stability at some α if $\sigma > \beta + 1$. By using an averaging procedure, one can show that there is a stable symmetric periodic orbit or limit cycle for all large enough α if $3\sigma > 2\beta + 1$. Note that the Lorenz equations are symmetric under the mapping $(x,y,z) \rightarrow (-x,-y,z)$. Also notice that the region $3\sigma > 2\beta + 1$ is larger than and includes the region of $\sigma > \beta + 1$. That is, whenever S_+ or S_- lose stability at a finite α, there is a stable periodic orbit at large enough α. In the case when $3\sigma > 2\beta + 1$ but $\sigma < \beta + 1$, it can be shown that there are two unstable periodic orbits, i.e., a symmetric pair of non-symmetric orbits which exist for all large enough α, in addition to the stable symmetric orbit. Various other regions and other limits of the equations can be analysed in different ways. For instance, the limit $\sigma \approx \alpha \rightarrow \infty$ and $\beta \approx 1$ has been analysed by Shimada and Nagashima [1978], Fowler and McGuinness [1982, 1984]. Again, the references cited earlier are urged to consult with.

If it was not for the complicated but beautiful results obtained by numerically integrating the Lorenz equations on a computer, the Lorenz equations would not have received such interest and attention. Indeed, a computer is necessary to proceed much beyond the qualitative description earlier. In the following we will give a few well-known figures just to entice the reader for further reading.

We will fix two parameters σ and β, and let α vary. It should be noted that the values used by Lorenz and most other researchers are $\sigma = 10$, $\beta = 8/3$, and similar behavior occurs for other values. With these parameters fixed, one find $\alpha_h \approx 24.74$. Lorenz then fixed $\alpha = 28$ and integrated Eqs.(6.7-1) numerically. Fig.6.7.1 shows a two-dimensional projection of a typical orbit calculated with the above set of parameters. The transients have been allowed to die away before plotting begins. The orbit appears to oscillate back and forth, rotating first on one side and then on the other and never closing up! Fig.6.7.2 show numerically calculated orbits at other parameter values. For the same σ and β, (a) $\alpha = 60$, resulted in a chaotic behavior; (b) $\alpha = 126.515$ resulted in a stable orbit; (c) $\alpha = 198$ resulted in an asymmetric chaotic attractor; and (d) $\alpha = 350$ resulted in a typical large-α stable symmetric periodic orbit.

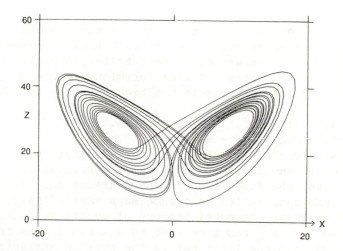

Fig.6.7.1 [Sparrow 1986] This is a plot of x vs z for σ =
10, β = 8/3, and α = 28.0.

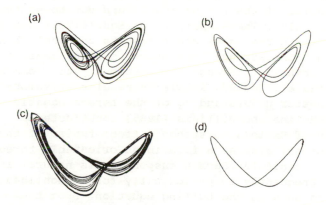

Fig.6.7.2 [Sparrow 1986] Same as Fig.6.7.1 for σ = 10, β =
8/3. (a) α = 60.0, chaotic behavior; (b) α = 126.515, a
stable orbit; (c) α = 198.0, an asymmetric chaotic
attractor; (d) α = 350.0, a large α stable symmetric
periodic orbit.

291

For any further details on the chaotic behavior of the
Lorenz equations as well as other well-known nonlinear
systems, see the references cited earlier. A prototype
equation to the Lorenz model of turbulence contains just one
nonlinearity in one variable has been proposed by Rossler
[1976]:

$$dx/dt = -(y + z), \quad dy/dt = x + 0.2y,$$
$$dz/dt = 0.2 + z(x - 5.7). \tag{6.7-2}$$

There is only a single nonlinear term, and of course there
is no longer any immediate physical interpretation.
Nonetheless, the flow in phase space allows for "folded"
diffeomorphisms, called horseshoe maps [Smale 1967], are
well-known in the geometric theory of dynamical systems. In
fact, each of them can give rise to a three-dimensional
suspension, and the limit set is the strange attractor and
whose cross-section is a two-dimensional Cantor set. The
flow is nonperiodic and structurally stable [Ruelle and
Takens 1971] even though all orbits are unstable. Thus, most
of the results of the Lorenz model turn out to be true for
Eqs.(6.7-2) also. The simplicity of Eqs.(6.7-2) has the
added attribute that some other results one would like to
obtain about the strange attractors, such as basin
structure, behavior through bifurcations, etc., can be
obtained easier. Indeed, Rossler's results stimulated and
helped a better understanding of the Lorenz equations.
Recently, Holmes and Williams [1985] constructed a
suspension of Smale's horseshoe diffeomorphism of the
two-dimensional disc as a flow in an orientable three
dimensional manifold. Such a suspension is natural in that
it occurs frequently in periodically forced nonlinear
oscillators such as the Duffing equation. From the
suspension, they constructed a know-holder or template in
such a way that the periodic orbits are isotopic to those in
the full three-dimentional flow. Theorems of existence,
uniqueness and nonexistence for families of torus knots, and
these families to resonant Hamiltonian bifurcations which
occur as horseshoes are created in a one-parameter family of

area preserving maps.

Saravanan et al [1985] have studied the effect of modulating the control parameter α with a frequency Ω, i.e., in the second equation of Eqs.(6.7-1), replace α by $\alpha[1 + \epsilon \cos(\Omega t)]$. They have found that this modulation is a very sensitive probe of the limit cycle of frequency Ω_o near the Hopf bifurcation point. For $\Omega = 2\Omega_o$, and low amplitudes, such modulation leads to a hastening of chaos. But for higher amplitudes, the destabilization ceases and beyond a critical amplitude the limit cycle of frequency Ω_o is stabilized. The existence of such a critical amplitude follows from a variant of the perturbation theory. They noted that the possibility of stabilizing the limit cycles is particularly interesting as it restores the correspondence with the real hydrodynamics up to the onset of chaos. The routes to chaos in this modulated systems are of interest for further study.

Shimizu and Morioka [1978] have shown that by changing variables, Lorenz equations can be written as a differential system describing a particle in some fourth order potential depending on a varying parameter. Coullet, Tresser and Arneodo [1979] have shown that such simple differential systems lead to the transition to stochasticity. Andrade [1982] used the Carleman embedding to obtain some results for the Liapunov exponents of the Lorenz model. Birman and Williams [1983a,b] in a series of papers have asked the questions such as can the periodic orbits of a dynamical system on S^3 or R^3 be knotted, and if so, what kinds of knots can occur, and what are the implications? Franks and Williams [1984] have further shown that with positive topological entropy, there are infinitely many distinct knot types. Results are obtained for the Lorenz model. Williams [1983] proved that Lorenz knots are prime. Holmes and Williams [1985] have constructed an suspension of Smale's horseshoe diffeomorphism (we have not discussed this topic due to its complicated construction, we refer the reader to Smale [1967], Irwin [1980], Guckenheimer and Holmes [1983])

293

of a two-dimensional disk as a flow in an orientable three dimensional manifold. Such a suspension occurs frequently in periodically forced nonlinear oscillators, such as the Duffing equation. From this suspension they constructed a template, which is a branched two-manifold with a semiflow, in such a way that the periodic orbits are isotopic to those in the full three-dimensional flow. Some existence, uniqueness, and nonexistence theorems for families of torus knots are obtained. They then were able to connect a countable subsequence of one-dimensional bifurcations with a subsequence of area-preserving bifurcations in a two parameter family of suspensions by using knot theory, kneading theory and Hamiltonian bifurcation theory. The moral of the results is that there are no universal routes to chaos! For more recent work on the classification of knotted periodic orbits in periodically forced nonlinear oscillators, as well as a review of earlier work, see Holmes [1988].

Recently, Agarwal, Banerjee and Bhattacharjee [1986] have shown that at the threshold of period-doubling chaos in a dynamical system such as the Lorenz system, the fractal dimension of the associated strange attractor assumes a universal value. Dekker [1986] using computer simulation demonstrated that the tunnelling orbits under a symmetric double-well potential can be assigned a fractal dimension.

The three-dimensional Lorenz flow is approximated by a two-dimensional flow with a branch curve with the use of the approximation of the Lorenz attractor by invariant two-dimensional manifolds [Dorfle and Graham 1983], and the probability density generated by the flow on the invariant manifolds in the steady state is obtained. It is also shown that the probability density arises in the Lorenz model subject to stochastic forces as a self-consistent approximation for very small but finite noise.

Recently, Lahiri and Nag [1989] have considered intermittency in inverted-pitchfork bifurcation of a 1D dissipative map and of a 2D conservative map. They have also considered the effect of noise and the scaling laws in the presence of noise. Furthermore, they have obtained the results for the saddle-node bifurcation in 2D area-preserving maps.

For a broader view of chaos, with its implications, see, for instance, Holden [1986], and Schuster [1984]. For a more current review of chaotic phenomena in nonlinear systems, see Hao [1988]. For chaotic phenomana in astrophysics, see for instance Buchler and Eichhorn [1987], Contopoulos [1985]. For nonlinear dynamical phenomena in chemical systems, see for instance, Vidal and Pacault [1984]. For fluid or plasma phenomena, see for instance, Marsden [1984], Sagdeev [1984], Rand and Young [1981]. For thoes readers who are interested in computer graphic presentations and numerical integration of nonlinear dynamical systems, there are several PC compatible software available for such purpose.

Although, we have mainly concentrated on nonlinear dynamical systems represented by ordinary differential equations, similar techniques can also be applied to nonlinear dynamical systems represented by partial differential equations. Birnir and Morrison [1987] have discussed structural stability and chaotic solutions of perturbed Benjamin-Ono equations for characterizing the turbulent motion. Bishop et al [1983, 1988] have discussed chaotic solutions to other perturbed nonlinear partial differential systems such as Sine-Gordon, and nonlinear Schrodinger equations. Olsen and Samuelsen [1987] have examined the one-dimensional sine-Gordon soliton equation in the presence of driving and damping numerically. They have found that by increasing the driving strength, the system exhibits an infinite sequence of period doubling leading to chaos. They noticed that the first bifurcation occured after a narrow regime characterized by intermittency-type of chaos and quasi-periodic oscillations. They further attributed the origin of the intermittency-type of chaos by the competition between two spatial patterns described by the presence of one or two breather-like modes respectively. Brandstater et al [1983] have presented evidence for low dimensional strange attractors in Couette-Taylor flow data. Doering et al [1987] have obtained an exact analytic computation of the Liapunov dimension of the universal attractor of the complex Ginzburg-Landau equation for a finite range of parameter values. Meanwhile, Sirovich and Rodriguez [1987] have used the Ginzburg-Landau equation to extract a complete set of uncorrelated coherent structures, which are then used as a basis for the dynamical description of coherent structures in the attractor set. Chate and Manneville [1987] discuss the transition to turbulence via spatiotemporal intermittency observed in a partial differential equation displays statistical features. The transition to chaos through intermittency has recently been observed in a steady-state plasma [Cheung, donovan and Wong 1988].

Even the traditional problems, such as the Navier-Stokes equations for an incompressible fluid on a two-dimensional torus, are rich in phenomenology. For instance, Franceschini [1983a,b] has numerically investigated two truncations of the above problem.

Over a decade of active research, a number of experiments have concluded that the production of complex, unpredictable turbulence for low-dimensional systems is due to few-dimensional attractors. Indeed, based on these and other observations and analysis, the "conventional wisdom" has concluded that the Newhouse-Ruelle-Takens route of few-dimensional chaos [Ruelle and Takens 1971, Newhouse, Ruelle and Takens 1978] is the proper explanation of the nature of turbulence. Nonetheless, there are several alternative explanations of turbulence which do not employ the attractive hypothesis, namely, the spin-glass relaxation [Walden and Ahlers 1980, Crutchfield 1984], spatial noise amplification [Deissler and Kaneko 1987], and transients [Crutchfield and Kaneko 1988]. Let us briefly describe three most common scenarios for turbulence. By the way, here chaos and weak turbulence are interchangeable.

The Newhouse-Ruelle-Takens (NRT) scenario is the oldest one in describing the route to turbulence. This can be stated in the following theorem [Newhouse, Ruelle and Takens 1978]:

Theorem 6.7.1 Let x be a constant vector field on the n-dim torus $T^n = R^n/Z^n$. If $n \geq 3$, every C^2 neighborhood of x contains a vector field x' with a strange Axiom A attractor. If $n \geq 4$, we may take C^∞ for C^2.

Under all these assumptions, the NRT scenario asserts that a strange attractor is likely to occur in the following sense. In the space of all differential equations, some equations have strange attractors, and others have none. Those which do have strange attractors form a set containing a subset which is open in the C^2 topology. The closure of this open set contains the constant vector fields on the torus T^3. The measurable consequences of the presence of strange attractor following the NRT scenario is the following: If a system starting from a stationary state solution undergoes three Hopf bifurcations when a parameter is varied, then it is very likely that the system possesses a strange attractor with sensitivity to initial conditions after the third bifurcation. And the power spectrum of such a system will exhibit first one, then two, and possibly three independent basic frequencies. If there is a strange attractor, when the third frequency is about to appear some broad-band noise will simultaneously also appear. When these happens, we consider the system chaotic or turbulent. Then it is natural to ask whether or not three-frequency quasiperiodic orbits are to be expected in typical nonlinear dynamical systems? This is the question raised by Grebogi, Ott and Yorke [1983]. Nonetheless, at that time the answer was positive but incomplete.

Another scenario leading to turbulence is the Feigenbaum scenario. These type of systems usually can be reduced to Poincaré return maps or area-preserving maps, and the bifurcations of the orbit structure are pitchfork bifurcations, i.e., a stable fixed point loses its stability and gives rise to a stable periodic orbit as a parameter is changed. This corresponds to a crossing of one eigenvalue of the tangent map through -1. In an experiment, if one observes subharmonic bifurcations at μ_1, μ_2, then according to the scenario, it is very probable for a further bifurcation to occur near $\mu_3 = \mu_2 - (\mu_1 - \mu_2)/\delta$, where $\delta = 4.66920...$ is universal [Feigenbaum 1978, 1979, 1980; Collet and Eckmann 1980, Greene, MacKay, Vivaldi and Feigenbaum 1981]. Some other universal fine structures in periodic doubling systems have also been discussed by Geisel and Nierwetberg [1981, 1982]. Furthermore, if one has seen three bifurcations, a fourth one becomes more probable then the third one after the first two, etc. And at the accumulation point, one will observe aperiodic behavior, but without broad-band spectrum. This scenario is extremely well tested both numerically and experimentally. The periodic doublings have been observed in most low dimensional dynamical systems. Giglio, Musazzi and Perini [1981] presented experimental results of a Rayleigh-Bénard cell. By appropriate preparation of the initial state, the system can be brought into a single frequency oscillatory regime. When a further increase of the temperature gradient makes the system undergo a reproducible sequence of period-doubling bifurcations, the Feigenbaum universal numbers are also determined. Gonzalez and Piro [1983] considered a nonlinear oscillator externally driven by an impulsive periodic force, and an exact analytical expression for the Poincaré map for all values of parameters is obtained. This model also displays period-doubling sequences of chaotic behavior and the convergence rate of these cascades is in good agreement with the Feigenbaum theory.

The third scenario is the Pomeau-Manneville scenario [Pomeau and Manneville 1980], which is also termed transition to turbulence through intermittency. The mathematical status of this scenario is less than satisfactory. Nonetheless, this scenario is associated with a saddle-node bifurcation, i.e., the amalgamation of a

stable and an unstable fixed point. One can state this
scenario for general dynamical systems as follows. Assume a
one- parameter family of dynamical systems has Poincaré maps
close to a one-parameter family of maps of the interval, and
that these maps have a stable and unstable fixed point which
amalgamate as the parameter is varied. As the parameter is
varied further to $\mu = 1.75$ from the critical parameter value
μ_c, one will see intermittently turbulent behavior of random
duration. The difficulty with this scenario is that it does
not have any clear- cut or visible precursors like the other
two scenarios. Recently, Keeler and Farmer [1987]
investigated a one dimensional lattice of coupled quadratic
maps. They found that the motion of the spatial domain walls
causes spatially localized changes from chaotic to almost
periodic behavior and the almost periodic phases have
eigenvalues very close to one and with a $1/f$ low frequency
spectrum. This behavior has some aspects of Pomeau-
Manneville intermittency, but is quite robust under changes
of parameters.

We would like to emphasize that since a given dynamical
system may have many attractors, several scenarios may
evolve concurrently in different regions of phase space.
Therefore, it is natural if several scenarios occur in a
given system depending on how the initial state of the
system is prepared. Furthermore, the relevant parameter
ranges may overlap, thus although the basins of attraction
for different scenarios must be disjoint, they maybe
interlaced. Recently, Grebogi, Kostelich, Ott and Yorke
[1986] using two examples to show that basin boundary
dimensions can be different in different regions of phase
space. They can be fractal or nonfractal depending on the
region. In addition, they have shown that these regions of
different dimension can be intertwined on an arbitrarily
fine scale. They further conjectured that a basin boundary
typically can have at most a finite number of possible
dimensions. Indeed, a series of papers by Gollub and
coworkers using laser-Doppler methods have identified four

distinct sequences of instabilities leading to turbulence at low Prandtl number (between 2.5 to 5.0) in fluid layers of small horizontal extent [Gollub and Benson 1980, Gollub and McCarriar 1982, Gollub, McCarriar and Steinman 1982]. McLaughlin and Martin [1975] proposed a mathematical theme of the transition to turbulence in statically stressed fluid systems. Systems are classified according to Hopf bifurcation theorem. They have found certain kind of flows obey the Boussinesq conditions exhibit hysteresis, finite-amplitude instabilities, and immediate transition to turbulence. They have also found another kind of flow, such as a model of fluid convection with a low-Prandtl number, in which as the stress increases, a time-periodic regime precedes turbulence. Nonetheless, the transition to nonperiodic behavior in this model is found to proceed in accordance with the NRT scenario.

Although we have not discussed stochastic influence on the nonlinear dynamical systems, there are many interesting results in this area. For instance, Horsthemke and Malek-Mansour [1976], using the method of Ito [1944, 1951] stochastic differential equations, have shown that in the vicinity of the bifurcation point the external noise, even they are characterized by a small variance, can influence profoundly the macroscopic behavior of the system and gives rise to new phenomena not predicted by the deterministic analysis. In Ruelle [1985], the ergodic theory of differentiable dynamical systems is reviewed, and it is applied to the Navier-Stokes equation. He also obtains upper bounds on characteristic exponents, entropy, and Haussdorff dimension of attracting sets. Recently, Machacek [1986] presented a general method for calculating the moments of invariant measure of multidimensional dissipative dynamical systems with noise. In particular, moments of the Lorenz model are calculated. Nicolis and Nicolis [1986] have casted the Lorenz equations in the form of a single stochastic differential equation where a deterministic part representing a bistable dynamical system is forced by a

300

noise process. An analytically derived fluctuation-dissipation-like relationship linking the variance of the noise to the system's parameters provides a satisfactory explanation of the numerical results. This leads the authors to suggest that the emergence of chaos in the Lorenz model is associated with the breakdown of the time scale separation between different variables. This gives some credence to the frequent assertion in the literature that the elimination of fast variables generates noise in the subset of the slow variables.

Horsthemke and Lefever [1977] have demonstrated on a chemical dynamics model that even though the system is above the critical point, phase transitions can still be induced solely by the effect of external noise. Jeffries and Wiesenfeld [1985] have measured the power spectra of a periodically driven p-n junction in the vicinity of a dynamical instability. They have found that the addition of external noise introduces new lines in the spectra, which become more prominent as a bifurcation (either period doubling or Hopf bifurcation) is approached. Furthermore, they have found that the scaling of the peak, width, area and lineshape of these lines are in excellent agreement with the predictions [Wiesenfeld 1985]. Experimentally, the onset of chaos in dynamical systems can often be analyzed in terms of a discrete-time map with a quadratic extremum [Collet and Eckmann 1980]. Kapral and Mandel [1985] investigated the bifurcation structure of a nonautonomous quadratic map. They found that alghough nontrivial fixed points do not exist for such system, a bifurcation diagram can be constructed provided that the sweep rate is not too large. Morris and Moss [1986] constructed an electronic circuit model of a nonautonomous quadratic map, different from the one by Kapral and Mandel [1985]. They observed that bifurcation points are postponed by amounts which depend on both the sweep velocity and the order of the bifurcation. The measured results obey two scaling laws predicted by Kapral and Mandel [1985] and provide evidence for the universality of these scaling laws.

Newell, Rand and Russell [1988] recently suggested that the transport properties and dissipation rates of a wide class of turbulent flows are determined by the random occurrence of coherent events that correspond to certain orbits in the non- compact phase space which are attracted to special orbits which connect saddle points in the finite region of phase space to infinity and represent coherent structures in the flow field.

Recently, Chernikov et al [1987] have proposed that chaotic web may be able to explain the origin of ultra high energy cosmic rays. Eckmann and Ruelle [1985] gave an interesting review of the mathematical ideas and their concrete implementation in analyzing experiments in ergodic theory of dynamical systems.

Recently, Ruelle [1987] suggested the study of time evolutions with adiabatically fluctuating parameters, i.e., evolutions of the form $x_{t+1} = f(x, \mu(t))$ for discrete time. Similar to the nonlinearly coupled oscillators, Yuan, Tung, Feng and Narducci [1983] analyzed some general features of coupled logistic equations. They have found that for selected values of the three control parameters, chaotic behavior may emerge out of an infinite sequence of period-doubling bifurcations. There also exist large regions of control parameter space, where two characteristic frequencies exist, and the approach to chaos follows the NRT scenario. Like the coupled oscillators, as the ratio of these frequencies is varied, phase locking and quasiperiodic orbits are observed enroute to chaos. Recently, Klosek-Dygas, Matkowsky and Schuss [1988] have considered the stochastic stability of a nonlinear oscillator parametrically excited by a stationary Markov process. Harikrishnan and Nandakumaran [1988] recently analyzed numerically the bifurcation structure of a two-dimensional noninvertible map of the form suggested by Ruelle and they showed that different periodic cycles are arranged in it in the same order as in the logistic map. Indeed, this map

satisfies the general criteria for the existence of
Sarkovskii [1964] ordering, which is simply the order in
which different period cycles are arranged along the
parameter axis. On the other hand, the Henon [1976] map can
be considered as the two-dimensional analogue of the
quadratic map, yet the Sarkovskii theorem does not hold
[Devancy 1986].

One may also expect that external noise can influence
the nature of chaotic systems. Surprisingly, the theorem due
to Kifer [1974] states that for a dynamical system with an
Axiom A attractor, the system is insensitive to small
external noise. This is experimentally demonstrated for
Rayleigh-Benard system by Gollub and Steinman [1980]. Ruelle
[1986a,b] expanded to consider the analytic properties of
the power spectrum for Axiom A systems near resonance.
Recently, Ciliberto and Rubio [1987] studied experimentally
the spatial patterns in temporal chaotic regimes of
Rayleigh-Benard convection.

Recently, Cumming and Linsay [1987], using a simple
operational-amplifier relaxation oscillator driven by a sine
wave which can be varied in frequency and amplitude, have
presented experimental evidence for deviations from
universality in the transition to chaos from
quasiperiodicity in a nonlinear dynamical system. In fact,
they have shown that the power spectrum, tongue convergence
rate, and spectrum of critical exponents all differ from the
theory. Meanwhile, in a different domain, Cvitanovic [1985]
argued that period doubling and mode lockings for circle
maps are characterized by universal scaling and can be
measured in a variety of nonlinear systems. Indeed, such
phenomena not only have been observed in physical phenomena,
but also have been observed in biological and physiological
situations. For instance, the spontaneous rhythmic activity
of aggregates of embryonic chick heart cells was perturbed
by the injection of single current pulses and periodic
trains of current pulses. The regular and irregular
dynamical behavior produced by periodic stimulation, as

period doubling bifurcations, were predicted theoretically and observed experimentally by Guevara, Glass and Shrier [1981].

The above paragraphs show that even though we have some understanding of low dimensional (one- or two-dimensional) chaos, but there is still much more we do not understand and also with lots of confusion, to say the least!

In the past several years, a great deal of interest has been focused on the temporal evolution and behavior of chaos, which arises in spatially constrained macroscopic systems while some control parameters are varied. Recently, it has become increasingly more popular to study the spatial pattern formation and the transition to turbulence in extended systems [Thyagaraja 1979, Wesfreid and Zaleski 1984, Bishop, Campbell and Channell 1984], and bifurcations in systems with symmetries [Stewart 1988 and references cited therein, Sattinger 1983], and nonlinear mode competition between instability waves in the forced free shear layer [Treiber and Kitney 1988]. Recently, Pismen [1987] has investigated the strong influence of periodic spatial forces on the selection of stationary patterns near a symmetry-breaking bifurcation. Using simple model equations of long-scale thermal convection, he demonstrated: (i) the transition between alternative patterns, (ii) emergence of spatially quasiperiodic patterns, and (iii) evolution of patterns due to rotating phases. The presence of symmetry usually can simplify the analysis, yet symmetry can also lead to more complicated dynamical behavior. Swift and Wiesenfeld [1984] have examined the role of symmetry in systems displaying period-doubling bifurcations. They have found that symmetric orbits usually will not undergo period-doubling, and those exceptional cases cannot occur in a large class of systems, including the sinusoidally driven damped oscillators, and the Lorenz model. Experimentally, conservation laws are manifested through the observation of restrictions on transport of chaotic orbits in phase space. Recently, Skiff et al [1988] have presented experiments

304

which demonstrated the conservation of certain integrals of
motion during Hamiltonian chaos. Recently, Coullet, Elphick
and Repaux [1987] presented two basic mechanisms leading to
spatial complexity in one-dimensional patterns, which are
related to Melnikov's theory of periodically driven one-
degree-of-freedom Hamiltonian systems, and Shilnikov's
theory of two-degrees-of-freedom conservative systems [see,
for instance, Guckenheimer and Holmes 1984]. It is
interesting to note that Coullet and Elphick [1987] have
used Melnikov's analysis to construct recurrence time maps
near homoclinic and heteroclinic bifurcations, and they have
shown that an elegent method developed by Kawasaki and Ohta
[1983] to study defect dynamics is equivalent to Melnikov's
theory. Recently, Newton [1988] considered the spatially
chaotic behavior of separable solutions of the perturbed
cubic Schrodinger equation in certain limits. Unfortunately,
the limits under consideration were not applicable to
nonlinear optics, otherwise, its applications would be much
broader.

On the subject of Hamiltonian systems, recently Moser
[1986] gave a very interesting and broad review of recent
developments in the theory of Hamiltonian systems. Although
many real world problems are dissipative, nonetheless, as we
have pointed out in Ch. 2, Hamiltonian systems can give us
some global information on the structure of the solutions.
Hopefully, we can treat the dissipative or dispersive
systems as perturbations of Hamiltonian systems and use KAM
theory to determine the characteristics of the solutions.

On the subject of KAM theory, it is known that the KAM
tori survive small but finite perturbation and are expected
to break at a critical value of the parameter which depends
upon the frequency. Recently, Farmer and Satija [1985] and
Umberger, Farmer and Satija [1986] studied the breakup of a
two-dimensional torus described by a critical circle map for
arbitrary winding numbers. They have demonstrated that such
a breakdown can be described by a chaotic renormalization
group where the chaotic renormalization orbits converge on a

strange attractor of low dimension. And they have found that the universal exponents characterizing this transition are global quantities which quantify the strange attractor. Recently, MacKay and van Zeijts [1988] have found universal scaling behavior for the period- doubling tree in two-parameter families of bimodal maps of the interval. A renormalization group explanation is given in terms of a horseshoe with a Cantor set of two-dimensional unstable manifolds instead of the usual fixed point with one unstable direction. Satija [1987] recently has found that similar results are also valid in Hamiltonian systems, i.e., the breakup of an arbitrary KAM orbit can be described by a universal strange attractor, and the global critical exponents can characterize the breakup of almost all KAM tori.

Numerically, Sanz-Serna and Vadillo [1987] considered the leap-frog (explicit mid-point) discretization of Hamiltonian systems, and it is proved that the discrete evolution preserves the symplectic structure of the phase space. Under suitable restrictions of the time step, the technique is applied to KAM theory to guarantee the boundedness of the computed points.

Dimension is one of the basic properties of an attractor. Farmer, Ott and Yorke [1983] discussed and reviewed different definitions of dimension, and computed their values for typical examples. They have found that there are two general types of definitions of dimension, those that depend only on metric properties and those that depend on the frequency with which a typical orbit visits different regions of the attractor. They have also found that the dimension depending on frequency is typically equal to the Liapunov dimension, which is defined by Liapunov numbers and easier to calculate. An algorithm which calculates attractor dimensions by the correlation integral of the experimental time series was proposed by Grassberger and Procaccia [1983]. Nonetheless, due to nonstationarity of the system, consistent results may require using a small data set and covering short time intervals during which the system can be approximately considered stationary. Under such circumstances, several sources of error in calculating dimensions may result. Recently, Smith [1988] has computed intrinsic limits on the minimum size of a data set required for calculation of dimensions. A lower bound on the number of points required for a reliable estimation of the correlation exponent is given in terms of the dimension of the object and the desired accuracy. A method of estimating the correlation integral computed from a finite sample of a white noise is also obtained. Havstad and Ehlers [1989] have proposed a method to resolve such problems. Eckmann, Ruelle and Ciliberto [1986] discussed in detail an algorithm for computing Liapunov exponents from an experimental time series, and a hydrodynamic experiment is investigated as an example. Recently, Ellner [1988] has come up with a maximum-likelihood method for estimating an attractor's generalized dimensions from time-series data. Meanwhile, Braun et al [1987], Ohe and Tanaka [1988] have investigated the ionization instability of weakly ionized positive columns of helium glow discharges, and the transition to the turbulent state is experimentally observed. By calculating the correlation integral, they have determined that the periodic instability has low dimensionality while turbulent instability has a higher, but finite, dimensionality. Nonetheless, they have not determined the dimensions of strange attractors in their system by computing the Liapunov exponents for the experimental time series. This can be an interesting problem to relate the dimensions obtained via two approaches. Earlier, Farmer [1982] studied the chaotic attractors of a delay differential equation and he has found that the dimension of several attractors computed directly from the definition of dimensions agrees to within experimental resolution of the dimension computed from the spectrum of Liapunov exponents according to a conjecture of Kaplan and Yorke [1979]. Farmer and Sidorowich [1987] have presented a technique which allows us to make short-term predictions of the future behavior of a chaotic time series using information from the past. Recently, Tsang [1986] proposed an analytical method to determine the dimensionality of strange attractors in two-dimensional dissipative maps. In this method, the geometric structures of an attractor are obtained from a procedure developed previously. From the geometric structures, the Hausdorff dimension for the Cantor set is determined first, then for the attractor. The results have compared well with numerical results. Recently Auerbach et al [1987] were able to approximate the fractal invariant measure of chaotic strange attractors by the set of unstable n-periodic orbits of increasing n, and they also presented algorithms for extracting the periodic orbits from a chaotic time series and for calculating their stabilities. The topological entropy and the Hausdorff dimension can also be calculated. For nonlinear and nonstationary time series analysis, see Priestley [1988].

On the other hand, it is important to realize that there are other types of invariant sets in dynamical systems that are not attracting, and these non-attracting invariant sets also play a fundamental role in the understanding of dynamics. For instance, these non-attracting invariant sets occur as chaotic transient sets [Grebogi, Ott and Yorke 1983, Kantz and Grassberger 1985, Szepfalusy and Tel 1986], fractal basin boundaries [McDonald, Grebogi, Ott and Yorke 1985], and adiabatic invariants [Brown, Ott and Grebogi 1987]. In many cases, these invariant sets have complicated, Cantor set-like geometric structures and almost all the points in the invariant sets are saddle points. Hsu, Ott and Yorke [1988] called such non-attracting sets strange saddles. They discussed and numerically tested formula relating the dimensions of strange saddles and their stable and unstable manifolds to the Liapunov exponents and found to be consistent with the conjecture formulated by Kaplan and Yorke [1979].

Recently, Lloyd and Lynch [1988] have considered the

maximum number of limit cycles for a dynamical system of Lienard type: $dx/dt = y - F(x)$, $dy/dt = - g(x)$, with F and g are polynomials. They have found that for several classes of such systems, the maximum number of limit cycles that can bifurcate out of a critical point under perturbation of the coefficients in F and g can be represented in terms of the degree of F and g.

In their classic paper Li and Yorke [1975] posed a question that for some nice class of functions whether the existence of an asymptotically stable periodic point implies that almost every point is asymptotically periodic. Recently, Nusse [1987] established the following two theorems which affirm the question posed by Li and Yorke in part.

Theorem 6.7.2 Let f be a mapping of $C^{1+\alpha}$ for some positive real number α, from a nontrivial interval X into itself. Assume that f satisfies the following conditions: (i) The set of asymptotically stable periodic points for f is compact (if this set is empty, then there exists at least one absorbing boundary point of X for f). (ii) The set of points, whose orbits do not converge to an asymptotically stable periodic orbit of f or converge to an absorbing boundary point of X for f, is a nonempty compact set, and f is an expanding map on this set. Then we have: (a) The set of points whose orbits do not converge to an asymptotically stable periodic orbit of f or to an absorbing boundary point of X for f, has Lebesgue measure zero. (b) There exists a positive integer p such that almost every point in X is asymptotically stable periodic with period p, provided that $f(X)$ is bounded.

As a consequence, the set of aperiodic points for f, or equivalently, the set on which the dynamical behavior of f is chaotic, has Lebesgue measure zero. Furthermore, one can show that the conditions in Theorem 6.7.2 are invariant under the conjugation with a diffeomorphism. Nonetheless, since condition (i) and part of (ii) can not be ascertained a priori, they are not very useful for practical purposes.

<u>Theorem 6.7.3</u> Assume that f is a chaotic C^3-mapping from a nontrivial interval X into itself satisfying the following conditions:

(i) f has a nonpositive Schwarzian derivative, i.e.,
$(d^3 f(x)/dx^3)/(df(x)/dx) - (3/2)[(d^2 f(x)/dx^2)/(df(x)/dx)]^2 \leq$
0 for all x ϵ X with df(x)/dx $\neq$ 0;

(ii) The set of points, whose orbits do not converge to an absorbing boundary point(s) of X for f, is a nonempty compact set;

(iii) The orbit of each critical point for f converges to an asymptotically stable periodic orbit of f or to an absorbing boundary point(s) of X for f;

(iv) The fixed points of f^2 are isolated. Then we have:

(a) The set of points whose orbits do not converge to an asymptotically stable periodic orbit of f or to an absorbing boundary point(s) of X for f, has Lebesgue measure zero;

(b) There exists a positive integer p such that almost every point in X is asymptotically periodic with period p, provided that f(X) is bounded.

In the following, we will give some simple examples:

(A) X = [-1,1], f: X → X is defined by f(x) = 3.701x³ - 2.701x. It can be shown that f has two asymptotically stable periodic orbits with period three. Since f has a negative Schwarzian derivative and df/dx(x=1) = df/dx(x=-1) > 1, by Theorem 6.7.3 we have that almost every point in X is asymptotically periodic with period three.

(B) Let f be a chaotic map of C^3 from a compact interval [a,b] into itself with the following properties [Collet and Eckmann 1980]: (i) f has one critical point c which is nondegenerate, f is strictly increasing on [a,c] and strictly decreasing on [c,b]; (ii) f has negative Schwarzian derivative; (iii) the orbit of c converges to an asymptotically stable periodic orbit of f with smallest period p, for some positive integer p. But since the existence of an asymptotically stable fixed point in [a,f^2(c)) has not been excluded, thus the following cases can occur: (1) f has an asymptotically stable fixed point in

310

[a,f^2 (c)) and f has an asymptotically stable periodic orbit
which contains the critical point in its direct domain of
attraction. (2) f has an asymptotically stable fixed point
in [a,f^2 (c)) and the orbit of the critical point converges
to this stable fixed point. Moreover f has 2^n unstable
periodic points with period n for each positive integer n.
(3) f has no asymptotically stable fixed point in [a,f^2 (c));
consequently, the critical point is in the direct domain of
attraction of the asymptotically stable periodic orbit.
Since the map f satisfies the conditions of Theorem 6.7.3
and we have that almost each point in the interval [a,b] is
asymptotically periodic with period p.

For one-dimensional maps $x_{n+1} = mF(x_n)$, there are four
kinds of bifurcation likely to take place in general,
namely: (a) regular period doubling, (b) regular period
halving, (c) reversed period doubling, and (d) reversed
period halving. It is known that if the Schwarzian
derivative is negative at the bifurcation point, then either
(a) or (b) can take place, (if the Schewarzian derivative is
positive, then either (c) or (d) will take place) [Singer
1978, Guckenheimer and Holmes 1983, Whitley 1983]. Thus,
having a negative Schwarzian derivative is a condition which
implies that at the parameter value where a period doubling
bifurcation occurs, the periodic orbit is stable due to
nonlinear terms and prevent (c) and (d) to occur. There were
questions raised as to whether or not having a negative
Schwarzian derivative ruled out regular period halving.
Recently, Nusse and Yorke [1988] presented an example of a
one-dimensional map where F(x) is unimodal and has a
negative Schwarzian derivative. They showed that for their
example, some regular period halving bifurcations do occur,
and the topological entropy can decrease as the parameter m
is increased.

So far we have been discussing bifurcations of nonlinear
continuous or discrete systems. George [1986] has
demonstrated that bifurcations do occur in a piecewise
linear system.

311

Before we end this section on chaos, we would like to point out that it is fairly simple to demonstrate deterministic chaos experimentally. Briggs [1987] gives five simple nonlinear physical systems to demonstrate the ideas of period doubling, subharmonics, noisy periodicity, intermittency, and chaos in a teaching laboratory, such as in senior or graduate laboratory courses.

So far we have only scratched the surface of chaos, nonetheless, we hope that we have provided some sense of the richness and diversity of the phenomenology of chaotic behavior of dynamical systems. It will remain to be an exciting field for sometime to come, because not only there will be new phenomena to be discovered, but more importantly, there will be many surprises along the way.

For instance, Grebogi, Ott and Yorke [1982] have investigated the chaotic behavior of a dynamical system at parameter values where an attractor collides with an unstable periodic orbit, specifically, the attractor is completely annihilated as well as its basin of attraction. They call such events crises, and they found sudden qualitative changes taken place and the chaotic region can suddenly widen or disappear. It is well-known that in periodically driven systems, the intersection of stable and unstable manifolds of saddle orbits forms two topologically distinct horseshoes, and the first one is associated with the destruction of a chaotic attractor, while the other one creates a new chaotic attractor. Abrupt annihilation of chaotic attractors has been observed experimentally in a driven CO_2 laser and p-n junction circuits. Recently, Schwartz [1988] used a driven CO_2 laser model to illustrate how sequential horseshoe formation controls the birth and death of chaotic attractors. Hilborn [1985] reported the quantitative measurements of the parameter dependence of the onset of an interior crisis for a priodically driven nonlinear diode-inductor circuit. The measurements are in reasonable agreement with the predictions of Grebogi, Ott and Yorke [1982]. Grebogi, Ott and Yorke [1986] recently

312

have also found a theory of the average lifetime of a crisis for two- dimensional maps. Recently, Yamaguchi and Sakai [1988] studied the structure of basin boundaries for a two-dimensional map analytically, and they have shown that basin boundaries do not play the role of a barrier when the crisis occurs as the two parameters are changed.

The studies of nonlinear dynamical systems or maps modeling these systems have been concerned with how chaos arises through successive series of instabilities. The converse questions are: is there a nonlinear system or map whereby at certain control parameters the dynamics of the system can lead from chaotic states to ordered states? or how can the interaction of chaotic attractors lead to ordered states? Recently, Phillipson [1988] has demonstrated the emergence of ordered states out of a chaotic background of states by suitably contrived one-dimensional maps characterized by multiple critical points. Certainly similar considerations of maps of higher dimensionality will be investigated. Bolotin, Gonchar, Tarasov and Chekanov [1989] recently have studied the correlations between the characteristic properties of the classical motion and statistical properties of energy spectra for two-dimensional Hamiltonian with a localized region having a negative Gaussian curvature of the potential energy surface. They have found that at such potential energy surface with negative Gaussian curvature, the transition of regularity-chaos-regularity occurs. Further studies of a more general nature will be of great interest. It should be pointed out that the establishment of order out of chaos implies the imposition of rules on an otherwise unspecified situation. This is the basis of evolutionary processes in nature [Farmer, Lapedes, Packard and Wendroff 1986].

One can easily show or construct nonlinear dynamical systems with multiple attractors and consequently require very accurate initial conditions for a reliable prediction of the final states. This is the issue raised by Grebogi, McDonald, Ott and Yorke [1983]. We will leave this section

313

with this issue as food for thought.

6.8   A new definition of stability
As we have discussed earlier, there are several notions
of stability, and structural stability is intuitively easy
to understand and has some nice attributes. Yet we have also
noted that structural stability is in some sense a failure
because structurally stable systems are not dense for
dimensions ≥ 3. It is true that Anosov systems, AS systems,
and Morse-Smale systems (or hyperbolic strange attractors)
are structurally stable, but they are rather special, and
most strange attractors and chaotic systems which appear in
applications (such as the Lorenz model) are not structurally
stable.

We have also pointed out that there have been several
attempts to define stability such that the stable systems
are dense, but they have failed. Recently, Zeeman [1988]
proposed a new definition of stability for dynamical systems
which is particularly aimed at nonlinear dissipative
systems. There are several advantages of this new definition
of stability than that of structural stability. For
instance, the stable systems (in this new definition) are
dense, therefore most strange attractors are stable,
including non-hyperbolic ones. This approach offers an
alternative to structural stability. As we shall see, to a
certain extent, it is a complement to structural stability.

In the following, we shall briefly discuss Zeeman's
stability (for simplicity, we shall simply call it
stability), its advantages over and its differences from
structural stability, some examples, and some difficulties.
To limit our discussion, we shall only discuss the stability
of vector field and flows, we shall leave the
diffeomorphisms for the readers to read the paper by Zeeman.

Let M be a smooth oriented n-dimensional Riemannian
manifold. Let R_+ denotes the non-negative reals. Given a
smooth vector field v on M, and given $\epsilon > 0$, the

314

Fokker-Planck equation for v with ϵ-diffusion is the partial differential equation on M

   $u_t = \epsilon u_{xx} - \text{div}(uv)$,   (6.8-1)

where $u: M \times R_+ \to R_+$, $u(x,t) \geq 0$, $x \in M$, $t \geq 0$ and the integral of u over M equals to unity. Here u_{xx} is the Laplacian of u, and the divergence are determined by the Riemannian structrue (see Section 2.4). The function u represents the smooth probability density of a population on M driven by v and subject to ϵ-small perturbation. For further details on Fokker-Planck equation, see any probability theory or statistical mechanics books, see also Risken [1984].

   Let $u^{v,\epsilon}: M \to R_+$ denote the steady-state solution of Eq.(6.8-1). If M is compact (for noncompact M a suitable boundary condition on v will be required for the existence of u), it has been established for the existence, uniqueness, and smoothness of the steady-state solution of Eq.(6.8-1). As before, next we shall define the equivalence relations. Two smooth functions u, u': $M \to R$ are said to be <u>equivalent</u>, u - u', if there exist diffeomorphisms α, β of M and R such that the following diagram commutes:

$$\begin{array}{ccc} M & \xrightarrow{u} & R \\ \alpha \downarrow & & \downarrow \beta \\ M & \xrightarrow{u'} & R. \end{array}$$

We say a function is <u>stable</u> if it has a neighborhood of equivalents in the space of all smooth functions on M with C^∞ topology. We define two smooth vector fields v, v' on M to be <u>ϵ- equivalent</u> if $u^{v,\epsilon}$ - $u^{v',\epsilon}$. A vector field is said to be <u>ϵ-stable</u> if it has a neighborhood of ϵ-equivalents in the space X(M) of all smooth vector fields on M with C^∞ topology. A vector field is <u>stable</u> if it is ϵ-stable for arbitrarily small $\epsilon > 0$. Zeeman [1988] has shown that ϵ-stable vector fields are open and dense, and stable vector fields are dense (but not necessarily open).

   As an example, let M = R, v(x) = -x. Then the

315

Fokker-Planck equation becomes $u_t = \epsilon u_{xx} + (ux)_x$. Then we can solve the steady- state equation directly and we obtain
$$u = (A\int e^{y^2/2\epsilon}dy + B)e^{-x^2/2\epsilon}$$
where A and B are constants. If $A < 0$ then the bracket tends to $-\infty$ as $x \to \infty$, and so $u < 0$ for sufficiently large values of x. And if $A > 0$ then $u < 0$ for sufficiently large negative values of x. Therefore, $u \geq 0$ implies $A = 0$. And the condition of unit probability determines the value of B, and we have $u = e^{-x^2/2\epsilon}/2\pi\epsilon$. Clearly, the steady-state solution is just the normal distribution. Intuitively we can see that if there were no diffusion ($\epsilon = 0$), then the whole population would be driven towards the origin, approaching the Dirac delta function as $t \to \infty$. When $\epsilon > 0$, the diffusion term pushes the population away from the origin, opposing the incoming drive term until they reached a balance in the normal distribution.

Since the circle is the simplest nontrivial compact manifold, let us look at some examples of vector fields on S^1. Here we shall not get into the detailed construction of the probability density u(x), but we shall consider some special cases qualitatively.

Let v be a vector field on S^1 given by v: $S^1 \to R$. If $v > 0$ the flow of v is a cycle. The invariant measure of the flow is k/v, where k is a constant, chosen in such a way that $k/v = 1$. Furthermore, if v has only two critical points, a minimum at p and a maximum at q. In the situation of the pendulum, p is the position where the pendulum stands vertically upward and nearly at rest (highest potential energy but lowest kinetic energy), and q is the position where the pendulum hangs vertically downward with maximum velocity (lowest potential energy but highest kinetic energy). Then the invariant measure will have a maximum at p where the flow lingers longest and a minimum at q where the flow is the most rapid as shown in Fig.6.8.1(a). The steady state of the Fokker-Planck equation is an ϵ-approximation of the measure and qualitatively the same, with maximum and minimum occurring slightly before the flow reaches p and q respectively. We have discussed this in Ch.1 in detail.

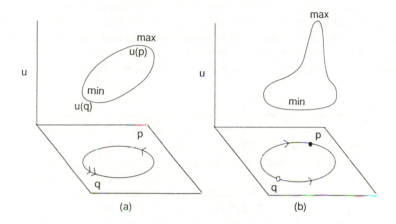

Fig.6.8.1 [Zeeman 1988]

Now consider the case where v has an attractor at p and
a repellor at q, as shown in Fig.6.8.1(b). The invariant
measure is the delta function at p, and the steady-state of
the Fokker-Planck equation will resemble a normal
distribution near p. If v is a gradient field, then the
maximum and minimum will be at p and q, but if v is
non-gradient, then the maximum will be near p but the
minimum may not be anywhere near q.

Note that the two steady-states shown in Fig.6.8.1 are
qualitatively similar, and by the (new) definition of
equivalence, they are equivalent. And indeed, they are ϵ-
equivalent. It is then nature to ask: if we take a
parametrized family of vector fields going from one to the
other, are all the members of this family equivalent? If so,
does this mean that the resulting bifurcation must be
stable? Surprisingly, both answers are affirmative.

Fig.6.8.2 shows two fold bifurcations, in each case a
sink (attractor) coalesces with a source (repellor) r at a
fold point f and disappears (see Section 6.3). Note that,

317

suppose one of the zeros p is non-hyperbolic, then a
C^∞-small perturbation of the vector field near p can alter p
to a source or to a sink. But since we leave all the other
original zeros of v unaltered, we have changed the number of
sources in either case. That was the reason behind Theorem
6.3.1. Thus, this type of system is not structurally stable.
Zeeman [1988] gives a slightly different argument.

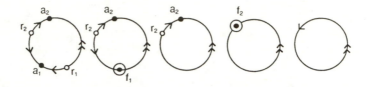

Fig.6.8.2 [Zeeman 1988]
     It appears that any two limit cycles in the plane are ϵ-
equivalent. In fact, the flows on any two limit cycles are
topologically equivalent. The following example will
illustrate that two limit cycles in the plane may not be
ϵ-equivalent.
     Fig.6.8.3(a) shows a limit cycle on which the flow is as
in Fig.6.8.1(a), with a source inside. This situation would
arise from the following dynamical equations in polar
coordinates [Zeeman 1981]:
     $dr/dt = r(1 - r)$, $d\theta/dt = 2 - r\cos\theta$.
The resulting steady-state resembles a volcano crater, with
the rim of the crater above the limit cycle, with a maximum
at p and a saddle at q on the lip of the crater and a
minimum at the source. Let us compare this with the
hysteresis cycle on a cusp catastrophe given by the
equations in Cartesian coordinates:
     $dx/dt = -y/k$, $dy/dt = k(x + Y - Y^3)$,   where $k \gg 0$.
This is the same as the van der Pol oscillator with large
damping: $d^2x/dt^2 + k(3x^2 - 1)dx/dt + x = 0$.
Here the limit cycle has two branches of slow manifold
separated by the catastrophic jumps between the two branches

318

of slow manifold as shown in Fig.6.8.3(b). The resulting
steady-state resembles two parallel mountain ridges with
maxima at p_1 and p_2 near the fold points of the
catastrophes, saddles at q_1 and q_2, and a minimum at the
source inside.

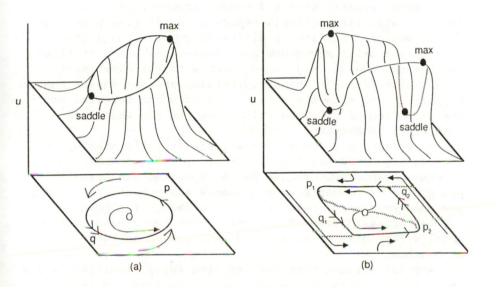

Fig.6.8.3 [Zeeman 1988]
    It is obvious that the two flows are topologically
equivalent but differ in character. But the two
steady-states are not equivalent, thus the two vector fields
are not ϵ-equivalent.
    The above three examples illustrate the difference
between stability and structural stability. Fig.6.8.1
illustrates two flows that are ϵ-equivalent but not
topologically equivalent, while Fig.6.8.3 illustrates two
flows that are topologically equivalent but not
ϵ-equivalent. The second bifurcation in Fig.6.8.2 is an

319

example of a flow that is stable but not structurally stable. Conversely, in a one-parameter family of flows joining those in Fig.6.8.3, there will be a bifurcation where the flow is structurally stable but not stable.

Roughly speaking, structural stability captures the topological properties of the system like basins of attraction, separatrices and saddle connections, while Zeeman's stability criteria captures qualitative properties of the measure. Putting it differently, structural stability retains the orbit structure but ignores the speed of flow and the smoothness, and loses touch with the experimental data, while Zeeman's stability retains the smoothness and keeps in touch with the data but ignores the dynamics and the direction of flow. In this sense, we say that they complement each other.

As we have mentioned earlier, structurally stable systems are not dense for dimensions ≥ 3 in general, while Zeeman's stable systems are dense. Thus the qualitative properties of Zeeman's stable models are robust, i.e., they are preserved under perturbations. Although the phenomenon may be a perturbation of the model, the robustness of the model will remain to be a valid description of the phenomenon.

Another serious criticism of structural stability is its lack of smoothness. As we have noticed before, in its definition, it is necessary that the equivalence relation be a homeomorphism (for topological conjugacy and equivalence, see, Section 4.2) rather than a diffeomorphism. For example, Fig.4.1.1(a) is topologically equivalent to it with a rotation so that it spirals out. Clearly the equivalence is only subject to homeomorphism but not diffeomorphism. The importance of smooth equivalence relations is that if we can prove a given situation is smoothly equivalent to a standard model, then all the smooth qualitative properties of the standard model are automatically carried over to the given situation. But if a given situation is only topologically equivalent to a standard model, then any smooth qualitative

properties of the given situation has to be stated and
proved separately.

It should be noted that Zeeman's stability can not
distinguish between volume-preserving flows. This is because
the theory is designed for dissipative systems, and is not
suitable for Hamiltonian systems. But nor is structural
stability useful to study Hamiltonian systems because all
Hamiltonian systems are structurally unstable.

Section 4 of Zeeman's paper develops an analogous theory
for the stability of diffeomorphisms. It has been pointed
out that this theory is not as elegant as that for flows.
The next two sections of Zeeman's paper prove the existence,
uniqueness, and classification theorems.

It is clear that these results are the beginning of a
very interesting and fruitful undertaking. At the end of the
paper, Zeeman suggested some open questions for further
research:
(a) Extend the results for flows to non-compcat manifolds
with suitable boundary conditions;
(b) Prove the analogous density and classification theorems
for diffeomorphisms;
(c) Develop a theory of unfoldings of unstable vector fields
of finite codimensions;
(d) Investigate the stability of specific strange
attractors, both Axiom A and non-Axiom A, such as Lorenz and
Henon;
(e) Extend the results from C^∞ to C^r, $r \in [1, \infty)$.

In these broad classes of problems, the last two classes
are of immediate impact on applications. Indeed, some
strange attractors, such as Lorenz, Cantor-like, Henon, van
der Pol, Duffing, coupled-attractors, etc. must be studied
with Zeeman's criteria for stability and classification.

Of course, with better understanding of this stability
criteria, a even more general and more satisfying definition
of stability may evolve. Examples and practical problems may
provide enough imputus for mathematicians to construct such
more satisfying criteria for stability.

321

Appendix

In this appendix, we shall use the Lorenz model to illustrate the center manifold theorem and its applications for bifurcation calculations. For more detail, see, e.g., Guckenheimer and Holmes [1983], Carr [1981].

We shall study the bifurcation of the Lorenz equation, Eq.6.7.1, occurring at the origin and $\alpha = 1$. The Jacobian matrix at the origin is

$$\begin{bmatrix} -\sigma & \sigma & 0 \\ \alpha & -1 & 0 \\ 0 & 0 & -\beta \end{bmatrix}.$$

When $\alpha = 1$, this matrix has eigenvalues 0, $-\sigma - 1$, and $-\beta$ with eigenvectors $(1, 1, 0)$, $(\sigma, -1, 0)$, and $(0, 0, 1)$. By using the eigenvectors as a basis for a new coordinate system, we have

$$\begin{bmatrix} u \\ v \\ w \end{bmatrix} = \begin{bmatrix} 1/(1 + \sigma) & \sigma/(1 + \sigma) & 0 \\ 1/(1 + \sigma) & -1/(1 + \sigma) & 0 \\ 0 & 0 & 1 \end{bmatrix} \begin{bmatrix} x \\ y \\ z \end{bmatrix}.$$

Then the Lorenz equation becomes

$$\begin{aligned}
du/dt &= -\sigma(u + \sigma v)w/(1 + \sigma), \\
dv/dt &= -(1 + \sigma)v + (u + \sigma v)w/(1 + \sigma), \\
dw/dt &= -\beta w + (u + \sigma v)(u - v),
\end{aligned} \tag{A-1}$$

and in matrix form we have

$$\begin{bmatrix} du/dt \\ dv/dt \\ dw/dt \end{bmatrix} = \begin{bmatrix} 0 & 0 & 0 \\ 0 & -(1+\sigma) & 0 \\ 0 & 0 & -\beta \end{bmatrix} \begin{bmatrix} u \\ v \\ w \end{bmatrix} + \begin{bmatrix} -\sigma(u + \sigma v)w/(1 + \sigma) \\ (u + \sigma v)w/(1 + \sigma) \\ (u + \sigma v)(u - v) \end{bmatrix}.$$

Now the linear part is in standard form. In the new coordinates, the center manifold is a curve tangent to the u-axis. The projection of the system onto the u-axis yields $du/dt = 0$. Nonetheless, the u-axis is not invariant because the equation for dw/dt contains u^2 term. If we make a nonlinear coordinate transformation by setting $w' = w -$

u^2/β, we have

$$dw'/dt = -\beta w' + (\sigma - 1)uv - \sigma v^2$$
$$+ 2\sigma u(u + \sigma v)(w' + u^2/\beta)/(\beta(1 + \sigma)).$$

Thus, in the new coordinate system (u, v, w'), we have

$$du/dt = -\sigma(u + \sigma v)(w' + u^2/\beta)/(1 + \sigma).$$

Now the projection of the system onto the u-axis in the new coordinate system gives $du/dt = (-\sigma/\beta(1 + \sigma))u^3$. Note also that there is no u^2 term in any of the equations for v and w'. Thus, the u-axis in the new coordinate system is invariant up to second order. Further efforts to find the center manifold can proceed by additional coordinate transformations which serve to make the u- axis invariant for the flow iteratively by changes in v and w'.

Chapter 7   Applications

7.1   Introduction

Following the discussion at the end of Section 6.5, we would like to point out the interaction of spatial and temporal patterns, then follow with specific illustrations. Suppose that the dynamics of a physical system is governed by the system of evolution equations

$$\partial v / \partial t = F(v, \mu) \qquad\qquad (7.1\text{-}1)$$

where v is an element of an appropriate Banach space, $\mu \in R$ is a real bifurcation parameter, and F is a nonlinear operator defined on a neighborhood of the origin satisfying $F(0,0) = 0$. We assume that the linearized operator

$$A = D_v F(0,0) \qquad\qquad (7.1\text{-}2)$$

has a simple zero eigenvalue and, addition, a simple pair of imaginary eigenvalues $\pm i\Omega_o$ (where $\Omega_o > 0$). The remaining spectrum of A is assumed to be to the left of the imaginary axis. Eq.(7.1- 1) has the stationary solution $v = 0$ for $\mu = 0$. When the externally controllable bifurcation parameter μ is varied away from zero, then due to the nonlinearity of F, two basic types of solutions bifurcate from the trivial one. They are the steady-state solutions associated with the zero eigenvalue, and time-periodic or Hopf solutions associated with the eigenvalues $\pm i\Omega_o$ of A. The nonlinearity of F causes these two solutions to interact and since they tend to the trivial solution $v = 0$ for $\mu \to 0$, F has a degenerate bifurcation at $(0,0)$. The degeneracy can be removed by subjecting F to small perturbations, representable by additional imperfection parameters σ in F itself, $F \to F_\sigma$. This is achieved by stably unfolding the algebraic bifurcation equations to which Eq.(7.1-1) will be reduced. Then, as the unfolding parameters (functions of σ) are varied, zero and imaginary eigenvalues occur for different values of μ and, with the degeneracy so removed, new bifurcation phenomena which are structurally stable spring up.

Since the linearization of Eq.(7.1-1) at $(0,0)$ has

$(2\pi/\Omega_0)$-periodic solutions, we seek periodic solutions of Eq.(7.1-1) near $(0,0)$ with period $2\pi/\Omega$ where Ω is close to Ω_0. Setting $s = \Omega t$, $u(s) = v(s/\Omega)$ so that u has period 2π in s. We can rewrite Eq.(7.1-1) as

$$N(u,\mu,\tau) = \tau du/ds + Lu - R(u,\mu) = 0, \qquad (7.1\text{-}3)$$

here $\tau = \Omega - \Omega_0$, $L = \Omega_0 d/ds - A$ and $R(u,\mu) = F(u,\mu) - Au$. In the space of (2π)-periodic vector-valued functions $u = u(s)$ the linear operator L has a three-dimensional nullspace spanned by the eigenfunctions $f = (f_1, f_2, f_3)$. We can reduce the bifurcation problem Eq.(7.1-3) to an algebraic one, i.e., the degenerate algebraic system of bifurcation equations

$$G(x,\mu,y) = \begin{bmatrix} a(x,\mu,y^2) \\ yb(x,\mu,y^2) \end{bmatrix} = 0 \qquad (7.1\text{-}4)$$

with $a(0,0,0) = b(0,0,0) = 0$, $a_x(0,0,0) = 0$. Eq.(7.1-4) describes the Z_2-covariant interaction between the Hopf and steady-state solutions of Eq.(7.1-1).

The multivalued solutions of Eq.(7.1-4) are the bifurcation diagrams in (x,μ,y)-space. We classify them together with their stable perturbations by means of imperfect bifurcation theory. Eq.(7.1-4) possess two coupled types of solutions, viz., pure steady-state solutions with amplitude x determined by $\{a(x,\mu,0) = 0, y = 0\}$, and periodic solutions with $y \neq 0$ obtained by the simultaneous solution of the equations $a(x,\mu,y^2) = 0$, $b(x,\mu,y^2) = 0$. The periodic solutions branch from the steady-state at a secondary Hopf bifurcation point and may further undergo tertiary bifurcations to tori [Armbruster 1983]. By changing coordinates so that the qualitative topology of the bifurcation diagram $G = 0$ is preserved, the special role of the externally controllable bifurcation parameter μ is respected, and G takes the simple polynomial forms given in Armbruster [1983], from which the solutions of $G = 0$ may easily be determined.

To classify all the possible stable and inequivalent, i.e., qualitatively different, bifurcation diagrams that may arise when a given $G(x,\mu,y)$ is subjected to small

perturbations which correspond to imperfections in F, the
universal unfoldings of G are determined. The unfolding
parameters of a universal unfolding of G are functions of
the imperfection parameters σ in $F_\sigma(v,\mu)$. Their number, the
contact codimension of G, is a measure of the degree of
complexity of the singularity. Hence, unfolding G displays
the effects of all imperfections. The result is a finite
list of generic perturbed bifurcation diagrams describing
interacting Hopf (H) and steady-state problems (S). Of major
interest for applications are special points in the
bifurcation diagrams, viz., limit points and secondary
bifurcation points (SB) which are here all Hopf bifurcation
points, and tertiary bifurcations (T) from the Hopf branch
to a torus. The stability properties are indicated by
assigning to each branch of a diagram its stability symbol
(-- is stable, etc.), i.e., the signs of the real parts of
the eigenvalues of the Jacobian DG. Fig.7.1.1 shows the
simplest secondary bifurcation of a Hopf branch (H) from a
steady-state in the (x,μ)-plane associated with the normal
from a $= x^2 + \epsilon_2 y^2 + \mu = 0$, b $= y(x - \alpha) = 0$. Fig.7.1.2
shows two Hopf branches bifurcating from a steady-state with
bistability. In Fig.7.1.3 a tertiary bifurcation point T
appears where a transition to a double-periodic solution
occurs. There are many more diagrams exhibiting a variety of
new phenomena such as gaps in Hopf branches, hysteresis
between Hopf and steady-state branches, periodic solutions
coming out of nowhere, i.e., not connected to steady-states,
and so on. As an application we shall discuss the problem of
optical bistability in Section 7.3.

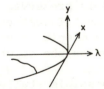

Fig.7.1.1. Secondary Hopf bifurcation (SB) emerging from a
steady state

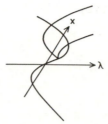

Fig.7.1.2. Simultaneous Hopf bifurcations in y-direction
           originating at a hystersis branch in the
           (x,λ)-plane

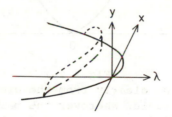

Fig.7.1.3. Hopf-steady-state interaction leading to a torus
           point T

7.2 Damped oscillators and simple laser theory

For a damped anharmonic oscillator with mass m, damping
constant Γ, restoring forces $-\alpha x - \beta x^3$, the equation of
motion (non-autonomous system) is

$$m\,d^2x/dt^2 + \Gamma\,dx/dt = -\alpha x - \beta x^3. \qquad (7.2-1)$$

The restoring force $k(x) \equiv -\alpha x - \beta x^3$ possesses a potential

$$k(x) = -\partial V/\partial x, \qquad (7.2-2)$$

where

$$V(x) = \alpha x^2/2 + \beta x^4/4. \qquad (7.2-3)$$

The potential can be plotted as a function of x for
different values of α and β. Fig.7.2.1 compares the usual
quadratic potential with the fourth order potential for a
given pair of α and β. Because of this potential, we can

easily discuss its stability.

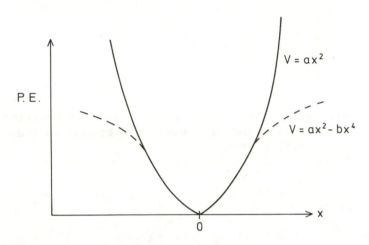

Fig.7.2.1

In the case $\beta > 0$, clearly from the graph of V vs x, the system is globally stable. Wherever the particle starts, it comes to rest at a finite value of x. On the other hand, we have global instability for $\beta < 0$. Now let us look at the local stability. It is suffice to consider only $\beta > 0$ case, as we notice the "symmetry" between $\beta > 0$ and $\beta < 0$. Let us first consider the steady state which is characterized by:

$$d^2 x/dt^2 = dx/dt = 0. \qquad (7.2-4)$$

The states of stable and unstable equilibria are defined by

$$\alpha x + \beta x^3 = 0. \qquad (7.2-5)$$

For $\alpha > 0$, we have the stable solution

$$x_o = 0. \qquad (7.2-6)$$

Mathematicians call this an "attractor". For $\alpha < 0$, the state $x_o = 0$ becomes unstable ("repeller"), and instead, we have two stable solutions

$$x_{1,2} = \pm(|\alpha/\beta|)^{\frac{1}{2}}. \qquad (7.2-7)$$

With the transition from $\alpha > 0$ to $\alpha < 0$ the system passes through an instability and the particle is now either at x_1 or x_2. In fluid dynamics, the change of one stability to

328

another is called an exchange of stabilities. In mathematics, the passing over from one stability to two new ones is known as Hopf bifurcation. The state where $\alpha = 0$, $x = 0$ is a marginal state, i.e., bifurcation point.

As one knows from experience in all areas of physical, biological, and social sciences, one finds that fluctuation must usually incorporate into the parameter equations. Let us assume the particle gets perturbative forces of equal magnitude but random in forward or backward directions. Let us add to Eq. (7.2- 1) the random force

$$\phi(t) = \eta \, \Sigma_j \, (-1)^n \, \delta(t - t_j), \qquad (7.2-8)$$

where η stands for the magnitude of the force, n_j is a random variable can either be 0 or 1 so that it gives the direction of the random force, t_j is a random time sequence, and δ is the usual Dirac δ-function. Thus Eq. (7.2-1) becomes

$$m d^2 x/dt^2 + \Gamma dx/dt = -\alpha x - \beta x^3 + \phi(t). \qquad (7.2-9)$$

As we have pointed out earlier, one can transform Eq. (7.2-9) into a system of two first order equations. To illustrate some of the features we would like to display, for simplicity we assume that the oscillator is heavily damped. In this case, we can formally set $m = 0$ in Eq. (7.2-9) which becomes

$$dx/dt = ax - bx^3 + F(t) \qquad (7.2-10)$$

where $a = \alpha/\Gamma$, $b = \beta/\Gamma$, $F(t) = \phi(t)/\Gamma$. One can discuss this equation from various viewpoints. If we assume the fluctuation is of the form Eq. (7.2-8), i.e.,

$$F(t) = F_0 \, \Sigma_j \, (-1)^{n_j} \delta(t - t_j), \qquad (7.2-11)$$

then we have $<F(t)> = 0$. One can also show that the correlation function

$$<F(t)F(t')> = (F_0^2/t_0) \, \delta(t-t') = C \, \delta(t-t'), \qquad (7.2-12)$$

where t_0 is the mean time between perturbative impulses. To analyze Eq. (7.2-12), we first discuss Eq. (7.2-10) with $F(t) = 0$. Then the time-dependent solutions of $dx/dt = -ax - bx^3$ are

$$x(t) = \pm \, a^{1/2} \{\exp[2a(t - t')] - b\}^{-1/2}, \quad \text{for } a > 0 \qquad (7.2-13)$$

and

$$x(t) = \pm |a|^{1/2} \{\exp[-2|a|(t - t')]\}^{-1/2}, \quad \text{for } a < 0. \qquad (7.2-14)$$

329

Here a has the meaning of an inverse relaxation time for the
system. In either cases $a > 0$ or $a < 0$, x tends to its
equilibrium value. Because an explicit solution can not be
found for more general cases of order-parameter equations,
as usual, we must discuss local stability by linearization.
Define the steady state coordinate x_s by $dx_s/dt = 0$ and
assume $x = x_s + \delta x$. First linearize Eq.(7.2-10) about $x_s = 0$
which yields

$$\delta(dx/dt) = -a\delta x, \qquad (7.2\text{-}15)$$

with the solution $\delta x = A \exp(-at)$.   (7.2-16)
For $a > 0$ the system is stable, $a < 0$ unstable, and for $a = 0$ it is a marginal state. When the relaxation constant a
approaches to 0, we have the critical slowing down
phenomena. If $a < 0$, the coordinate of the stable point is

$$|x|_s = (|a|/b)^{\frac{1}{2}}. \qquad (7.2\text{-}17)$$

By inserting $x = x_s + \delta x$ (7.2-18)
into $dx/dt = -ax - bx^3$, we have

$$dx/dt = dx_s/dt + d(\delta x)/dt$$
$$= - a\delta x - 3b|a|\delta x/b - x_s(a + b|a|/b)$$
$$= - 2|a|\delta x, \qquad (7.2\text{-}19)$$

this is because $a < 0$, $a + |a| = 0$. Thus yields the
relaxation time

$$\tau = (2|a|)^{-1}. \qquad (7.2\text{-}20)$$

Now consider the fluctuation of x in the linearized theory
and solve the equation

$$d(\delta x)/dt + a\delta x = F(t) \qquad (7.2\text{-}21)$$

which yields

$$\delta x = \exp(-at)\int_{t_0}^{t} \exp(a\tau)F(\tau)\, d\tau. \qquad (7.2\text{-}22)$$

The correlation function of the coordinate gives a measure
for the temporal behavior of the system. Substituting
Eq.(7.2-22) into $<\delta x(t)\delta x(t')>$ yields (with Eq.(7.2-12))

$$<\delta x(t)\delta x(t')> = C \exp[-a(t - t')]/2a, \quad t \geq t' \qquad (7.2\text{-}23)$$

for $t_0 \rightarrow -\infty$. From Eq.(7.2-23) it is clear that as $a \rightarrow 0$ not
only the relaxation time τ , but also the coordinate
fluctuation become infinite. It is important to point out
that <u>the divergence of Eq.(7.2-23) for $a \rightarrow 0$ is caused by
linearization procedure</u>. In other words, while the

330

fluctuation δx for $a < 0$ or $a > 0$ are finite and can be neglected in most cases, the linearization procedure breaks down near the point $a = 0$. However, in the exact theory, it is still true that at the critical point $a = 0$ the fluctuation in δx becomes very large.

At any rate, the concepts of critical fluctuation, critical slowing down, symmetry breaking, etc. are part of the standard repertoire of phase-transition theory.

In statistical mechanics, Eq.(7.2-10) can be considered as an extension of the Langevin equation of Brownian motion
$$dv/dt = - av + F(t). \tag{7.2-24}$$
It is quite simple to solve the linearized form of Eq.(7.2-10), i.e., Eq.(7.2-24), but the solution of Eq.(7.2-10) becomes a formidable task even for this simple case with nonlinearity kept. One may want to proceed to the Fokker-Planck equation given by:
$$df(q,t)/dt = -\partial(K(q)f)/\partial q + \tfrac{1}{2}\partial^2 (Q(q)f)/\partial q^2 . \tag{7.2-25}$$
Here $f(q,t)dq$ is the probability of finding the particle in the interval $(q, q+dq)$ at a time t. The drift coefficient $K(q)$ and diffusion coefficient $Q(q)$ are defined by:
$$K(q) = \lim_{t \to 0} (1/t)<q(t) - q(0)> \tag{7.2-26}$$
and $Q(q) = \lim_{t \to 0} (1/t)<(q(t) - q(0))^2>. \tag{7.2-27}$
One has to imagine that Eq.(7.2-10) in this context is solved for a time interval which comprises many pushes of $F(t)$ but small compared with the overall motion of the system. In the present case one readily finds that $K(q)$ is identical with the force $k(x)$ in Eq.(7.2-2), i.e.,
$$K(q) = - aq - bq^3 = - \partial V/\partial q \tag{7.2-28}$$
and $Q(q) = C, \tag{7.2-29}$
where C is defined as the coefficient in the correlation function Eq.(7.2-12). The Fokker-Planck Eq.(7.2-25) then reads
$$df(q,t)/dt = - \partial [(- \partial V/\partial q)f - \tfrac{1}{2}C \, \partial f/\partial q]/\partial q. \tag{7.2-30}$$
This equation has the form of a conservative law. Let the probability current be denoted by j, then we have
$$df/dt = - \partial j/\partial q. \tag{7.2-31}$$
In the stationary case, $f = 0$ and we find the solution

$f = \eta \exp(-2V/C) = \eta \exp[-(2/L)(aq^2/2 + bq^4/4)]$,   (7.2-32)
here we have taken into account that f vanishes at infinity. The distribution function f is of great importance because it governs the stability, the fluctuation, and the dynamics of the system.

(a) For stability, a simple comparision with our previous considerations reveals that these systems are globally stable whereby f is normalizable. V in Eq.(7.2-32) serves as a Liapunov function V_L (let $V_L = V(q) - V(q_0)$) which satisfies the following conditions:

(i) $V_L(q)$ is C^1 in a region surrounding $q_0 \in D$,
(ii) $V_L(q_0) = 0$,
(iii) $V_L(q) > 0$ in D,
(iv) $dV_L/dt = k(q)\, \partial V/\partial q \le 0$.

Liapunov's theorem then states that: If there exists such a Liapunov function V_L in D, then $q_0 \in D$ is stable.

(b) For instance, by expanding the exponent $V(q)$ in f in Eq.(7.2-32) about the steady state, using $q = q_s + \delta q$, one get an expression for the probability of finding a fluctuation of the size δq,

$f(q) = \eta \exp\{-2[V(q_s)/C + \partial^2 V(q)/\partial q^2 |_q (\delta q)^2/2C]$.   (7.2-33)

(c) With $V(q)$, one can develop the dynamics of the system. We shall not go into this in detail.

With proper replacement of variables, one can study the thermodynamics of the system, for instance, one can obtain the relation between fluctuation and dissipation, Einstein's relation for the probability $W(q)$ for a fluctuation of size δq for Brownian motion, as well as demonstrate the Landau theory of phase transitions. So far we have discussed a simple system in "thermal" equilibrium. One can also demonstrate by explicit examples which lead to equations of the form Eq.(7.2-10) or their generalizations to many degrees of freedom for systems which are far from thermal equilibrium.

In different cases, x (or generalized coordinates q) may represent very different quantities, e.g., the laser light field, electric current, velocity field for fluids,

concentrations of chemical reactions, etc. Now the steady
states $q = q_s \neq 0$ are now maintained by a balance between
energy input and dissipation. It should be emphasized that a
great deal of the analysis illustrated above applies equally
well to those more general cases in diverse fields of
science, engineering, and social science.

In the following, we shall briefly discuss a simple
laser theory which is a system far from thermal equilibrium.
This allows one to study cooperative effects in greater
detail.

Simple laser theory

We shall describe the laser field either quantum
mechanically or classically, and our formulation will be in
such a way that the equations can be understood as
classical ones. Nonetheless, in most cases they possess
exact quantum mechanical analogue.

The usual treatment of lasers consider the electric
field strength $\mathbf{E}(\mathbf{x},t)$ and decomposes it into spatial modes
$\mathbf{b}_i(\mathbf{x})$, and $\mathbf{b}_i(\mathbf{x})$ are determined by the usual resonator
theory. Thus we assume that these spatial modes are
determined completely and they form an orthonormal set. Then
the expansion of $\mathbf{E}(\mathbf{x},t)$ can be written as

$$\mathbf{E}(\mathbf{x},t) = \Sigma_i \, E_i(t)\mathbf{b}_i(\mathbf{x}), \tag{7.2-34}$$

where $E_i(t)$ are time-dependent amplitudes. One can decompose
these amplitudes into positive and negative frequency parts
according to

$$E_i(t) = E_i^+(t)e^{-i\Omega_i t} + E_i^-(t)e^{i\Omega_i t}, \tag{7.2-35}$$

where Ω_i is the frequency of mode i in the unloaded cavity,
i.e., without the presence of laser active atoms, and $E_i^{\pm}$
are slowly varying amplitudes. We want to derive equations
of motion for these slowly varying amplitudes except for
ultrashort pulses. We shall use dimensionless units for $E_i^{\pm}$
by putting

$$E_i^- = -i(\hbar\Omega_i/2\epsilon_0)^{\frac{1}{2}}a_i^+,$$
$$E_i^+ = i(\hbar\Omega_i/2\epsilon_0)^{\frac{1}{2}}a_i. \tag{7.2-36}$$

Quantum mechanically, a_i^+, a_i are creation and annihilation

333

operators of photons of the mode i, and in classical treatment they are merely c-number time-dependent complex amplitudes. Of course, in some cases, it is preferable to use the field instead of the mode decomposition. Also, due to the quantum-classical correspondence, one can replace the quantum correlation by classical averages.

For simplicity, let us consider single mode lasers, so that we can drop the index i. In a region not too far above and below the laser threshold, the mode amplitude $a^+(t)$ obeys the simple equation

$$da^+(t)/dt = - Ka^+(t) + Ga^+(t) + F^+(t), \qquad (7.2-37)$$

where K accounts for the losses by the mirrors, refraction, absorption due to impurities, etc., G describes the gain by the stimulated emission, and $F^+(t)$ represents the fluctuation or noise of the amplitude. This noise term can stem from the spontaneous emission of the atoms into all modes, the interaction of the atoms with lattice vibrations, the pumplight, etc. For more detail, see Haken [1970] or Sargent, Scully and Lamb [1974].

As before, we assume that the statistical average of the fluctuation vanishes, i.e.,

$$<F^+(t)> = <F^-(t)> = 0 \qquad (7.2-38)$$

and they have correlation function

$$<F^+(t)F^-(t')> = C \ \delta(t-t'), \qquad (7.2-39)$$

as in Eq. (7.2-12). Eq. (7.2-39) expresses the fact that the fluctuations have a very short "memory" compared to other time constants in the systems. The constant C depends on the cavity width, the number of thermal photons, damping constants, occupation numbers of the individual atoms, etc.

The gain function G is proportional to the number of excited atoms N_2 minus the number of atoms in the ground state N_1. Furthermore, the gain depends on the line shape of the atoms. The closer the laser frequency Ω to the atomic resonance v, the larger the gain. For the homogeneous Lorentzian linewidth with half-width Γ, the real part of the gain is [Haken 1970; V. Arzt et al 1966; Fleck 1966]:

$$R_eG = (N_2 - N_1)\Gamma|g|^2/[\Gamma^2 + (\Omega - v)^2], \qquad (7.2-40)$$

334

here constant g contains the optical matrix element. It is important to notice that the inversion $N_2 - N_1$ is lowered due to the process of stimulated emission. That is, for not too high laser amplitudes, we have the instantaneous inversion

$$N_2 - N_1 = (N_2 - N_1)_0 - \text{constant} \cdot a^+(t) a(t),$$
$$(N_2 - N_1)_0 = D_0. \tag{7.2-41}$$

The first term on the right hand side, $(N_2 - N_1)_0$, is the unsaturated inversion, the second term describes the cowering of the inversion due to laser action. In Eq. (7.2-41) we assume that the atomic inversion responds immediately to the field. Substituting Eq. (7.2-41) into Eq. (7.2-40) we obtain the saturated gain:

$$G^s = G^u - \beta a^+ a. \tag{7.2-42}$$

Introducing Eq. (7.2-42) into Eq. (7.2-37) we have the basic laser equation derived previously [Haken & Sanermann 1963; Lamb 1964; Haken 1964].

$$da^+/dt = - Ka^+ + G^u a^+ - \beta a^+ (a^+ a) + F^+. \tag{7.2-43}$$

This is the familiar equation of Eq. (7.2-10). It should also be mentioned that without the noise term F^+, Eq. (7.2-43) is similar to the Duffing equation which describes the behavior of the hardened spring with cubic stiffness term. As we have pointed out in Chapter 4, this system has a unique closed orbit which is the Ω-set of all the orbits except the fixed point. Furthermore, this system is auto-oscillatory since all solutions (except one) tend to become periodic as time increases.

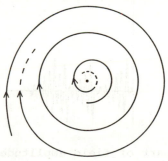

Once again, we may interpret Eq.(7.2-43) as the equation
of a strongly overdamped motion of a particle in the
potential field $V = (K - G^u)|a^+|^2/2 + \beta|a^+|^4/4$. In addition,
the particle experiences random pushes by the fluctuating
force $F^+(t)$.

Let us first discuss the situation below threshold.
Fig.7.2.2a shows the potential V in one dimension for $G^u - K$
≤ 0. After each push excerted by the fluctuating force on
the particle, it falls down the slope of potential hill.
When we multiply E (or a^+) by $\exp(i\Omega t)$ to obtain E(t) of
Eq.(7.2-34) and consider a sequence of random pushes,
Fig.7.2.2b results. As was shown by Mandel & Wolf [1961],
the field amplitude is Gaussian distributed as is
represented by Fig.7.2.2c.

Fig.7.2.2 Laser below threshold, G - K < 0.

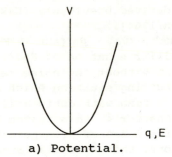

a) Potential.

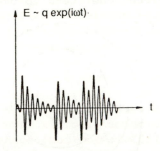

b) Real part of field amplitude vs. time.

336

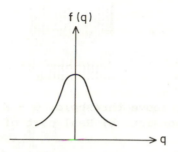

f(q)

q

c) Gaussian distribution of field amplitude.

For above laser threshold, $G^u - K > 0$, the potential curve of Fig.7.2.3a applies. The state $E = 0$ has become unstable and is replaced by a new stable state $E_0 \neq 0$. For the moment, if we ignore the fluctuations, a coherent wave emerges as in Fig.7.2.3b. When we take into account the impact of fluctuations, we must resort to higher than one-dimensional potential, Fig.7.2.3c. The random pushes in the central direction will result the same effect as in the one-dimensional situation, but the random pushes in the tangential direction will cause phase diffusion. Indeed, this is the basis leading to the prediction of Haken [1964] that laser light is amplitude stablized with small superimposed amplitude fluctuations and a phase diffusion. Furthermore, if the corresponding Langevin-type equation, Eq.(7.2-10 or 7.2-43) is converted into a Fokker-Planck equation, Eq.(7.2-25), the steady-state distribution function can be easily obtained by using the adiabatic elimination method [Risken 1965]. These and further properties of laser light were further studied both theoretically and experimentally by a number of authors, for instance the references cited in Haken [1970], Sargent, Scully and Lamb [1974], and the recent book by Milonni and Eberly [1988].

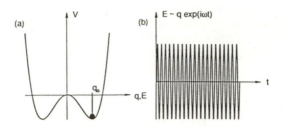

Fig.7.2.3 Laser above threshold, G - K > 0. a) Potential in one dimension. b) Real part of field without fluctuations.

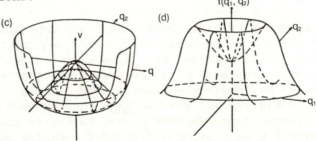

Fig.7.2.3 c) Potential in two dimensions. d) Laser light distribution function. [Haken 1986]

It has to be said that the laser was recognized as the first example of a nonequilibrium phase transition and a perfect analogy to the Landau theory of phase transition could be established [Graham and Haken 1968,1970; Haken 1983]. When we plot the stable amplitude E_o vs. $(G^u - K)$, we obtain the bifurcation diagram of laser light, Fig.7.2.4.

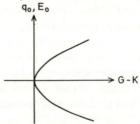

Fig. 7.2.4 Bifurcation diagram of laser light

338

It should also be pointed out that if one considers the classical dispersion theory of the electromagnetic waves, then the electric field satisfies the wave equation,

$$-\nabla^2 E + (1/c^2)\partial^2 E/\partial t^2 + 4\pi\sigma_c \partial E/\partial t/c^2$$
$$= -4\pi\partial^2 P/\partial t^2/c^2, \qquad (7.2-44)$$

where σ_c is the conductivity which describes the damping of the field, and P is the macroscopic polarization. One can think of atoms dispersed in a medium and we may represent the polarization as a sum over the individual atomic contributions at x_i by

$$P(x,t) = \Sigma_i \, \delta(x - x_i)p_i(t), \qquad (7.2-45)$$

where p_i is the dipole moment of the i-th atom. Then the field equation (7.2-44) is supplemented by the equation of the atom i,

$$\partial^2 p_i/\partial t^2 + 2\alpha\partial p_i/\partial t + \beta^2 p_i = e^2 E(x,t)/m, \qquad (7.2-46)$$

where α is the damping constant of the atoms. Once again, we can see that Eq.(7.2-46) is a special case of Eq.(7.2-9). Indeed, if one pursued this further by decomposing the dipole moments in terms of raising and lowering operator for atomic levels, and using the interacting Hamiltonian, one can get field equations, the equation for the atomic dipole moments, and the equation for the atomic inversion. We shall not get into any more details here. Interested readers please see, e.g., Haken [1970], Sargent, Scully and Lamb [1974].

Statistical concept of physical processes, not too far from equilibrium, was first established in the theory of Brownian motion [Langevin 1908]. The theory of Fox and Uhlenbeck [1970], resulted in a general stochastic theory for the linear dynamical behavior of thermodynamical systems close to equilibrium, which includes the Langevin theory of Brownian motion, the Onsager and Machlup theory [1953] for irreversible processes, the linearized fluctuating equations of Landau and Lifshitz for hydrodynamics [1959], and the linearized fluctuating Boltzmann equation as special cases. For more detailed discussions of the general theory of stochastical processes, see, for instance, Gardiner [1985].

As has been pointed out by Fox [1972], in the general theory
as well as in each of the above special cases, the
mathematical descriptions involve either linear partial
integro- differential equations or set of linear
inhomogeneous equations. These inhomogeneous terms are the
usual stochastic (or fluctuating) driving forces of the
processes frequently termed the Langevin fluctuations. These
terms are being referred to as additive fluctuations for
additive stochastic processes. Fox [1972] has since
systematically introduced the stochastic driving forces for
homogeneous equations in a multiplicative way, and such
processes are called multiplicative stochastic processes.

Multiplicative stochastic processes arise naturally in
many disciplines of science. Fox [1972] has established some
mathematical groundwork for multiplicative stochastic
processes (MSP), and pointed out the relevance of MSP in
nonequilibrium statistical mechanics. Shortly after lasers
were invented, it became clear that both lasers and
nonlinear optical processes were examples of dynamical
processes far from thermodynamical equilibrium. Later,
Schenzle and Brand [1979], motivated by laser theory and
nonlinear processes, demonstrated that: (i) an ensemble of
two-level atoms interacting with plane electromagnetic
waves, described by the well-known Maxwell-Block equations;
(ii) parametric down conversion as well as parametric up
conversion; (iii) stimulated Raman scattering; and (iv)
autocatalytic reactions in biochemistry, are examples of
nonlinear processes whose stochastic fluctuations are
multiplicative. They have obtained some very interesting
results. It is natural to expect that many real systems have
both types of fluctuations, that is, there may be a mixture
of both processes. Indeed, they have discussed such systems
with mixture of both fluctuations [1979]. Some further
discussions on multiplicative fluctuations in nonlinear
optics and the reduction of phase noise in coherent
anti-Stokes Raman scattering have been discussed by Lee
[1990]. There are many unanswered questions on the

340

fluctuations of nonlinear phenomena are awaiting for further study.

In general, relaxation oscillations characterized by two quite different time scales can be described by $x_t = f(x,y,\mu,\epsilon)$ and $y_t = \epsilon g(x,y,\mu,\epsilon)$, where a_t is the partial derivative of a with respect to t, $\mu \ll 1$ and μ is the control parameter. Baer and Erneux [1986] have shown how the harmonic oscillations near the bifurcation point progressively change to become pulsed, triangular oscillations.

There are several classics which deal with nonlinear oscillations. One of them is by Minorsky [1962]. It has a great deal of information and results worked out. Indeed, many recent results for applications of nonlinear oscillations can be found in Minorsky [1962]. Another book by Krasnosel'skii, Burd and Kolesov [1973] is also a very useful source. Another classic is the one by Nayfeh and Mook [1979]. This one provides many details of low dimensional nonlinear oscillations and their dynamical evolutions. In the 1950' and 1960's, Lefschetz edited a seriers of volumes on the contributions to the theory of nonlinear oscillations [1950, 1950, 1956, 1958, 1960].

7.3   Optical instabilities

One of the unexpected features of early development of laser systems was the presence of output pulsations even under steady pumping conditions. It did not take long to recognize this result as an important aspect of lasers. In fact, spiking was observed in some masers even before the discovery of lasers [Makhov et al, 1958; Kikuchi et al, 1959; Makhov et al, 1960]. As early as 1958, Khaldre and Khokhlov [1958], Gurtovnick [1958], and Oraevskii [1959] linked the output pulsations to the emergence of dynamical instabilities. Uspenskii [1963; 1964], Korobkin and Uspenskii [1964] already in the early 1960's studied the instability phenomena for homogeneously broadened laser

341

systems. The current efforts in optical instabilites seek to answer the same overall questions as the ones which manifested themselves in other disciplines, namely: What are the origins and the functions of the evolutionary structures of the dynamical systems? Are there universal laws which demand the growth of certain instabilities and structures? What differentiates among the many possible routes of a system to certain macroscopic behavior?

The earliest theoretical models of laser action were based on the description of the energy exchanges between a collection of inverted two-level atoms and the cavity field. Later on, we have found that the rate equation approach is inadequate to provide a faithful description of the observed instabilities [Hofelich-Abate and Hofelich, 1968]. A very significant advance in the field of laser instabilities was the one by Haken [1975], who established the homeomorphism between the single-mode laser model and the Lorenz equations. Such a homeomorphism between the single-mode laser and the Lorenz equations establishes that deterministic chaos is also a part of chaotic laser behavior as long as the single-mode approximation is sufficiently accurate for the laser system. More importantly, such a homeomorphism unifies seemingly different phenomena from different disciplines, and Synergetics provides the motivation and the guidance for an organized approach to the problem of dynamical systems and chaotic behavior in general, and laser instabilities in particular.

In the following, we shall discuss briefly the Maxwell-Bloch equations for a simple ring cavity and demonstrate some interesting features of a ring cavity, such as: (i) if the detuning of the incident light with the absorber introduced, in a stationary situation the transmitted field becomes a multi-valued function of the incident field; (ii) the stationary solution is not always stable even when it belongs to the branch with positive differential gain, in fact, in some cases the transmitted field exhibits a chaotic behavior.

Let the simple ring cavity be the following:

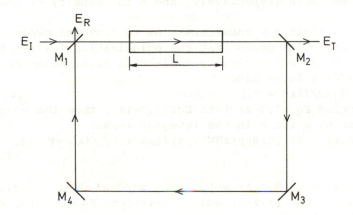

where E_I, E_T and E_R are the incident, transmitted, and reflected field respectively. L is the length of the sample cell containing a two level absorber (for simplicity, homogeneously broadened) and L_T the total length of the optical path in the ring cavity. Also assume the reflectivity of M_1 and M_2 be R and 1 for M_3 and M_4. Let E(t,z) be the complex envelope of the electric field, then we have the following boundary conditions

$$E(t,0) = T\ E_I(t) + R\ E(t - l/c,L)\exp(ikL_T),$$
$$E_T(t) = T\ E(t,L)\exp(ikL),$$

where $T \equiv 1 - R$ and $l \equiv L_T - L$. The propagation of the electric field in the non-linear absorber can be described by the Maxwell-Block equations [Sargent, Scully and Lamb 1974]:

$$\partial E/\partial z = 4\pi i n\mu k\sigma, \qquad (7.3\text{-}1a)$$
$$\partial N/\partial \tau = -\Gamma_{\parallel}(N + 1/z) + i\mu(\sigma^*E - \sigma E^*)/z, \qquad (7.3\text{-}1b)$$
$$\partial \sigma/\partial \tau = (i\Delta\Omega - \Gamma_{\perp})\sigma - i\mu NE, \qquad (7.3\text{-}1c)$$

here $T \equiv t - z/c$ is the retarded time, σ is the dimensionless polarization and $N \equiv \frac{1}{2}(N_1 - N_2)$, μ the transition dipole moment, and $\Delta\Omega = \omega - \Omega$ (where Ω is the transition frequency of the two level atom) is the detuning

343

frequency, $\Gamma_\perp$ and $\Gamma_\parallel$ are the transverse and longitudinal relaxation rates respectively, and n the density of the atoms.

Let us limit our considerations to fast transverse relaxation, which means that the polarization follows the electric field adiabatically (adiabatic elimination method) i.e., $\partial\sigma/\partial\tau \approx 0$, we have

$$\sigma = i\mu NE/(i\Delta\Omega - \Gamma_\perp). \qquad (7.3\text{-}2)$$

Substituting Eq.(7.3-2) into Eq.(7.3-1a), then the electric field can be written in the integral form.

$$E(\tau+z/c,z) = E(\tau,0)\exp[2\theta W(\tau,z)(i\Delta\Omega + \Gamma_\perp)/(\Delta\Omega^2+\Gamma_\perp^2)], \qquad (7.3\text{-}3)$$

where

$$\theta \equiv 2\pi nk\mu^2 \quad \text{and} \quad W(\tau,z) \equiv \int_o^z N(\tau + z'/c,z')dz'. \qquad (7.3\text{-}4)$$

Substituting Eqs.(7.3-3) and (7.3-4) into Eq.(7.3-1b) and then integrating over z, one has

$$\partial W(\tau,z)/\partial\tau = -\Gamma_\parallel(W + z/2)$$
$$- \mu^2 |E(\tau,0)|^2 \{\exp[4\theta\Gamma_\perp W/(\Delta\Omega^2 + \Gamma_\perp^2)] -1\}/4\theta. \qquad (7.3\text{-}5)$$

Introducing the dimensionless quantities

$$\epsilon(t,z) \equiv \mu\, E(t,z)/z\sqrt{\Gamma_\perp\Gamma_\parallel(1 + \Delta^2)},$$
$$x \equiv t\Gamma_\parallel,$$
$$\phi(t) \equiv W(t - L_T/c,L)/L, \qquad (7.3\text{-}6)$$
$$\Delta \equiv \Delta\Omega/\Gamma_\perp.$$

Combining Eqs.(7.3-3) and (7.3-5) together with the boundary conditions with the dimensionless quantities in (7.3-6) we have the following set of equations which do not involve the optical coordinates:

$$\epsilon(x,0) = \sqrt{T}\epsilon_I(x) +$$
$$R\epsilon(x-k,0)\exp(\alpha L\phi(x))\exp\{i(\alpha L\Delta(\phi(x)-\tfrac{1}{2})-\delta_o\}, \qquad (7.3\text{-}7a)$$
$$d\phi(x)/dx = -(\phi(x)+\tfrac{1}{2}) - 2|\epsilon(x-k,0)|^2[\exp(2\alpha L\phi(x))-1]/\alpha L, \qquad (7.3\text{-}7b)$$
$$\epsilon_T = \sqrt{T}\epsilon(x-k,0)\exp(\alpha L\phi(x))\exp\{i(\alpha L\Delta(\phi(x)+\tfrac{1}{2})-(\delta_o+kl)\}, \qquad (7.3\text{-}8)$$

where $\epsilon_T(x) \equiv \mu E_T(t - 1/c)/2\{\Gamma_\perp\Gamma_\parallel(1 + \Delta^2)\}^{\frac{1}{2}}$,
$\epsilon_I(x) \equiv \mu E_I(t)/2\{\Gamma_\perp\Gamma_\parallel(1 + \Delta^2)\}^{\frac{1}{2}}$,
and $\alpha \equiv 2\theta\Gamma_\perp/(\Delta\Omega^2 + \Gamma_\perp^2)$,
is the effective absorption coefficient, and

$$\delta_0 = - k(\sqrt{\epsilon_\iota'}L + 1) + 2\pi M,$$

(where $\epsilon_\iota' \equiv 1 - 4\pi n\mu^2 \Delta\Omega/(\Delta\Omega^2 + \Gamma_\bot^2)$ is the linear dielectric constant, and $2\pi M$ is the multiple of 2π nearest to $k(\sqrt{\epsilon_\iota^T}L+1)$) being the mistuning parameter of the ring cavity and $k \equiv \Gamma_\parallel L_T/c$ being the dimensionless round trip time. Eqs.(7.3-7a,b) and (7.3- 8) can be interpreted as difference - differential equations and whose solutions are uniquely determined by the initial value $\phi(0)$ and the boundary condition $\epsilon(x,0)$ for $- k \leq x < 0$.

For the stationary case, we can set $d\phi(x)/dx = 0$ and $\epsilon(x,0) = $ constant. By eliminating $\epsilon(x,0)$ from Eqs.(7.3-7a,b) and (7.3-8), one obtains the relationship between the stationary solution of the transmitted field and the incident field:

$$|\epsilon_I|^2 = |\hat{\epsilon}_T|^2 \{ \exp(-\alpha L\hat{\phi}) - R]^2$$
$$+ 4R\exp(-\alpha L\hat{\phi})\sin^2[\delta(|\hat{\epsilon}_T|^2)/2]/T^2 \}. \qquad (7.3-9)$$

where $\delta(|\hat{\epsilon}_T|^2) \equiv \delta_0 - \alpha L \quad (\phi + 1/2),$ \qquad (7.3-10)

and $\wedge$ denotes the stationary solution and $\hat{\phi}$ is related to $|\hat{\epsilon}_T|^2$ by

$$(\hat{\phi} + 1/2)/[\exp(-2\alpha L\hat{\phi}) - 1] = 2|\hat{\epsilon}_T|^2/T\alpha L. \qquad (7.3-11)$$

Here $\hat{\phi}$ is a monotonic increasing function of $|\hat{\epsilon}_T|^2$ from $-1/2$ to zero. Also, note that $\delta(|\hat{\epsilon}_T|^2)$ denotes the intensity dependent mistuning parameter, which is originated from the nonlinear shift of wavenumber (or frequency) in the absorber.

In the limit of $\alpha L_\Delta \rightarrow 0$ and $\delta_0 \rightarrow 0$, Eqs.(7.3-9) and (7.3-11) reduce to the absorptive bistability obtained by Bonifacio and Lugiato [1978]. On the other hand, if the conditions $\alpha L_\Delta \ll 1$, $\alpha L \ll 1$, $|\delta_0| \ll 1$ and $|\hat{\epsilon}_T|^2/T \ll 1$ are satisfied, then Eq.(7.3-9) reduces to

$$|\epsilon_I|^2 = |\hat{\epsilon}_T|^2 [1 + 4R(\alpha L_\Delta|\hat{\epsilon}_T|^2/T - \delta_0/2)^2/T^2], \qquad (7.3-12)$$

by the approximation $\delta(|\hat{\epsilon}_T|^2) \approx \delta_0 - 2\alpha L_\Delta|\hat{\epsilon}_T|^2/T$. Eq.(7.3-12) agrees with the relation obtained by Gibbs et al [1976] exhibiting the dispersive bistability experimentally.

If the parameter αL_Δ is sufficiently large, the transmitted field oscillates as a function of the incident field due to the factor $\sin^2[\delta(|\hat{\epsilon}_T|^2)/2]$. Consequently, the

345

ordinary bistable behavior will be drastrically modified. For fixed parameter αL and reflectivity R to be 4.0 and 0.95 respectively, the relations between the transmitted field and the incident field are shown in Fig.7.3.2 for various values of αL_A.

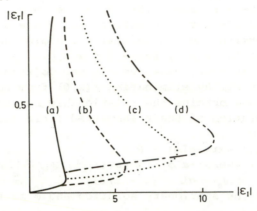

Fig. 7.3.2. Relations between the transmitted field and the incident field for (a) αL = 0.0, (b) αL_A = 2 , (c) αL_A = 4, and (d) αL_A = 6   [Ikeda 1979].

As αL_A increases, the "ordinary" bistable relation (a) for αL = 0 changes to the ones typified by (c) and (d), i.e., new branches appear in the lower tensity of $|\hat{\epsilon}_T|$ and their number increases with αL_A. Indeed, when the magnitude of αL is small enough, then a pair of new branches with negative and positive differential gains is generated when αL_A is increased by 2π. Such multiple-valuedness is due to the intensity dependent mistuning of the cavity with the incident light. The possibility of multiple-valued response of the transmitted light has also been discussed by Felber and Manburger [1976] for F-P cavity system containing a Kerr (cubic) medium. They have also pointed out that the physical origin of the bistable behavior is the nonlinear increase of optical path length at high cavity energy density which brings the initially detuned cavity into resonance with the field. Once the resonance is achieved the transmissivity is high.

For the discussion of the stability of the multiple-valued stationary state, we shall consider the limiting case k << 1, i.e., very short cavity round trip time. In this limit, we may set $d\phi(x)/dx$ in Eq.(7.3-7b) equal to zero. Then Eqs.(7.3-7a,b) and (7.3-8) reduce to the following difference equations:

$$\epsilon_{o,n} = \sqrt{T}\epsilon_{In} + R\epsilon_{o,n-1}\exp(\alpha L\phi_n)\exp\{i(\alpha L_\Delta(\phi_n+\tfrac{1}{2}) - \delta_o)\},$$
(7.3-13a)

$$\epsilon_{Tn} = \sqrt{T}\epsilon_{o,n-1}\exp(\alpha L\phi_n)\exp\{i[\alpha L_\Delta(\phi_n+\tfrac{1}{2}) - (\delta_o + kl)]\},$$
(7.3-13b)

where $\epsilon_{o,n}$, ϵ_{Tn}, and ϵ_{In} denote $\epsilon(x_o+nk,0)$, $\epsilon_T(x_o+nk)$ and $\epsilon_I(x_o+nk)$ respectively, and ϕ_n relates with $\epsilon_{o,n-1}$ by

$$(\phi_n + 1/2)/[1 - \exp(\alpha L\phi_n)] = 2|\epsilon_{o,n-1}|^2/\alpha \qquad (7.3-14)$$

It should be noted that when $\alpha L = \delta_o = 0$ one can choose $\epsilon_{o,n}$ as real, thus Eq.(7.3-13a) can be solved by graphical method. It can be shown that in the case of pure absorptive bistability, the branch with the positive differential gain is always stable. The stability problem in general has not been properly analyzed as yet. Note that as we have discussed earlier in Chapter 6, in particular, the last section, the proper mathematical setting and their techniques are readily available. See elsewhere in these Notes and references therein.

For more detail discussion of the Maxwell-Bloch theory of ring laser cavity, see, for instance, Lugiato et al [1987], Risken [1986; 1988], Milonni et al [1987]. It suffices to say that from Eqs.(7.3-1) one can identify, formally and intuitively, that the Maxwell-Bloch equations are closely related to the Lorenz equations, and consequently the unstable region of the parameter space. Some detailed derivations of this relationship can be found in Miloni et al [1987]. The Lorenz model has been well-studied, and we shall not go into any detail here. One would expect that the projection of the phase space trajectory in the E-N-plane will produce the "butterfly"

347

figures as in the Lorenz model. Indeed, one would.

Since the electric field amplitude is not directly observable, it is interesting to know how to distinguish a symmetry breaking transformation using traditional measurements. The solution is the application of heterodyne techniques using a stable reference source. For heterodyne techniques in infrared and optical frequencies, see for instance, Kingston [1978]. It suffices to say that the main differnce between a symmetric and an asymmetric electric fields is that the symmetric one has a zero average electric field, while the asymmetric one has non-zero average electric field. Thus, the heterodyne spectrum of a symmetric solution has symmetric frequency components around $|\Omega - \Omega_o|$ but with zero spectral power, while an asymmetric solution displays a distinguishing signature of having a line at the beat frequency $|\Omega - \Omega_o|$ and with symmetric sidebands.

It should be noted that a survey of the available experiments indicates that significant areas of disagreement still exist between the theoretical predictions of the Maxwell-Bloch model and the experimental data. Indeed, nobody really knows how to model the active medium with sufficient accuracy, in particular, for solid state lasers. Even for gas lasers, which can operate with a homogeneous broadened gain and yet show behavior which is not compatible with the usual descriptions. In fact, Lippi et al [1986] have shown that the CO_2 laser whose unstable behavior near threshold is in striking difference with the theoretical descriptions.

Hendow and Sargent [1982; 1985] and Narducci et al [1986] have discovered a new type of instability, the phase instability, and its dynamical origin can be attributed by the loss of phase stability instead of its amplitude. The experimental confirmation showed good qualitative agreement between the theoretical predictions and the experimental observations [Treducci et al 1986]. Experiments have also shown that frequent appearance of regular and chaotic pulsations in inhomogeneous broadened lasers [Casperson

348

1978; 1980; 1981; 1985; Bentley and Abraham 1982; Maeda and
Abraham 1982; Abraham et al 1983; Gioggia and Abraham 1983;
1984].

An important warning about the adequacy of the
plane-wave approximation came from the lack of quantitative
agreement between the predictions of the plane-wave
stationary theory of optical bistability [Bonifacio and
Lugiato 1976; 1978; Lugiato 1984], and the failure of the
time-dependent plane-wave calculations to match the observed
pulsing pattern [Lugiato et al 1982]. Indeed, a growing
number of experimental and theoretical results support the
view that transverse effects play a significant role and
maybe even more influential when the optical resonator
contains an active medium [LeBerre et al 1984; Valley et al
1986; Derstine et al 1986; McLaughlin et al 1985; Moloney et
al 1982; Moloney 1984].

There is another type of optical instability which is
very distinct from the optical chaos mentioned above
relating to the Lorenz model of a nonlinear dynamical
system. This other type of optical instability is due to
nonlinear delay feedback, which is related to the iterative
maps.

In the following, the stability results of linearized
equations due to Ikeda [1979] is presented. The linear
motion of $\epsilon_{o,n}$ around its stationary solution is
characterized by two eigenvalues of a 2x2 evolution matrix,
and the stationary solution is stable only if each of the
eigenvalues has an absolute value less than unity. The
stability has also been studied numerically. As expected,
the branches with negative differential gain $d|\epsilon_T|/d|\epsilon_I| < 0$
are always unstable. It is interesting to note that even the
branch with positive differential gain is not always stable.
The stationary solution becomes stable when $|\epsilon_I|$ is set in
the vicinity of the supremum or infimum of the branch, as
illustrated in the following figure.

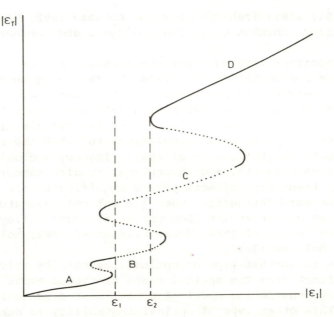

Fig.7.3.3 ——— Stable, ... unstable. A & D branches are stable. Note that all stationary solutions in the region $\epsilon_1 < |\epsilon_I| < \epsilon_2$ are unstable.

Unstable positive differential gain region can lead to regenerative oscillation, for more details, see e.g., Goldstone & Garmire [1983]. As we have noticed that in the region $\epsilon_1 < |\epsilon_I| < \epsilon_2$ all stationary solutions are unstable. What happens in this region? From an appropriate initial value $\epsilon_{o,o}$, one can iterate Eqs.(7.3-13a,b) and it results in an erratic behavior of the transmitted field. It has been found that as the iterated step is advanced the plotted point tends to be "attracted" into a figure which appears to consist of an infinite set of one-dim curves. It is also interesting to know that almost identical figures are obtained when the initial value $\epsilon_{o,o}$ is changed over a wide range. This suggests that the figure represents the "strange attractor" of the difference equation (7.3-13a,b). For

350

details, see, Ikeda [1979]; Ikeda, Daido and Akimoto [1980, 1982].

For more general discussion on iterate maps, see, Collet and Eckmann [1980]. For more detailed discussions and further references on optical bistability, see, Bowden, Ciftan and Robl [1981]; Bowden, Gibbs and McCall [1984]; Gibbs [1985]; Gibbs, Mandel, Peyghambarian and Smith [1986]; Goldstone [1985]; Lugiato [1984]; Zhang and Lee [1988]. Gibbs [1985] provides the most comprehensive treatment of optical bistability up to 1984-5, including references. Goldstone's review article is also recommended. For recent results, see those conference proceedings. There are some review articles on optical instabilities, e.g., Boyd, Raymer and Narducci [1986]; Narducci and Abraham [1988]; Orayevskiy [1988]; Chrostowski and Abraham [1986], Milonni et al [1987], and forthcoming conference proceedings.

7.4   Chemical reaction-diffusion equations

The usual study of chemical processes is concentrated in that several chemical reactants are put together at a certain time and then processes taking place are then studied. In equilibrium thermodynamics, one usually compares only the reactants and the final products and observes the rate as well as the direction of a process taking place. Here we would like to consider the following situation which can be served as a model for biochemical processes. Let us suppose several reactants are continuously put into a reactor vessel where new chemicals are continuously produced and the end products are then removed in such a way that they satisfy steady state conditions. Certainly, these processes can be maintained for any finite amount of time only under conditions far from thermal equilibrium. Indeed, a number of interesting questions, such as, under what conditions can we get certain products in a well-controlled large concentration? Can such processes produce some spatial or temporal patterns? Some of the answers may have some

351

bearing on the formation of structures in biological systems and on the theories of evolution.

Just to illustrate the applications, in the following we shall discuss deterministic reaction equations with diffusion, leaving the stochastic point of view as an "after thought". Before we get into any specific reaction processes, it suffices to say that the time rate of change of the concentration of a given species can be related to the rate constants, concentrations of other species, reverse reaction rate constants, losses, and diffusion terms, which can be written as

$$dn/dt = D_n \nabla^2 n + g(n). \qquad (7.4-1)$$

For some situations, $g(n)$ may be derived from a potential V, i.e., $g(n) = - \partial V / \partial n.$ $\qquad (7.4-2)$

When studying the steady state, $dn/dt = 0$, we want to derive a criterium for the coexistence of two phases, i.e., we consider a situation in which we have a change of concentration within a certain layer. To study the steady state equation, we may invoke an analogy with an oscillator, or more generally, with a particle under the influence of the potential $V(n)$ by noting the following correspondence (just for one-dimension):

$x \rightarrow$ time, $\qquad V \rightarrow$ potential, $\qquad n \rightarrow$ coordinate.

Now let us consider a reaction-diffusion model with two or three variables:

$A \rightarrow X, \qquad B + X \rightarrow Y + D, \qquad 2X + Y \rightarrow 3X, \qquad X \rightarrow E,$

between molecules of the species A, B, D, E, X, Y. Only the following corresponding concentrations enter into the equations of the chemical reaction, a, b, n_1, and n_2 for A, B, X, and Y respectively. We shall treat a and b fixed, whereas n_1 and n_2 are treated as variables. Thus the reaction-diffusion equations read as:

$$\partial n_1 / \partial t = a - (b + 1) n_1 + n_1^2 n_2 + D_1 \nabla^2 n_1, \qquad (7.4-3)$$
$$\partial n_2 / \partial t = b n_1 - n_1^2 n_2 + D_2 \nabla^2 n_2, \qquad (7.4-4)$$

where D_1 and D_2 are diffusion coefficients. For simplicity, let us only consider one spatial dimension. We shall subject the concentrations n_1 and n_2 to two kinds of boundary conditions; either

$n_1(0,t) = n_1(1,t) = a$, $n_2(0,t) = n_2(1,t) = b/a$,   (7.4-5)

or n_1 and n_2 remain finite for $x \to \pm \infty$.   (7.4-6)

One can easily verifies that the stationary state of Eqs. (7.4- 3,4) is given by

$n_{1s} = a$, $n_{2s} = b/a$.   (7.4-7)

In order to see whether any new solution classes can occur, i.e., if any new spatial and/or temporal structure may arise, we can perform a linear stability analysis of the reaction-diffusion equations, Eqs. (7.4-3,4). Let

$n_1 = n_{1s} + q_1$; $n_2 = n_{2s} + q_2$,   (7.4-8)

and linearize Eqs. (7.4-3,4) with respect to q_i's. It is straight forward that the linearized equations are

$\partial q_1/\partial t = (b - 1)q_1 + a^2 q_2 + D_1 \partial^2 q_1/\partial x^2$,   (7.4-9)

$\partial q_2/\partial t = - bq_1 - a^2 q_2 + D_2 \partial^2 q_2/\partial x^2$.   (7.4-10)

The boundary conditions (7.4-5,6) require that:

$q_1(0,t) = q_1(1,t) = q_2(0,t) = q_2(1,t) = 0$,   (7.4-11)

and q_i's remains finite for $x \to \pm \infty$.

Treating q as a column matrix, then Eqs. (7.4-9,10) can be written as

$\partial q/\partial t = Lq$,   (7.4-12)

where $L = \begin{bmatrix} D_1\partial^2/\partial x^2 + b - 1 & a^2 \\ -b & D_2\partial^2/\partial x^2 - a^2 \end{bmatrix}$ (7.4-13)

One can satisfy the boundary conditions (7.4-11) by setting

$q(x,t) = q_o \exp(\mu_l t) \sin(l\pi x)$,   with $l = 1,2,\ldots$ (7.4-14)

As usual, solving the characteristic equation

$\mu^2 - \alpha\mu + \beta = 0$,   (7.4-15)

where $\alpha = (-D_1' + b - 1 D_2' - a^2)$,

$\beta = (-D_1' + b - 1)(-D_2' - a^2) + ba^2$,

and $D_i' = D_i l^2 \pi^2$, $i = 1, 2$.

One can easily show that an instability occurs if $Re(\mu) > 0$. For fixed a but changing the concentration b beyond the critical value b_c, one can find oscillating solutions as well as bifurcating solutions.

The above equations may serve as a model for a number of

biochemical reactions and can also provide some understanding, at least qualitatively, the Belousov-Zhabotinski (B-Z) reactions where both temporal and spatial oscillations have been observed [Hastings and Murray 1975]. Kuramoto and Yamada [1976] were first to propose the possibility of turbulence-like behavior of reactant concentrations in oscillatory chemical reactions. Their discussion was based on the reduced form of B-Z equations. Schimitz, Graziani, and Hudson [1977] recently published experimental data obtained with the B-Z reaction which showed evidence of chaotic states. Olson and Degn [1977] presented results on chaos in a biochemical system, the horseradish peroxidase reaction. Indeed, these studies were guided by the pioneer work of Rossler who had shown that a simple set of three ordinary differential equations could produce chaos [1976]. Consequently, chaos is most likely to be found in laboratory chemical reactors.

Subsequently, Hudson et al [1979] have shown that an entire sequence of states, some periodic and some chaotic, could be obtained by varying a single parameter, the flow rate or residence time. Similar behavior has since been confirmed by Turner et al [1981]. Experimental studies on the B-Z reactions continue, and the investigation of chaotic behavior in chemical reactors is quite active. Indeed, even higher forms of chaos, i.e., more than one positive Liapunov characteristic exponent, is likely to be found in chemical systems [Rossler and Hudson 1983]. There have been several studies in heterogeneous systems governed by partial differential equations.

The experiments were carried out in an isothermal continuous stirred tank reactor (CSTR). The reactants are fed more recently by means of precise constant volume pumps. And data are taken with a platinum wire electrode and a bromide ion electrode which are connected to a digital computer. Data have been obtained as a function of flow rate, temperature, and feed concentration. We shall limit our discussion to results obtained at a single temperature

354

and mixed feed concentrations. A portion of the series of
oscillations with the bromide electrode is shown in
Fig.7.4.1. The oscillations in 1a (with the residence time τ
= 6.76 min) are alternating single and double peaks and
those in 1b (τ = 6.26 min) are double peaks. Both
oscillations are periodic and stable. Chaotic behavior is
observed at τ = 5.89 min in 1c. This behavior is
reproducible and continues until the external conditions are
changed. This chaos is primarily an irregular mixture of two
and three peaks. Two other regions of chaotic behavior were
found in 1e (τ = 5.63 min.) and 1g(τ = 5.34 min.). It should
be pointed out that the ability of realistic B-Z reaction
models to generate chaos is not yet completely clear.
Indeed, Noyes and his coworkers [1978] have seen only
periodic solutions eventhough they have analyzed carefully
at such models. On the other hand, by modifying these
equations, Tomita and Tsuda [1979] and Turner et al [1981]
have obtained chaos. Recently Hudson et al [1984] have
presented simulated results of chaos in two single,
irreversible, exothermic reaction whose reactors coupled
through the heat transport. Such a single reaction has been
shown to produce sustained oscillations in a non-adiabatic
continuous stirred reactor by both experiments and
simulations. It is shown that for two almost identical
reactors, if only one parameter, the heat transfer
coefficient governing heat flow between the two reactors, is
varied, the system changes from periodic, through
quasiperiodic, and finally becomes chaotic. It
is also interesting to note that the flow appears to have
the topology of a folded torus such as that found with the
driven van der Pol osicillations [see, e.g., Guckenheimer
and Holmes 1983]. The fact that chaos is found in two
coupled tanks indicates that complex behavior may be
prevalent in many other systems involving
reaction–diffusion.

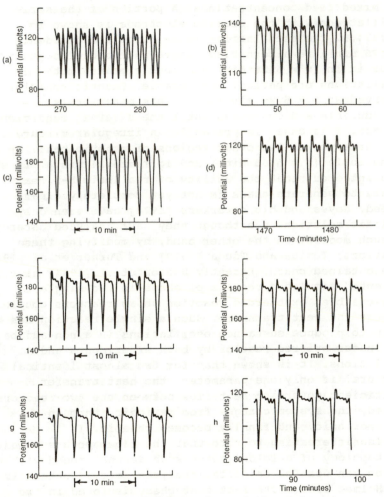

Fig. 7.4.1 Summary of the behavior of the Belousov-Zhabotinski reaction with variation of a single parameter; T = 25°C; (a) τ = 6.76 min; (b) τ = 6.26 min; (c) τ = 5.89 min; (d) τ = 5.85 min; (e) τ = 5.63 min; (f) τ = 5.50 min; (g) τ = 5.34 min; (h) τ = 5.28 min. [Hudson et al 1979].

Recently, Argoul et al [1987] have done experiments on the B-Z reaction in a continuous flow reactor which reveal a spiraling strange attractor arises from the interaction of a local subcritical Hopf bifurcation with a global homoclinic bifurcation. They further point out that the proximity of these two bifurcations justifies the application of a theorem by Sil'nikov [1965,1970] which ensures the existence of chaos. (See Section 6.3). Tam et al [1988] reported the first experimental observation of a new type of spatially extended open chemical system, the Couette reactor. This is an effectively one- dimensional reaction-diffusion system with well-defined boundary conditions. The experiment reveals steady, periodic, quasiperiodic, frequency-locked, period-doubled, and chaotic spatiotemporal states, and qualitatively agrees with the model, and provides some insight into the physical mechanism for the observed behavior.

In a series of papers, Gray and Roberts [1988a,b,c,d] and Gray [1988] have developed a complete analysis of chemical kinetic systems describable by two coupled ordinary differential equations and contain at most three independent parameters. They considered the thermally coupled kinetic oscillators studied by Sal'nikov.

It should be pointed out that Ohtsuki and Keyes [1987] have utilized a field-theoretic renormalization-group method to investigate crossover behavior in nonequilibrium multicritical phenomena of one-component reaction-diffusion systems. An expression for crossover exponents is derived and mean-field values of them are obtained as a function of n.

As an application to the reaction-diffusion equation, recently Conrad and Yebari [1989] studied a simple model of the dissolution-growth process of a solid particle in an aqueous medium in the stationary case. The resulting nonlinear eigenvalue problem consists of a reaction-diffusion equation in the aqueous medium limited by an unknown interface. The various types of bifurcation

357

diagrams depending on the nonlinear reaction term are described, and it is also found that more than one solution exist in general. And chemical dissolution of a two-dimensional porous medium by a reactive fluid which produces a fractal pattern is studied by Daccord [1987]. An interpretation of the evolution of the injection pressure with time which yields the fractal dimension is also presented.

A reaction-diffusion equation related to gasless combustion of solid fuel has been studied. A formal bifurcation analysis by Matkowsky and Sivashinsky [1978] has shown that solutions demonstrate behavior typical for the Hopf bifurcation. A regorous tretment of this problem is developed by Roytburd [1985]. In order to circumvent difficulties involving a possible resonance with the continuous spectrum, appropriate weighted norms are introduced. A suitable version of the Hopf bifurcation theorem is developed and the existence of time periodic solutions is proven.

Under general assumptions, Kopell and Ruelle [1986] studied the temporal and spatial complexity of solutions to systems of reaction-diffusion equations. The time averaged versions of complexity give upper bounds on entropy and Hausdorff dimension of any attracting set.

Recently, Parra and Vega [1988] have considered a first-order, irreversible, exothermic reaction in a bounded porous catalyst with a smooth boundary of one, two or three dimensions. They considered the cases for small Prater and Nusselt numbers, and a large Sherwood number; two isothermal models are derived. Linear stability analysis of the steady states of such models shows that oscillatory instabilities appear for appropriate values of the parameters. They have also carried out a local Hopf bifurcation analysis to ascertain whether such bifurcation is subcritical or supercritical.

For those readers who are interested in the dynamics of shock waves and/or reaction-diffusion equations, Smoller

358

[1983] provides a comprehensive study of these subjects. A more chemically oriented discussion can be found in Vidal and Pacault [1984].

Numerically, a variety of time-linearization, quasi-linearization, operator-splitting, and implicit techniques which use compact or Hermitian operators has been developed for and applied to one-dimensional reaction-diffusion equations by Ramos [1987]. It is shown that time-linearization, quasi-linearization, and implicit techniques which use compact operators are less accurate than second-order accurate spatial discretizations if first-order approximations are employed to evaluate the time derivatives. Furthermore, quasi-linearization methods are found to be more accurate than time-linearization schemes. Nonetheless, quali-linearization methods are less efficient because they require the inversion of block tridiagonal matrices at each iteration. Comparisons among the methods are shown in terms of the L^2-norm errors. Some improvements in accuracy are also indicated.

7.5   Competitive interacting populations, autocatalysis, and permanence

As we have seen in Section 1.1, the predator-prey model of interacting populations, in terms of Lotka-Volterra equations, is rich in structures. In this section, we shall first discuss the effect of crowding, then we shall discuss in some detail autocatalysis and permanence. Multispecies and their applications in biochemical reactions will also be discussed briefly.

Before we discuss any specific situations, it is interesting to note that many systems of nonlinear differential equations in various fields are naturally imbedded in a new family of differential equations. Each equation belonging to that family can be brought into a factorized canonical form for which integrable cases can be identified and solutions can be found by quadratures.

359

Recently, Brenig [1988] has developed such a technique, and generalized multi-dimensional (multi-species) Volterra equations are used as examples to illustrate the power of such an approach.

(i) The effect of crowding
Let us introduce a term representing retardation of growth due to crowding in the predator-prey problem of Votka-Volterra equations in Section 1.1. In particular, we consider the equations

$$dN_1/dt = aN_1 - bN_1N_2 - eN_1^2 \qquad\qquad (7.5\text{-}1)$$
$$dN_2/dt = -cN_2 + dN_1N_2.$$

Again, all the constants a, b, c, d, e are positive.

One equilibrium point is $N_1 = N_2 = 0$. Another one is $N_1 = a/e$, $N_2 = 0$, corresponding to the equilibrium of the logistic growth of the prey in the absence of predators. Any equilibrium point with nonzero values of both N_1 and N_2 must satisfy:

$$a - bN_2 - eN_1 = 0, \qquad\qquad - c + dN_1 = 0.$$

Thus the unique equilibrium solution is

$$N_1 = c/d, \qquad N_2 = (da - ec)/bd.$$

If $a/e > c/d$, this is a positive equilibrium.

Applying the same procedure as in Section 1.1, we can change the variables

$$x_1 = dN_1/c, \qquad x_2 = bdN_2/(ad - ce). \qquad\qquad (7.5\text{-}2)$$

Then it converts Eq.(7.5-1) to

$$dx_1/dt = \alpha x_1(1 - x_2) + \beta x_1(1 - x_1),$$
$$dx_2/dt = - cx_2(1 - x_1),$$

where $\alpha = a - ce/d$, $\beta = ec/d$. For $x_1 > 0$, $x_2 > 0$, let us define $V(x_1,x_2) = cx_1 - c \log x_1 + \alpha x_2 - \alpha \log x_2$.

It has a minimum at the equilibrium point $(1,1)$. Indeed, we find that $dV(x_1,x_2)/dt = - c \beta(1 - x_1)^2 \leq 0$.

Thus, V is a Liapunov function of the system. Using Corollary 5.2.15, one can show that $(1,1)$ is asymptotically stable over the interior of the positive quadrant.

Hainzl [1988] studied the predator-prey system a la Bazykin [1976] whcih depends on several parameters, including the stability of equilibria, the Hopf bifurcation, the global existence of limit cycles, the global attractivity of equilibria, and the codimension two bifurcations.

Recently, Hardin et al [1988] analyzed a discrete-time model of populations that grow and disperse in separate phases, where the growth phase is a nonlinear process that allows for the effects of local crowding, while the dispersion phase is a linear process that distributes the population throughout its spatial habitat. The issues of survival and extinction, the existence and stability of nontrivial steady states, and the comparison of various dispersion strategies are discussed, and the results have shown that all of these issues are tied to the global nature of various model parameters.

Recently, Tucker and Zimmerman [1988] have studied the dynamics of a population in which each individual is characterized by its chronological age and by an arbirary finite number of additional structural variables, and the nonlinearities are introduced by assuming that the birth and loss processes, as well as the maturation rates of individuals, are controlled by a functional of the population density. The model is a generalization of the classical Sharpe-Lotka-McKendrick model of age-structured population growth [Sharpe and Lotka 1911, McKendrick 1926], the nonlinear age-structured model of Gurtin and MacCamy [1974], and the age-size-structured cell population model of Bell and Anderson [1967]. Weinstock and Rorres [1987] investigated the local stability of an equilibrium population configuration of a nonlinear, continuous, age-structured model with fertility and mortality dependent on total population size. They introduce the marginal birth and death rates, which measure the sensitivities of the fertility and mortality of the equilibrium population configuration to changes in population size. They have found that in certain cases the values of these two parameters completely determine the stability classification.

(ii) Autocatalytic reaction:
    Recently, some experimental investigations were
undertaken in studying competition, selection and permanence
in biological evoluation and molecular systems far from
equilibrium. Polynucleotides, such as RNA and enzymes of
simple bacterio- phages are used in these experiments
[Biebricher 1983]. Polynucleotides, strangely enough, have
an intrinsic capability to act as autocatalysts built into
their molecular structures. A combination of an
autocatalytic reaction, a degradation reaction and a
recycling process was found to be adequate in representing
an appropriate mechanism for modeling such system. We
consider the open system
$$A + X \underset{c'}{\overset{c}{\rightleftharpoons}} 2X, \qquad X \underset{d'}{\overset{d}{\rightleftharpoons}} B, \qquad B \overset{r(E)}{\rightarrow} A. \qquad (7.5-3)$$
The rate constant of the recycling reaction r(E) is
determined by an external energy source E. A simple example
of such a process is a photochemical reaction using a light
source. The dynamics of the mechanism (7.5-3) can be
described by
$$da/dt = rb + c'x^2 - cax, \qquad (7.5-4a)$$
$$db/dt = dx - (d' + r)b, \qquad (7.5-4b)$$
$$dx/dt = cax + d'b - c'x^2 - dx, \qquad (7.5-4c)$$
where we denote the concentrations of A, B and X by a, b, x
respectively. A trivial constant of motion is the total
concentration of all substance, i.e.,
$$a + b + x = c_0 = \text{constant}. \qquad (7.5-5)$$
Thus we are left with a two-dimensional system defined on
the state space:
$$S = \{(a,b,x) \in R_+^3 : a + b + x = c_0\}.$$
Clearly the fixed points of Eqs.(7.5-4) are
$$P_0 = (b_0, x_0) = (0,0), \qquad (7.5-6a)$$
$$P_1 = (b_1, x_1)$$
$$= [cc_0(d'+r)-dr]/[(c+c')(d'+r)+cd] \cdot (d/(d'+r),1). \qquad (7.5-6b)$$
If $c_0 < c_{crit} \equiv dr/[c(d'+r)]$, then $P_1 \notin R_+^3$, i.e., outside the
physically relevant state space S. At $c_0 = c_{crit}$, $P_1 = P_0$ and
enters S for $c_0 > c_{crit}$. Local stability analysis shows that

362

P_1 is asymptotically stable and P_0 is a saddle for $c_0 > c_{crit}$, and P_0 is a sink for $c_0 < c_{crit}$.

In order to find global stability, we apply the Dulac-function, x^{-1}, to the vector field, Eqs.(7.5-4), and obtain:

$$x^{-1} db/dt \equiv (db/dt) = d - (d'+r)b/x, \qquad (7.5-7a)$$
$$x^{-1} dx/dt \equiv (dx/dt) = c(c_0 - b - x) + d'b/x - c'x - d. \qquad (7.5-7b)$$

These equations (or vector field), Eqs.(7.5-7), have a strictly negative divergence on S. Thus the flow (7.5-7) is area contracting, consequently, periodic orbits are not possible. It is appropriate to recall and restate the famous theorem of Poincaré-Bendixson (Theorem 4.3.1): A nonempty compact Ω- or α- limit set of a planar flow, which contains no fixed points, is a closed orbit. Thus, the Poincaré-Bendixson theorem implies that the stable stationary solutions P_0 or P_1 are indeed globally stable; every solution starting in S converges to the stable fixed point.

If the degradation is irreversible, i.e., $d' = 0$, global stability of P_1 can also be proved by means of the Liapunov function:

$$V = c(b - b_1)^2 + 2d(x - x_1 \log x), \qquad (7.5-8)$$

where (b_1, x_1) are the coordinates of P_1, given by (7.5-6b).

The autocatalytic reaction mechanism described by (7.5-3) admits only two qualitatively different types of dynamical behavior. The plane of external parameters (r, c_0) is split into two regions corresponding to the stability of either P_0 or P_1 (Fig. 7.5.1)

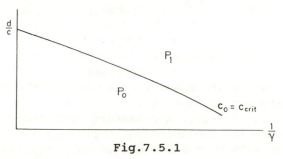

Fig.7.5.1

363

If all reactions are irreversible ($c' = d' = 0$), a condition
which is approximately true for most biochemical and
biological systems, the concentration of A at the stationary
solution P_1 and the critical value c_{crit} coincide and they are
independent of r. The irreversible autocatalytic processes
consume the reactant A to the ultimate limit, i.e., P_1 would
become unstable if the concentration of A went below c_{crit}.

(iii) Multispecies - competition between autocatalytic
      reactions
    Now consider a network of 2n+1 reactions: n first-order
autocatalytic reactions followed by n degradation processes
and coupled to a common irreversible recycling reaction,
which is again controlled from the outside, which allows
driving the system far from equilibrium:

$$A + X_i \; \underset{c_i'}{\overset{c_i}{\rightleftharpoons}} \; 2X_i, \qquad i = 1, 2,\ldots, n \qquad (7.5\text{-}9a)$$
$$X_i \; \underset{d_i'}{\overset{d_i}{\rightleftharpoons}} \; B, \qquad\qquad\qquad (7.5\text{-}9b)$$
$$B \; \overset{r(E)}{\longrightarrow} \; A. \qquad\qquad\qquad (7.5\text{-}9c)$$

One may consider this reaction scheme as competition between
n autocatalysts for the common source of material A for
synthesis. For the reversible reactions (7.5-9a,b), there is
a thermodynamic restriction of the choice of rate constants
due to the uniqueness of the thermodynamic equilibrium:

$$\bar{b}/\bar{a} = c_i d_i / c_i' d_i' \qquad\qquad\qquad \text{for all} \qquad i = 1,\ldots.n,$$

and $\bar{a}$ and $\bar{b}$ are equilibrium concentrations.
    The dynamics of mechanism (7.5-9) again can be described
by:

$$da/dt = rb + \Sigma_i \; c_i' x_i^2 - \Sigma_i \; ac_i x_i, \qquad (7.5\text{-}10a)$$
$$db/dt = \Sigma_i \; d_i x_i - (r + \Sigma_i \; d_i')b, \qquad (7.5\text{-}10b)$$
$$dx_i/dt = x_i(c_i a - c_i' x_i - d_i) + d_i' b. \qquad (7.5\text{-}10c)$$

Of course, we still have the constant of motion,

$$a + b + \Sigma_i \; x_i = c_o = \text{constant}.$$

The fixed points are readily computed for the case of
irreversible degradation, i.e., $d_i' = 0$. There are 2^n fixed
points of Eqs.(7.5-10). A convenient notation which
correspond to the nonvanishing components is: P_i is the
fixed point with $\bar{x}_i \neq 0$ and $\bar{x}_k = 0$ for $k \neq i$; P_{ij} is the

fixed point with $\bar{x}_i \neq 0$, $\bar{x}_j \neq 0$, and $\bar{x}_k = 0$ for $k \neq i, j$;
etc. See Fig.7.5.2.

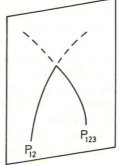

Fig.7.5.2

Without a loss of generality, we can arrange the indices
in such a way that $d_1/c_1 < d_2/c_2 < d_3/c_3 < \ldots < d_n/c_n$. It is
straightforward to show from Eqs.(7.5-10) that the presence
of species X_i at a stable stationary point implies also the
presence of species X_j with $j < i$. Consequently, all fixed
points different from P_0, $P_1, \ldots, P_n$, $P_{12}, \ldots, P_{n-1,n}$,
$P_{123}, \ldots, P_{12\ldots n}$ are unstable irrespective of the values of the
external parameters r and c_0. Indeed, only one of the fixed
points listed above is stable depending on r and c_0: if c_i'
> 0, then a sequence of double point bifurcations of the
type $P_{12\ldots i} \to P_{12\ldots i,i+1}$ with increasing value of c_0. At the
point of coincidence, a change in stability takes place and
a new species is introduced into the system at the stable
stationary state. The larger the environment, which can
sustain a larger total concentration c_0, the more species
can coexist. A similar phenomenon has been found for the
competition of self-replicating macro-molecules in
Lotka-Volterra food chain [So 1979; Gard and Hallman 1979],
or based on Michaelis-Menten type kinetics [Epstein 1979].

When the autocatalytic reactions are also irreversible,
i.e., $c_i' = 0$, a condition usually met by ecological
systems, and if the values of the quotients d_i/c_i are
distinct, then only P_0 and the one species fixed points, P_1,

365

$P_2, \ldots, P_n$ exist at finite values of c_0. P_1, the steady state corresponding to the lowest value of d_i/c_i, is stable for $c_0 > d_1/c_1$. In this case we observe the selection of the fittest species, which is characterized by the smallest quotient of degradation rate over the replication rate.

Theorem 7.5.1 For an irreversible degradating dynamical system (7.5-10) ($d_i' = 0$, for all $i \in [1,n]$), if there is no stationary point in the interior of the state space, i.e., $P_{12\ldots n}$ does not exist, then at least one species dies out.

Proof: Suppose the system of linear equations
$$x_i^{-1} dx_i/dt = c_i a - c_i' x_i - d_i = 0, \qquad i \in [1,n]$$
$$db/dt = \Sigma_i \, d_i x_i - rb = 0,$$
has no solution $(\bar{a}, \bar{b}, \bar{x}_1, \ldots, \bar{x}_n)$ with positive entries, then a well-known convexity theorem ensures the existence of real numbers q_i ($i \in [0,n]$), such that
$$q_0(\Sigma_i \, d_i x_i - rb) + \Sigma_i \, q_i(c_i a - c_i' x_i - d_i) > 0,$$
for all a, b, $x_i > 0$, $i \in [1,n]$. A Liapunov function
$$V = q_0 b + \Sigma_i \, q_i \log x_i, \qquad \text{for } x_i > 0,$$
satisfies $dV/dt = q_0 db/dt + \Sigma_i \, x_i^{-1} q_i \, dx_i/dt > 0$.
Thus, all orbits converge to points where at least one of the concentrations x_i vanishes.

May and Leonard [1975] have shown that for three competitors, the classic Lotka-Volterra equations possess a special class of periodic limit cycle solutions, and a general class of solutions in which the system exhbits nonperiodic population oscillations of bounded amplitude but ever increasing cycle time. As another example of an application of a theorem by Sil'nikov [1965], a criterion is obtained which allows one to construct one-parameter families of generalized Volterra equations displaying chaos [Arneodo, Coullet and Tresser 1980].

Another example of a Lotka-Volterra time dependent system is the parametric decay instability cascading with one wave heavily damped [Picard and Johnston 1982]. This can display Hamiltonian chaos even without ensemble averaging. It seems likely that many such systems can be nearly ergodic without the necessity of ensemble averaging.

366

Recently, Gardini et al [1987] have studied the Hopf bifurcation of the three population equilibrium point and the related dynamics in a bounded invariant domain. And the transitions to chaotic attractors via sequences of Hopf bifurcations and period doublings are also discussed.

Recently, Beretta and Solimano [1988] have introduced a generalization of Volterra models with continuous time delay by adding a nonnegative linear vector function of the species and they have found sufficient conditions for boundedness of solutions and global asymptotic stability of an equilibrium. They have considered a predator-prey Volterra model with prey-refuges and a continuous time delay and also a Volterra system with currents of immigration for some species. Furthermore, a simple model in which two predators are competing for one prey which can take shelter is also presented.

Goel et al [1971] studied many species systems and considered the population growth of a species by assuming that the effect of other species is to introduce a random function of time in the growth equation. They found that the resulting Fokker-Planck equation has the same form as the Schrodinger and Block equations. They have also shown that for a large number of species, statistical mechanical treatment of the population growth is desirable. They have developed such treatment.

(iv) Permanence:

In 1975, Gilpin [1975] showed that competitive systems with three or more species may have stable limit cycles as attractor manifolds, and he argued that it would be reasonable to expect to find this in nature.

The question of whether all species in a multispecies community governed by differential equations can persist for all time is one of the fundamental ones in theoretical ecology. Nonetheless, various criteria for this property vary widely, where asymptotic stabilty and global asympototic stability are two of the criteria most widely

367

used. But neither of these criteria appears to reflect intuitive concepts of persistence in a satisfactory manner. The asymptotic stability is only a local condition, while the global asympototic stability rules out cyclic behavior. Hutson and Vickers [1983] argued that a more realistic criterion is that of permanent coexistence, which essentially requires that there should be a region separated from the boundary (corresponding to a zero value of the population of at least one species) which all orbits enter and remain within.

Putting it differently, an interesting population dynamical system is called permanent if all species survive, provided, of course, they are present initially. More precisely, a system is permanent if there exists some level $k > 0$ such that if $x_i(0) > 0$ for all $i = 1,..,n$, then $x_i(t) > k$ for all $t > T > 0$. This property was called cooperativity [e.g., Schuster, Sigmund, and R. Wolff 1979]. Schuster et al [1980] has also discussed self replication and cooperation in autocatalytic systems. It was used in the context of molecular evolution where polynucleotides effectively helped each other through catalytic interactions as we have just discussed. The notation applies equally well in ecology, but semantically it is awkward to speak of cooperative predator-prey communities. Moreover, Hirsch [1982] has defined a dynamical system $dx/dt = f(x)$ as cooperative if $\partial f_i/\partial x_j > 0$ for all $j \neq i$, which seems to be more an appropriate usage of this term.

Another equivalent way of defining a permanent system is to postulate the existence of compact set C in the interior of the state space such that all orbits in the interior end up in C. Since the distance from C to the boundary of the state space is strictly positive, this implies that the system is stable against fluctuations, provided that these fluctuations are sufficiently small and rare; indeed, the effect of a small fluctuation upon a state in C would not be enough to send it all the way to the boundary and would be compensated by the dynamics which would lead the orbit of the perturbed state back into C. The details of what happens in C is of no relevance in the context of permanence. In a more mathematical term, one can also define a permanent system as the one admitting a compact set C in the interior of the state space such that C is a globally stable attractor.

In ecology or biology, there is a difference between the following two terms. Roughly speaking, a community of interacting populations is permanent if internal strife cannot destroy it, while it is uninvadable if it is protected against disturbance from without. Certainly, uninvadability is not a property per se, one has to specify which invaders the community is protected against.

Let $x_1, \ldots, x_m$ be the densities of established populations and $x_{m+1}, \ldots, x_n$ be the densities of potential invaders. Then the established community is uninvadable if (a) it is permanent, and (b) small invading populations will also persist. Mathematically, the $(x_1, \ldots, x_m)$ community is uninvadable inthe $(x_1, \ldots, x_n)$ state space (where $n > m$), if there exists a compact set C in the interior of the $(x_1, \ldots, x_m)$ subspace, which is a stable attractor in the $(x_1, \ldots, x_n)$ space, and even a globally stable attractor in the $(x_1, \ldots, x_m)$ subspace. There are several results on permanence.

Theorem 7.5.2 A necessary and sufficient condition for an interacting dynamical system to be permanent is the existence of a Liapunov function [Gregorius 1979].

The above theorem was obtained by introducing the concept of repulsivity of certain sets with respect to dynamical systems defined on metric spaces.

An important necessary condition for permanence is the existence of an equilibrium in the interior of the state space. This is just a consequence of the Brouwer's fixed point theorem (Corollary 2.4.18). Hutson and Moran [1982] have proved the necessary conditions for the discrete systems, and Sieveking has proved for the continuous systems.

Another useful condition of the permanence of continuous systems of the type $dx_i/dt = x_i F_i(x)$ was given by Hofbauer [1981]. If there exists a function L on the state space S which vanishes on the boundary ∂S and is strictly positive in the interior, and if the time derivative of L along the orbits satisfies $dL/dt = Lf$, where f is a continuous function with the property that for any x on ∂S, there is T > 1 with

$$1/T\int_0^T f(x(t))dt > 0, \qquad\qquad (7.5\text{-}11)$$

then the system is permanent. Note that, if f in Eq.(7.5-11) is strictly positive on ∂S, then Eq.(7.5-11) is always satisfied and L is just a Liapunov function. Then Eq.(7.5-11) means that L is an "average Liapunov function".

Hutson and Moran [1982] showed that a discrete system $(T\mathbf{x})_i = x_i F_i(\mathbf{x})$ is permanent if there exists a non-negative function P on S which vanishes exactly on ∂S and satisfies

$$\sup_{k>0}\ \lim_{y\to x}\ {}_{y\in S}^{\circ}\ \inf P(T^k\mathbf{y})/P(\mathbf{y}) > 1 \quad \text{for all } \mathbf{x} \in \partial S.$$

Hutson and Moran [1982] have studied the discrete equations given by an analogue of the Lotka-Volterra model,

$$(T\mathbf{x})_1 = x_1 \exp(b_1 - a_{11}x_1 - a_{12}x_2),$$
$$(T\mathbf{x})_2 = x_2 \exp(-b_2 + a_{21}x_1 - a_{22}x_2),$$

with all b_i and $a_{ij} > 0$. They have shown that the system is permanent iff it admits a fixed point in the interior of S. This is similar to the continuous Lotka-Volterra case. Nonetheless, for the continuous case, permanence implies global stability, but the discrete case allows for more complicated asymptotic behavior. Sigmund and Schuster [1984] offers an exposition of some general results on permanence and uninvadability for deterministic population models.

In a closely related discussion, Gard and Hallam [1979] have discussed the persistence-extinction phenomena and have shown that the existence of persistence or extinction functions for Lotka-Volterra systems, which are Liapunov functions, then the systems are persistent or extinct respectively.

Hofbauer and Schuster [1984], Hofbauer et al [1979,1981] have shown that in a flow reactor, hypercyclic coupling of self- reproducing species leads to cooperation, i.e., none of the concentrations will vanish. Yet autocatalytic self-reproducing macromolecules usually compete, and the number of surving species increases with the total concentration.

Before we end this subsection and move on to higher order autocatalytic systems, let us briefly mention the exclusion principle, which states roughly that n species can not coexist on m resources if m < n. This can easily be proved for the case that the growth rates are linearly dependent on the resources. But for more general types of dependence, this principle is not valid. As an exercise, prove this statement.

For example, Koch [1974] presented a computer simulation showing that two predator species can coexist on a single prey species in a spatially homogeneous and temporally invariant environment. This result was then interpreted in the context of Levin's extended exclusion principle [Levin 1970]. Armstrong and McGehee [1976] provided a different interpretation of the relation of Koch's work to Levin's.

(v) Second-order autocatalysis:

In order to demonstrate the enormous richness of the dynamics of higher order catalytic systems, we now study the trimolecular reaction coupled to degradation and recycling in analogy to Eqs.(7.5-3):

$$A + 2X \underset{c'}{\overset{c}{\rightleftharpoons}} 3X \tag{7.5-12a}$$
$$X \underset{d'}{\overset{d}{\rightleftharpoons}} B \tag{7.5-12b}$$
$$B \overset{r}{\rightarrow} A. \tag{7.5-12c}$$

The corresponding differential equations are:
$$da/dt = rb + c'x^3 - cax^2, \tag{7.5-13a}$$
$$db/dt = dx - (d' + r)b, \tag{7.5-13b}$$
$$dx/dt = cax^2 + d'b - c'x^3 - dx, \tag{7.5-13c}$$
and together with the conservation equation:
$$a + b + x = c_0 = \text{constant.}$$

For simplicity, we only consider the irreversible degradation case, i.e., d' = 0, nonetheless, the results obtained are also valid for d' > 0. Note that the fixed point $P_0(\bar{a} = c_0, \bar{b} = \bar{x} = 0)$ is always stable. The two other fixed points P and Q are positioned at:
$(p,q) = \{c_0 c \pm [c_0^2 c^2 - 4d(c + c' + cd/r)]^{\frac{1}{2}}\}/2(c + c' + cd/r)$.
Note that p > q, as p grows and q shrinks with increasing c_0. It is easy to show that Q is always a saddle point, and P either a sink or a source. So, for $c_0 < c_{sad} \equiv [4d(c + c' + cd/r)]^{\frac{1}{2}}/c$, P_0 is the only fixed point, and as c_0 crosses c_{sad}, by a saddle-node bifurcation, P and Q emerge into the interior of S.

To determine the stability of P, we evaluate the trace of the Jacobian, A, at P:
$$trA(x=p) = d - r - p^2(c + c'), \qquad (7.5-14)$$
which is a decreasing function of c_0. If at $c_0 = c_{sad}$, Eq.(7.5- 14) is negative, then P is stable for all $c_0 > c_{sad}$. This corresponds to a direct transition from region I to region V and it always happens if d ≤ r.

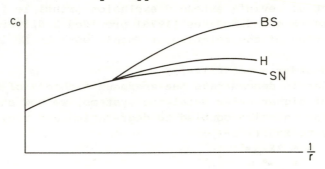

Fig.7.5.3.

Analogously to the earlier discussion, now using x^{-2} as the Dulac function, and using the Poincaré-Bendixson theorem to exclude the existence of periodic orbits, one concludes that all solutions have to converge to one of the fixed points P_0, P, or Q. If the numerical value of trA of Eq.(7.5-14) at $c_0 = c_{sad}$ is positive, the fixed point P is an unstable node.

372

With increasing c_o, the eigenvalues of A become complex. But since trace A is decreasing, the eigenvalues have to cross the imaginary axis at $c_o = c_H$ where a Hopf bifurcation takes place. The value of c_H can be obtained from Eq.(7.5-14) by setting trace $A(x = p) = 0$. Marsden and McCracken [1976] showed that this Hopf bifurcation is always subcritical. This means that the bifurcating periodic orbit is always unstable and occurs for $c_o > c_H$, at such value of c_o, P is stable. A further increase of c_o leads to growth of the periodic orbit. At $c_o = c_{BS}$, the periodic orbit includes the saddle point Q. Thus, the periodic orbit changes to a homoclinic orbit and disappears for $c_o > c_{BS}$. This phenomenon is called "blue sky bifurcation" [Abraham and Marsden 1978].

When $c_o > c_{BS}$, orbits starting from the boundary of S have a chance to converge to P. At values close to c_{BS}, but larger than c_{BS}, the admissible range of initial concentration $x(0)$ for final survival of the autocatalyst is very small. Indeed, if $x(0)$ is too small or too large , the orbit will converge to P_o and the autocatalyst will die out. For much larger values of c_o, the basin of attraction of P becomes almost the entire S because the saddle Q tends toward P_o for $c_o \to \infty$.

Recently, Kay and Scott [1988] have studied the behavior of a third order autocatalytic reaction diffusion system. They have found that for indefinitely stable catalysts, the model exhibits ignition, extinction and hysteresis, and the range of conditions over which multiple stationary states are found to decrease as the concentration of the autocatalyst in the reservoir increases. They have also found that the final ignition and extinction points merge in a cusp catastrophe with the consequent loss of multiplicity. On the other hand, with a finite catalyst lifetime, the dependence of the stationary state composition on the diffusion rate or the size of the reaction zone shows more complex patterns. The stationary state profile for the distribution of the autocatalytic species now allows multiple internal extrema, that is the onset of dissipative structures. And the cubic autocatalytic system also provides the simplest, yet chemically consistent, example of temporal and spatial oscillations in a reaction-diffusion system.

We would like to mention that the existence of two stable solutions on S is a common and well-known feature of higher order autocatalyic reactions. It is also striking to note that higher order autocatalytic reaction networks are very sensitive on initial conditions. In the following paragraphs, we shall briefly discuss competition between higher order autocatalytic reactions. As in Eq.(7.5-9), we have

$$A + 2X_i \underset{c_i'}{\overset{c_i}{\rightleftharpoons}} 3X_i, \qquad I_i \underset{d_i'}{\overset{d_i}{\rightleftharpoons}} B, \qquad B \overset{r}{\rightarrow} A, \qquad i \in [1,n].$$

For simplicity, let us just discuss the case with competitors, i.e., $n = 2$, and irreversible degradation ($d_i' = 0$). Then the dynamical equations are:

$$db/dt = d_1 x_1 + d_2 x_2 - rb, \qquad (7.5\text{-}15a)$$
$$dx_1/dt = x_1(c_1 a x_1 - c_1' x_1^2 - d_1), \qquad (7.5\text{-}15b)$$
$$dx_2/dt = x_2(c_2 a x_2 - c_2' x_2^2 - d_2), \qquad (7.5\text{-}15c)$$

with $a = c_o - x_1 - x_2 - b$.

Again, the fixed points $P_o(\overline{x}_1 = \overline{x}_2 = \overline{b} = 0)$ is stable. Furthermore, any x_i which is very small is characterized by $dx_i/dt < 0$, thus will tend toward zero. Contrary to the competition between linear autocatalytic system Eq.(7.5-10), all invariant surfaces of S are attracting. On both planes, i.e., $x_1 = 0$ and $x_2 = 0$, we observe all the behavior discussed earlier. For instance, we obtain two stable fixed points $P_1(\overline{x}_1 > 0, \overline{x}_2 = 0)$ and $P_2(\overline{x}_1 = 0, \overline{x}_2 > 0)$. Increasing c_o further for $c_i' > 0$, we observe the creation of two pairs of fixed points in the interior of S. If c_o is large enough, one of the fixed point $P_{12}(\overline{x}_1 > 0, \overline{x}_2 > 0)$ will be stable. Nonetheless, the various bifurcations occurring with increasing c_o are diffcult to analyze and one may expect chaotic behavior near the Hopf and homoclinic bifurcations. In the case of $c_i' = 0$, only one pair of fixed points emerges in the interior of S. These are always a source and a saddle point. There is no stable fixed point in S even for large c_o. This indicates that in the irreversible case, all

374

orbits converge to one of the planes $x_1 = 0$ or $x_2 = 0$. Thus, at least one of the competitors has to die out, just as we have found in the linear case, there is no region of coexistence of the two competitors when the autocatalytic processes are irreversible.

For further discussion on autocatalysis, hypercycles, and permanence, see for instance, the papers in Schuster [1984], Freedman and Waltman [1977], Epstein [1979], Eigen et al [1980]. For more quantitative results concerning competitions between three species, see the classical paper by May and Leonard [1975].

Even though the predator-prey model of interacting populations in terms of Lotka-Volterra equations and the autocatalysis equations in terms of reaction-diffusion equations are mainly for the study of population ecology and biochemical reactions, it should be pointed out that the laser equations for a two-level system have the exactly same set of differential equations as the original Lotka-Volterra equations. Likewise, some problems in semiconductor physics have very similar differential equations as the Lotka-Volterra equations or the autocatalysis equations we shall see in the next section.

7.6 Examples in semiconductor physics and semiconductor lasers

By looking at the dynamical equations of the predator-prey problem with crowding and autocatalysis, one can easily recognize their formal similarities or equivalence to some of the problems in semiconductor physics or semiconductor laser physics. In this section, we want to show that the formalism and results obtained in Section 5 can readily be applied to some situations in semiconductor physics and semiconductor lasers.

(i) Limit cycle for excitons:
There has been some work on the possibility of

375

spontaneous oscillations of the electron concentration and temperature due to nonlinear dependence of the Auger coefficient on temperature [Degtyarenko, Elesin and Furmenov 1974]. Nonetheless, this model does not contain an autocatalytic process, here we consider such process following Landsberg and Pimpale [1976] and Pimpale et al [1981] in a model with the following three steps: stimulated production of excitons: $e + h + X \xrightarrow{c} 2x$, radioactivedecay of excitons at recombination center: $X \xrightarrow{d} r$, photogeneration of carriers: $r \xrightarrow{g} e + h$. It is very clear that these processes are irreversible first order autocatalytic processes as in Eq.(7.5-3) where $(e + h) \equiv A$, $r \equiv B$, $c'= d'= 0$. The irreversibility is because the semiconductor is assumed to be far from equilibrium. Here we assume that the semiconductor to be nondegenerate and under high excitation so that the electron and hole concentrations are taken to be equal. Then the differential equations describing the dynamical system are (one can obtain from Eq.(7.5-4) too):

$$dn/dt = gb - Cn^2 x, \qquad\qquad (7.6-1a)$$
$$dx/dt = Cn^2 x - dx. \qquad\qquad (7.6-1b)$$

The conservation law of total concentration and the other dynamic equation (reduced form of Eq.(7.5-4b)) are taken into account. Note that, the exciton decay process has interesting nonlinear features: the decay rate is proportional to the number of recombination centers available where there is a large number of excitons available for decay, i.e., for $x \gg 1/q$, $dx/dt \approx -1/q$ where $1/q$ is concentration of recombination centers. But when the number of excitons is much lower than the number of recombination centers, the exciton decay rate is proportional to the exciton concentration, i.e., for $x \ll 1/q$, $dx/dt \approx -x$. Since the Michaelis-Menten S-shaped law seems to be a good interpolation law for the exciton decay rate, we then replace Eq.(7.6-1b) by

$$dx/dt = cn^2 x - dx/(1 + qx). \qquad\qquad (7.6-1b')$$

The exciton decay rate on recombination centers, given by

the second term of Eq.(7.6-1b'), is shown qualitatively in Fig.7.6.1.

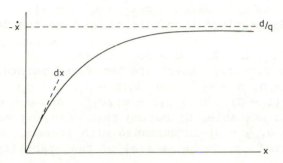

Fig.7.6.1

For small x, (xq <<1), one has the usual exponential decay. But for large x, there is a departure from the usual exponential decay. Experimentally, such a departure has been observed in qualitative agreement with Fig. 7.6.2 [Klingenstein and Schmod 1979]. A generalization to an n-th order reaction is also possible [Ibanez and Velarde 1977], but we shall not discuss it any further. Note, here we have also ignored electron-hole, electron-impurity recombinations and other excitonic processes.

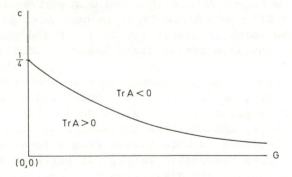

Fig.7.6.2

We can rescale the dimensions in such a way, (as an exercise) such that Eq.(7.6-1a,b') becomes

$$dn/dt = G - cn^2 x, \quad dx/dt = cnx^2 - x/(1 + x). \quad (7.6-2)$$

For physically meaningful solutions, we have the following constraints:

$$x \geq 0, \quad n \geq 0, \quad G > 0, \quad c > 0.$$

As in section 7.5-(ii), there are two fixed points:

$$P_0(x = 0, n = \infty) \quad \text{and} \quad P_1(x = x_0, n = n_0)$$

where $x_0 \equiv G/(1 - G)$, $n_0 \equiv |(1 - G)/c|^{\frac{1}{2}}$ and $0 < G < 1$. Clearly, P_0 is unstable, by noting that changing variable $y = 1/n$, $P_0(x = 0, y = 0)$ is unstable with respect to even linear perturbation. To get a feel of the stability of P_1, let

$$x(t) = x_0 + u(t), \quad n(t) = n_0 + v(t),$$

then Eq.(7.6-2) becomes

$$d/dt \begin{bmatrix} u \\ v \end{bmatrix} = A \begin{bmatrix} u \\ v \end{bmatrix}, \quad (7.6-3)$$

where

$$A = \begin{bmatrix} G(1 - G) & 2G(c/(1 - G))^{\frac{1}{2}} \\ -(1 - G) & -2G(c/(1 - G))^{\frac{1}{2}} \end{bmatrix}.$$

Note that $\det A = 2G(1 - G)^{3/2}c^{\frac{1}{2}} > 0$ for physically meaningful values of G and c. And, the Poincare index of the fixed points is +1. Thus the stability of P_1 is determined by the value of Tr A, where Tr $A = G(1 - G) - 2[c/(1 - G)]^{\frac{1}{2}}$. The bifurcation values of G and c are given by Tr $A = 0$, i.e., $(1 - G)^3 = 4c$. Since $(0,0)$ is unstable, it appears that the fixed point is stable for Tr $A < 0$ and unstable for Tr $A > 0$. The unstable steady state leads to oscillation. See Fig.7.6.2.

Another interesting phenomenon is the Gunn effect, or Gunn instability [see for instance, Seeger 1982] in a semiconductor having a two-valley conduction band, such as GaAs, is caused by the negative resistance arising from the transfer of electrons in high fields from a lower light-mass valley to an upper heavy-mass valley. It has been determined that the Gunn instability gives rise to anomalously large low-frequency current noise over a wide range of fields beyond the threshold of the instability. Recently, Nakamura

378

[1988] has performed carrier dynamics simulation to investigate the low-frequency current noise mechanism in a two-valley semiconductor. It has been noted that the current noise is caused by the chaotic transfer of carriers in the accumulation layer of a high-field domain.

7.7   Control systems with delayed feedback

Roughly speaking, the purpose of control is to manipulate the available inputs of a dynamical system to cause the system to behave in a more desirable manner than the one without control. Almost every movement our body makes in our daily life is under control. In fact, most of them involve feedback of various means. This can also easily generalize to our physical world.

There are two major types of control. In open-loop control, the input is generated by some processes external to the system itself, and then is applied to the system. A specific input may be generated by analysis such as developing the yearly production plan for a company, or may be generated repeatedly by a physical device when directing a physical parocess or machine such as programmable switch on and off of a heating system. At any rate, the underlying feature of open-loop control is that the input function is determined completely by an external process. In closed-loop control, the input is determined on a continuing basis by the behavior of the system itself as expressed by the behavior of the outputs. This is also called feedback control because the outputs are fed back to the input. There are many reasons why feedback control is often preferable to open-loop control. One is that a feedback rule is often simpler then a comparable open-loop scheme because it may require a fair amount of computation and complex implementation. It is very important that feedback can automatically adjust to unforeseen system changes or to unanticipated input disturbances. One of the other important reasons is that much feedback can rapidly adjust to changes

379

leads to the fact that feedback can increase the stability of the system. In this respect, mathematically, a basic feature of feedback is that it can influence the characteristic polynomial of the system.

Let us consider the effect of feedback applied to a very simple system, $dx(t)/dt = u(t)$. One can easily show that this system is marginally stable. Now suppose the input is controlled by feedback and suppose $u(t) = \alpha(x_o - x(t))$. This system certainly yields $x(t) \to x_o$. Note that by defining $y = x - x_o$, the feedback system is governed by the simple equation $dy(t)/dt = -\alpha y(t)$, and the variable $y(t)$ is the error. Thus, feedback has converted the original marginally stable linear system into an asymptotically stable system. As stated before, an important objective of feedback is to make the system more stable than it would otherwise be. Therefore, it is natural to ask how much influence feedback can have on the eigenvalues of a system. An important result is the <u>eigenvalue placement theorem</u>. This theorem states that if a system is completely controllable and if all state variables are available as outputs, then by suitable direct feedback it is possible for the feedback system to have any desired characteristic polynomial.

Before we continue on nonlinear feedback control systems, we would like to introduce the result of Sontag [1982] which deals with the necessary and sufficient conditions for asymptotic controllability of a controllable system. The system under consideration is of the following type:

$$dx(t)/dt = f(x(t),u(t)), \qquad\qquad (7.7-1)$$

where the states $x(t) \in R^n$, control parameter $u(t)$ is in a metric space X, and the map f is assumed locally Liptchitz to ensure the existence and uniqueness of solutions for small time intervals. Furthermore, admissible control functions $u(t)$ are all measurable and locally bounded functions, and X is topologized by the weak topology. The system Eq.(7.7-1) is <u>asymptotically controllable</u> if the following conditions are satisfied: (i) for each state x

380

there exists a control for which the corresponding solution
$\phi(x,u,t)$ is defined for all t, and converges to zero; (ii)
if for a given $\epsilon > 0$, there is a $\delta > 0$ such that for any x
of norm less than δ, there is a u as in (i) with the ensuing
orbit having $\|x(t)\| < \epsilon$ for all t; (iii) there is a
neighborhood U of the zero state, and a compact subset K of
X such that if x(0) is in U, there exists an input as in
(ii) with values in K almost everywhere. Then we have the
following theorem:

   <u>Theorem 7.7.1</u>   The system, Eq.(7.7-1), is asymptotically
controllable iff there is a Liapunov-like function for it.

   Here a <u>Liapunov-like function</u> for Eq.(7.7-1) is a real
function V defined on R^n such that:
(i)   V is continuous;
(ii) $V(x) > 0$ for $x \neq 0$, $V(0) = 0$;
(iii) $\{x|\ V(x) < a\}$ is bounded for all a;
(iv) for each $x \neq 0$ there is a relaxed control u with
$V'(x,u) < 0$;
(v)   there is a neighborhood L of the zero state and a
compact subset K of U such that, for states x in L, the
control u can be chosen with values in K (more precisely,
the measures u(t) are supported in K almost everywhere).
The derivative along the chosen trajectory is defined as:
$V'(x,u) = \lim_{t\to 0}\ \inf[(V(\phi(t,x,u)) - V(x))/t]$.

   In practice, the value of u is determined after a
measurement on the state x(t) at t was made. If this
measurement is in error, say x(t) + e(t) is measured rather
than x(t), the governing equation of motion will have the
form
$$dx(t)/dt = X(x(t) + e(t)). \qquad (7.7-2)$$
It is this concept which leads to the definition of
stability with respect to measurement. Intuitively, X is
stable with respect to measurement if any solutions of
Eq.(7.7-2) and $dx(t)/dt = X(x(t))$, satisfying the same
initial conditions, remain arbitrarily close over any finite
positive time interval whenever the supremum of $|e(t)|$ over
this time interval is restricted to be sufficiently small.

In general, the initial value problem for such a vector field X may not have a solution. Nonetheless, if there is an absolutely continuous function $\phi(t)$ which satisfies the initial condition and $d\phi(t)/dt = X(\phi(t))$ almost everywhere, then $\phi(t)$ is called a classical solution.

Consider a control system of the form

$$dx/dt = g(x, u(x)), \quad x = (x_1, x_2, \ldots, x_n), \quad u = (u_1, \ldots, u_n),$$
$$(7.7-3)$$

with values $u(x)$ to be chosen from a control set U. Let the target manifold S be contained in $[0,\infty) \times R^n$. If g is bounded and Lipschitzian and u is a given Lipschitzian control, then an initial value problem of Eq.(7.7-3) with data $x(0) = x^0$ has a unique solution, with value at time t denoted by $\phi(t,0,x^0)$. Assume that $\phi(t,0,x^0) \in S$. Then the question is: If S has dimension less than n in R^{n+1}, is it possible that, for each x in some neighborhood $N(x^0) \subset R^n$ of x^0, there exists a value $t(x)$, and $0 \leq t(x) < \infty$, such that $\phi(t(x), 0, x) \in S$?

Before we address this question, we shall briefly discuss the generalized concept of solutions for such equations as given by Filippov [1964]. Let X be a measurable function defined almost everywhere in a domain $Q \subset R^n$ with values in a bounded set in R^n. With $X(x)$ associate the convex set

$K\{X(x)\} = \cap_{\delta > 0} \cap_{\mu(N)=0}$ closed convex hull$\{X(U(x,\delta) - N\}$,

where $U(x,\delta)$ is a closed δ neighborhood of x, N an arbitrary set in R^n and μ is n dimensional Lebesgue measure. Then an absolutely continuous vector valued function ϕ, defined on $[0,T]$, is called a <u>Filippov solution</u> of $dx/dt = X(x)$ if for almost all t, $d\phi(t)/dt \in K\{X(\phi(t))\}$. In Filippov [1964], it has been shown that such solutions will always exist, and many of their properties are discussed. In particular, if X is continuous, then $K\{X(x)\} = X(x)$.

Let $U(x,\delta)$ denote a compact, spherical neighborhood of radius δ, about the point x in R^n, and coA denotes the convex hull of a set A. A vector field X, for which a classical solution ϕ of $dx/dt = X(x)$ with arbitrary initial

data x^o exists, is said to be <u>stable with respect to</u>
<u>measurement</u> if given $\epsilon > 0$ and finite $T > 0$, there exists a
$\delta > 0$ such that whenever e is a measurable function on $[0,T]$
with values in R^n and norm less than δ for which a
corresponding solution μ of $dx/dt = X(x(t) + e(t))$, $x(0) =$
x^o, exists on $[0,T]$, then $\|\phi - \mu\| < \epsilon$. Hermes [1967] has
established the following theorem:

   <u>Theorem 7.7.2</u>   If X is stable with respect to
measurement, then every classical solution is a Filippov
solution.

   If X is stable with respect to measurement, solutions
for $t \geq 0$ of the initial value problem for the corresponding
differential equation are unique, and such a solution, when
evaluated at a fixed positive time, varies continuously with
the initial data. Thus, with increasing time, solutions may
join but not branch. Thus, it is felt that feedback controls
which are meaningful from the standpoint of applications
should lead to vector fields which are stable with respect
to measurement. Nonetheless, to characterize such vector
fields directly is not a easy task. In the next section, we
shall illustrate such a system in feedback control of
semiconductor lasers and phased arrays.

   For linear difference-differential equations of retarded
delay feedback, Infante [1982] has outlined a
characterization of a broad class of such difference-
differential systems which have the property that their
asymptotic stability and hyperbolicity characteristics are
not affected by the value of the delays involved.

   Recently, an der Heiden and Walther [1983] and an der
Heiden [1983] have shown that under centain conditions, the
difference-differential equation

   $dx(t)/dt = f(x(t-1)) - ax(t)$,   (7.7-4)
describing delayed feedback control systems, admits chaotic
solutions.

   Recurrent synaptic feedback is an important mechanism
for regulating synchronous discharges of a population of
neurons. It can be found in the brain, the the spinal chord

and in the sensory systems. Recurrent feedback occurs when activity in a population of neurons excites, via axon collaterals, a second population, which in turn excites or inhibits the first. Models that include recurrent feedback fall into two main categories. The first consists of models of a specific part of the nervous system where recurrent feedback is important, while the models in the second category are developed specifically to describe resurrent feedback without reference to a particular part of nerve system. Nonetheless, both types include time delays due to the conduction times and synaptic delays in the feedback circuit. The model of Mackey and an der Heiden [1984] incorporates the stoichiometry of synaptic transmitter-receptor interactions and time delays and leads to a nonlinear term for recurrent feedback. An even simpler model for recurrent feedback is that of Plant [1981], which is a modification of the FitzHugh-Nagumo equations for a nerve cell [FitzHugh 1961] emphasizes the behavior of an individual cell membrance. Recently, Castelfranco and Stech [1987] provide a more comprehensive understanding of the structure of periodic orbits in Plant's model and viewed the system as a two-parameter Hopf bifurcation problem associated with delayed feedback in the FitzHugh model.

Ikeda et al [1982] have studied delayed feedback systems in detail and they have found that higher-harmonic oscillating states apear successively in the transition to chaos. It has been pointed out that the first order transitions between these states account for the frequency-locked anomaly observed by Hopf et al [1982] in a hybrid optical bistable device.

An inverse Liapunov problem was studied by Liu and Leake [1967]. They studied the controllable dynamical system of Eq.(7.7-1) and established means to determine the system function f or the control function u so that Eq.(7.7-1) behaves in an "acceptable" manner such as uniform asymptotic stability in the whole (u.a.s.w.), which is equivalent to asymptotic stability in the whole if the function f is autonomous or f is periodic in t, and equivalent to asymptotic stability if, f is also linear. Indeed, such stability is probably the most desirable type of stability for practical systems. They have shown the following theorem:

Theorem 7.7.3   Let $f \in C^0$ (local Lipschitz condition) and let x = 0 be the equilibrium is u.a.s.w. iff there exists a scalar function V(x,t) such that:
(i) V(x,t) is of class C^1, positive definite, decrescent, and radially unbounded; and
(ii) whose Eulerian derivative dV(x,t)/dt is negative definite. Here the Eulerian derivative is defined by dV/dt = <f, grad V> + V_t, where <> denotes the inner product, V_t = $\partial V/\partial t$, and grad V the gradient of V.

Unfortunately, there are no general means available whereby one can actually find V. The inverse problem is that, given a pair of scalar functions V(x,t) and dV(x,t)/dt, one must find the function f (or u) so that the Eulerian derivative is satisfied. The solution of the inverse problem together with the known results of Liapunov's direct method can be applied directly to the synthesis (or design) of nonlinear control systems. They also have to find the necessary and sufficient conditions for multi- loop feedback control systems and their complete asymptotic controllability conditions.

For the feedback system of dx/dt = Ax + bf(cx), a very interesting stability criterion for the asymptotic stability in the large is the Popov criterion [Popov 1962].

In a different vein, Westcott [1986] gave a brief review of the application of feedback control theory to macro-economic models.

7.8   Semiconductor laser linewidth reduction by feedback control and phased arrays

A narrow-linewidth, frequency stablized laser is a very useful device in many aspects of pure and applied research such as spectroscopy, laser communication, lidar and remote sensing, heterodyne detection, etc. Unlike conventional lasers, the field power spectra of semiconductor lasers are not dominated by mechanical cavity fluctuations, but instead, external noise sources such as ambient temperature fluctuations, injection current fluctuations due to the noises from a power supply, and so on, in addition to the spontaneous emission, contribute to the linewidth. Electrical or optical feedback techniques have been very successful in reducing linewidth significantly. Simple electrooptic feedback systems have been utilized for diode lasers to correct for amplitude and/or frequency fluctuations due to variations in drive current and ambient temperature. For a recent review of frequency stable semiconductor lasers, see, e.g., Lee and Chen [1990]. Simultaneous intensity and frequency stabilization can be achieved by detecting the fluctuations in these characteristics and to control both the diode operating temperature and drive current [Yamaguchi and Suzuki 1983].

One may have noticed that the huge difference, over nine orders of magnitude between the limiting linewidth and the cavity linewidth for conventional gas or semiconductor lasers, is mainly due to the very short photon lifetime, t_c, which is proportional to $\Delta v_{\frac{1}{2}}^{-1}$ in the resonantors. Clearly, to increase t_c, i.e., decrease the linewidth, in particular for semiconductor lasers, is to increase the cavity length L by placing the laser in an external resonator and by using high reflectance mirrors. Indeed, semiconductor laser linewidths in kHz regime in an external cavity have been demonstrated [Shay 1987].

There are several papers which provide simplified analytical discussions of external cavities with some restrictions [Fleming and Mooradian 1981, Henry, 1986]. These analyses can provide guidelines as to the amount of linewidth reduction which can be realized with external cavity. It should be pointed out that rigorous analyses of

386

the external cavity of a diode laser are quite complex and
not thoroughly understood. It is known that the spectral
behavior is strongly dependent on the level of optical
feedback [Patzak et al 1983a,b, Lenstra et al 1985, Tkach
and Chraplyvy 1985, Henry and Kazarinov 1986, Henry 1986]
and the phase relationship between the diode field and the
feedback field [Chen 1984]. As in injection locking, in
general, the higher the feedback amount is, the greater the
linewidth reduction can be achieved. Nonetheless, it should
be cautioned that at a varying feedback level at a certain
high level of feedback, the linewidth reduction tends to
saturate, and is then followed by an abrupt increase in
linewidth by several orders of magnitude. This effect was
observed for instance by Goldberg et al [1982] and Favre et
al [1982]. It has been pointed out that both the saturation
phenomenon and the sudden increase in linewidth are
consequences of the nonlinear dynamics, for instance,
[Olesen et al 1986, Otsuka and Kawaguchi 1984]. As has been
pointed out by Lenstra et al [1985], the state of increased
linewidth has the distinct signature of a chaotic attractor.
Indeed, the coherence collapse is due to optical delay
feedback [Dente et al 1989]. Due to their high gains,
semiconductor lasers are very sensitive to feedbacks.
Feedback not only induces phase (i.e., frequency)
fluctuations, but also induces amplitude fluctuations, and
may eventually lead to chaos.

Very recently, Li and Abraham [1989] have shown that
the semiconductor laser with optical feedback functions as a
laser with an optical servoloop can be considered a
self-injection locking laser, and they also determine the
effect of various parameters on the noise reduction and on
the stability of single- mode operation of the laser.

In order to achieve high power, semiconductor lasers
have to be phase locked together to form a phased array.
Such phase locked semiconductor phased arrays have been
demonstrated [Teneya et al 1985, Hohimer 1986, Leger et al
1988, Carlson et al, 1988]. Earlier studies of semiconductor

phased arrays are mainly focused on their spatial mode structures under steady state operating conditions. These spatial mode patterns in terms of supermodes are observed and well understood [Scifres et al 1979, Botez 1985, Butlet et al 1984, Peoli et al 1984]. Nonetheless, little has been said about the temporal or time evolutionary behavior of such phase locked semiconductor arrays. Only very recently, there have been a few papers concerning the dynamical behavior of the phased arrays [Elliott et al 1985, Wang and Winful 1988]. What we would like to know are: (a) under what operating conditions will the semiconductor laser array be phase locked, (b) how long will it take for locking to take place, (c) will the locking be stable, i.e., be persistent for a very long time, (d) if it can only be locked for a "short" time, then what happens after they become unlocked, (e) what are the details and routes to instabilities, (f) can we classify the various routes to instabilities from the variations of various operating parameters, (g) and above all, with reasonable operating conditions, can one prove a general statement concerning the existence or non-existence of a stable region for a semiconductor laser array.

There are many ways to initiate the phase locking of the semiconductor phased array. One way is the injection locking, which has been demonstrated for gain guided semiconductor laser array [Hohimer et al 1986]. But for index guided laser array, some phase shifters may be necessary for phase locking. Another method for phase locking is using a common external cavity for the laser array, whereby the phase locking is accomplished by mutual feedback, i.e., self-organization. Of course, in order to increase the amount of mutual feedback, the self-imaging technique can be employed [Leger et al 1988]. Here we shall not debate the relative merit of either method, or other generic methods, it suffices to say that in the following formulations and discussion, the general conclusion will be the same. Even though we shall only discuss the formulation of an semiconductor laser array undergoing mutual feedback,

the case of injection locking can be modified trivially as
we shall indicate.

For a M element diode-laser array oscillating in a
common cavity with mutual feedback, the system of the
dynamical equations is the following:

$$\frac{dN_i}{dt} = \frac{J_i(t)}{ed} - \frac{N_i}{\tau_{s_i}} - G(N_i, I_i)\{I_i(t) + \sum_{j=1}^{M} \xi_{ij}^{1/2} I_i(t-\tau_{ji}) \cos(\omega_j \tau_{ji} + \phi_{ij})\}$$
$$+ BN_i^2 + AN_i^3, \tag{7.8-1}$$

$$\frac{dI_i}{dt} = [G(N_i, I_i) - \frac{1}{\tau_1}] I_i(t) \cos(\phi_i(t) - \phi_i(t-\tau) + \omega_i \tau + \phi_{oi})$$
$$+ \sum_{j=1}^{M} \{[I_i(t)\xi_{ij} I_j(t-\tau_{ji})]^{1/2}$$
$$\times \cos[(\Delta\omega_{ij})t + \phi_i(t) - \phi_j(t-\tau_{ji}) + \omega_j \tau_{ji} + \phi_{oj}]\}, \tag{7.8-2}$$

$$\frac{d\phi_i}{dt} = \frac{\alpha}{2}[G(N_i, I_i) - \frac{1}{\tau}] - [\omega_i(N_i) - \Omega] - \xi_{ii}^{1/2}[\frac{I_i(t-\tau)}{I_i(t)}]^{1/2}$$
$$\times \sin[\phi_i(t) - \phi_i(t-\tau) + \omega_i \tau + \phi_{oi}]. \tag{7.8-3}$$

Here $i \in [1,M]$, τ_{s_i} is the semiconductor carrier diffusion
time, B is the carrier spontaneous recombination
coefficient, A is the Augur coefficient, ξ_{ij} is the coupling
coefficient and τ_{ij} is the roundtrip time between the i-th
and the j-th elements, τ_1 is the photon lifetime, $\tau = \tau_{ii}$ is
the self-feedback delay time, α is the absorption
coefficient, and Ω is the outcoupling frequency of the
coupled phased-array. For "identical" lasers, we assume that
$\tau_{s_i} = \tau_s$. The first square bracket on the right hand side of
Eq. (7.8-3) depicts the hole burning of homogeneously

389

the laser frequencies $\omega_i(N_i)$ depend on the carrier density by

$$G(N_i) = G(N_{th}) + (\frac{\partial G}{\partial N_i})\,(N_i - N_{th})\,, \qquad (7.8-4)$$

$$\omega_i(N_i) = \omega_i(N_{th}) = (\frac{\partial \omega_i}{\partial N_i})\,(N_i - N_{th})\,, \qquad (7.8-5)$$

where N_{th} is the threshold carrier density, $N_{th} \rightarrow N_{tr}$. Here the coupling coefficients ξ_{ij} consist only the external mirror or grating reflectivity, transmission and diffraction losses, and any intra-cavity loss. The product of ξ_{ij} with the appropriate phase terms correspond to the coupling coefficients in Lang and Yariv [1986, 1987]. Furthermore, the injected current J_i can be an explicit fuction of time, this will be of particular interest for applications utilizing FM. The mechanical analogue is a system of damped, nonlinearly coupled oscillators with time varying applied force. To simplify the problem further and to avoid delay feedback instability, such as Ikeda instability [1979], and allow us to use the slow varying envelope approximation (SVEA), one can assume a short external cavity. In particular, the self-image cavity can satisfy this requirement.

For a constant current injection, the system is an autonomous one. It is well-known [Lefschetz 1962, Arnold 1973] that for an autonomous nonlinear dynamical system $dx/dt = f(x)$, where x is a M-vector, the stability criteria of the system is equivalent to the corresponding linearized system $dx/dt = Ax + g(x)$, where A is the coefficient matrix and $g(x)$ is the nonlinear part, provided $\partial g/\partial x_j$ approach zero uniformly in t as the equilibrium solution $\|x\|$ approaches zero. This is always the case if each component

of g(**x**) is a polymonial in x_i which begins with terms of
order 2 or higher. Put it differently, if each component of
g(**x**) is differentiable, it will also satisfy the condition.
It is also well-established that the linearized system is
stable if all the eigenvalues of the system is negative, or
the real part of the eigenvalues be negative. From these
established results, we have the following algorithm for
determining the stability region of the coupled
semiconductor phased array. We first transform the coupled
dynamical system into standard form, i.e., first find the
equilibrium (steady state) solutions and transform the
origin to the equilibrium points by setting **z** = **x** - **x**$_o$. We
then linearize the system and write f(**x**$_o$ + **z**) in the form A**z**
+ g(**z**) where A is the linear part of f(**x**$_o$ + **z**). We then
compute the eigenvalues of the linearized system, i.e., A.
By varying all the operating parameters of the system, and
finding the values where all the eigenvalues are negative,
one can then "map" out the stability region of the system.

Clearly, the system, Eqs.(7.8-1-3), is very complex and
nearly intractable even numerically. Nonetheless, we shall
discuss its qualitative properties in a series of full
length papers elsewhere. The most simple case which still
preserves both the self-feedback and cross-feedback is the
coupling of two semiconductor lasers in an external cavity.
As we have just mentioned, in order to avoid delay feedback
instability, and allow us to use SVEA, we assume a short
external cavity, such that $\tau < \tau_s$. The typical carrier
diffusion time is of the order of 1 nsec, then the external
cavity length is assumed to be 2 cm. Thus we have $\tau_{12} = \tau_{21} = \tau_{11} = \tau_{22} = \tau$. From SVEA, we have $I_i(t - \tau) = I_i(t)$.
Furthermore, from the symmetry of ξ_{ij} we have, $\xi_{12} = \xi_{21} = \kappa$.
We than have the following set of six first order
nonlinearly coupled ordinary differential equations. Notice
the symmetry between 1 and 2, in later discussions this
symmetry should be kept in mind.

$$dN_1/dt = J/ed - N_1/\tau_s - G[I_1(t) + \kappa^{\frac{1}{2}} I_2(t - \tau)$$

$$\cdot \ \cos(\omega_2\tau + \phi_{o2})] + BN_1{}^2 + AN_1{}^3, \tag{7.8-6a}$$

$$dN_2/dt = J/ed - N_2/\tau_s - G[I_2(t) + \kappa^{\frac{1}{2}}I_1(t - \tau)$$
$$\cdot \ \cos(\omega_1\tau + \phi_{o1})] + BN_2{}^2 + AN_2{}^3, \tag{7.8-6b}$$

$$dI_1/dt = [G - 1/\tau_l]I_1(t) + \kappa^{\frac{1}{2}}[I_1(t)I_2(t - \tau)]^{\frac{1}{2}}$$
$$\cdot \ \cos[\phi_1(t) - \phi_2(t - \tau) + \omega_2\tau + \phi_{o2}], \tag{7.8-6c}$$

$$dI_2/dt = [G - 1/\tau_l]I_2(t) + \kappa^{\frac{1}{2}}[I_2(t)I_1(t - \tau)]^{\frac{1}{2}}$$
$$\cdot \ \cos[\phi_2(t) - \phi_1(t - \tau) + \omega_1\tau + \phi_{o1}], \tag{7.8-6d}$$

$$d\phi_1/dt = \alpha[G - 1/\tau]/2 - [\omega_1(N_1) - \omega] - \kappa^{\frac{1}{2}}[I_1(t - \tau)/I_1(t)]^{\frac{1}{2}}$$
$$\cdot \ \sin[\phi_1(t) - \phi_1(t - \tau) + \omega_1\tau + \phi_{o1}], \tag{7.8-6e}$$

$$d\phi_2/dt = \alpha[G - 1/\tau]/2 - [\omega_2(N_2) - \omega] - \kappa^{\frac{1}{2}}[I_2(t - \tau)/I_2(t)]^{\frac{1}{2}}$$
$$\cdot \ \sin[\phi_2(t) - \phi_2(t - \tau) + \omega_2\tau + \phi_{o2}], \tag{7.8-6e}$$

As we have prescribed earlier, we can transform the "origin" of the system to $(\bar{N}_1, \bar{N}_2, \bar{I}_1, \bar{I}_2, \bar{\phi}_1, \bar{\phi}_2)$, then linearize the system and obtain the matrix A. The eigenvalues of A are then computed for each set of operating parameters. We shall not go into any detail in doing so [Lee 1988]. It suffices to say that with a large size computer, one can find out within what range of operating parameters the system of two semiconductor lasers coupled by a short external cavity will become phase locked and remain locked indefinitely. Of course, one can also find the stability boundary of the operating parameters. For applications, Lee [1988] has found that the stability of these two coupled semiconductor lasers is sensitive to the coupling coefficient (or feedback amount), but most sensitive to the frequency detuning between the two lasers and their relative initial phases.

It is more interesting to know when those operating (control) parameters cross the stability boundary, in other words, what are the routes of chaos will this system follow?

There have been many experimental attempts to map out the dynamics of this "simple" system, yet none have been successful. The reason is that the detectors and oscilloscopes are too slow (usually of the order of one nanosecond) while the coupling of two diodes is taking place at about 30 - 50 psec. Thus, the oscilloscope displays a big blob of hundreds of return maps. Certainly, this situation is not satisfactory to say the least. Another approach is to use a streak camera [Elliott 1985]. But that is not satisfactory either due to the fact that it can not take sufficient samples for a short time span, i.e., to map out the time evolution of the dynamics. Consequently, one has to find some way to overcome the difficulty posed by the fast coupling of diode lasers.

We would like to point out from Eqs.(7.8-6) that if the two coupled diode lasers are operated at small signal gain regime, i.e., $G(N_i) = aN_i + b$, where a and b are constants, the only nonlinear coupling terms between N_i and E_i (here i = 1,2) are of the form of direct product. In this case, the phase space of this system is a product space. In other words, the phase space of this system can be decomposed into two "orthogonal" subspaces, the recombination rate subspace, N_1 and N_2, a two-dimensional subspace, and the subspace of amplitudes and phases of the two lasers, a four-dimensional subspace. Due to the product nature of the phase space, one can determine the dynamics of the system both experimentally and theoretically by studying its two orthogonal subspaces. Simply put, it is analogous to the understanding of the helical motion of an electron under the influence of a constant magnetic field. It is the product of the circular motion of an electron under a constant magnetic field, and the motion of the electron along z-axis. The composite motion is the product action.

The trajectories "projected" on the (N_1, N_2)-subspace may be complicated, nonetheless tractable. Furthermore, one can solve the dynamics of N_1 and N_2 rate equations fairly easily by keeping the rest constants. Moreover, from the time scale, the carrier recombination rate is about one nanosecond, longer than the field couplings, which are of the order of 30 - 50 psec, or the round trip time of the external cavity. Nonetheless, in order to study the dynamics of the two semiconductor lasers, without mixing the instabilities and chaos induced by the nonlinear delayed feedback [Ikeda 1979], the external cavity has to be very short. Indeed, in order to aviod the complexity of the nonlinear delayed feedback instability, the external cavity round trip time should not be greater than 0.1 nsec. So, the main effort of the study should be on the coupling dynamics of the field amplitudes and phases, that is, the study of the four-dimensional subspace of the field amplitudes and phases.

Theoretical study of the coupled six equations, Eqs.(7.8-6), is straight forward in the sense that with appropriate numerical schemes and large enough computer, one can have detailed numerical simulations and beautiful graphics for the dynamics of the system. And indeed, we can theoretically answer the questions we have posed earlier. What we are interested in, concerned with, and proposing to do, instead, is to find an experimental approach to study the system, which will be more satisfying. As we have pointed out earlier, direct measurements are hopeless due to the extremely fast coupling of the system. This is true even with the help of the very recent silicon photodetectors invented by Seigmann [1989], which have detector response time of a few psec. This is because the detectors can not have a sampling rate of 100 GHz or higher. Indeed, it is unlikely such a device will be available in the near future. It is then clear that some means of direct detection with a much slower sampling rate is necessary.

In the following, we will propose a set of experiments to be performed, which can unambiguously map out the coupling dynamics of the four-dimensional subspace of the fields and phases of two lasers experimentally. The idea is to use a laser medium which has a long radiative decay rate, such as those solid-state lasers, e.g., Nd:YAG laser, which has decay rate of 0.23 msec, or gas lasers, such as CO_2 laser, which has decay rate of 0.4 msec. If the external cavity round trip time is about 1 or 2 nsec, then the coupling time can only be longer than this, which can be resolved by the state of art detectors with time resolution of 0.1 nsec. In fact, we may not need such fast detectors. This is because even though the external cavity round trip time is about 2 nsec, due to relatively "low" gain of these laser media, effective coupling may take several round trips, thus the coupling time is on the order of tens of nsec. At any rate, this is not very important to our discussion, we only want to point out that relatively fast yet inexpensive detectors will be sufficient.

Even for these slow decay rate laser media, we still have a set of six coupled, nonlinear differential equations to deal with, which included two laser rate equations. But since the coupling time is stretched rather long, compared with semiconductor lasers, to a few to tens of nsecs, to accommodate the slow detector response time, it is still far too short compared with the radiative decay time of the laser which is of the order of 0.2 to 0.4 msec. Thus, the two laser rate equations can be considered as steady-state. Thus, we can study the nonlinear coupling of the fields and phases of the two lasers using existing technology.

Here we shall briefly discuss the four control parameters for the study of the dynamics of nonlinear coupling of two (say solid-state) lasers. For the sake of illustration, let us here consider the coupling of two small Nd:YAG lasers. Let these two Nd:YAG lasers be monolithic ones, such as the one for frequency stable application consatructed by Zhou et al [1985], but with somewhat larger gain volume. And they are temperature controlled with electric coolers. Let laser A be the "reference" one, that is, except the output power, no other parameters will be varied. Thus, this laser can be truely monolithic in design. The other laser, laser B, has PZT mirror as the back mirror,

thus the laser is slightly tunable. The power reflectivity
of the outcoupling mirror is the same as the "reference"
laser so that they all have the same amount of feedback.
Outside the outcoupler of laser B, there is an electro-optic
phase modulator, this can change the relative phase of these
two lasers. And in front of the coupling mirror, we can put
a calibrated neutral density filter to attenuate and to
control the mutual coupling of these two lasers. By
controlling the pump power of laser B, we can also vary the
output power density of laser B. This in turn will change
the relative power density of the two lasers, and
consequently, the relative feedback amount of these two
lasers. Of course, there are other control parameters one
can vary, but in order to closely simulate a diode phased
array, these four control parameters for the
four-dimensional subspace are the only relevant ones. For
diodes, the inject current densities are the other relevant
parameters.

In solving Eqs.(7.8-6c,d,e,f), we can restate these
equations by subtracting these two set of equations and
obtaining the set of two nonlinear differential equations of
the difference of field amplitude, ΔE, the difference of
resonance frequencies, $\Delta \omega$, and the difference of the phases
of these two lasers, $\Delta \phi$. Together with the other control
parameter, the feedback coupling coefficient κ, we have the
completely deterministic system. We have also simplified not
only the system, but also made the experiment simpler
because we only have to measure all the variables or control
parameters in their difference (or relative values).
Nonetheless, this simplification should not overshadow the
crucial importance of "slowing down" the dynamical coupling
of two lasers by using lasers of very slow radiative decay
rates.

We believe only if this four-dimensional subspace is
well understood experimentally, then do we have a chance to
understand experimentally the full system, i.e., the
coupling of two semiconductor lasers.

Now let us comment on the semiconductor laser phased array. Of course, there are many ways to phase lock semiconductor laser array, but the most esthetic as well as the simplest from the system point of view is the mutual feedback approach such as the self-imaging of the Talbot cavity [Jansen et al 1989, D'Amato et al 1989]. In principle, if the two-dimensional array has a very large number of "identical" elements (in both the geometric and physical sense), then with a very short time the array will self- adjust to the external cavity and a stable phase locked two- dimensional array can be formed. Nonetheless, the physical characteristics of each of the elements are critical to the phase locking process as well as the stability of such locking. As we have pointed out earlier, the locking is less sensitive to the feedback amount and the field intensity and gain inside each semiconductor lasers, and the locking is very sensitive to the frequency detuning and the phase. But these two most sensitive variables are very much determined by the semiconductor laser material processing. Indeed, in studying this 3N-dimensional system, where N is the number of semiconductor lasers and N is very large, one has to resort to nonlinear stochastic differential equations. The solution, if it exists, can give the statistical properties or requirements of the semiconductor material processing. Which in turn can give the material tolerance of the semiconductor laser phased array. Now we are getting into the real application, namely, the material engineering.

An interesting problem closely related to the problem discussed above is the random neural networks. Recently, Sompolinsky and Crisanti [1988] have studied a continuous-time dynamical model of a network of N nonlinear elements interacting via random asymmetric couplings. They have found that by using a self-consistent mean-field one can predict a transition from a stationary pahse to a chaotic phase occurring at a critical value of the gain parameter. In fact, for most region of the gain parameters, the only

stable solution is the chaotic solution.

References

N. B. Abraham, in **Far From Equilibrium Phase Transitions**, ed. by L. Garrido, Springer-Verlag, 1988.

N. B. Abraham, in 1989 Yearbook of Encyclopedia of Physical Science and Technology, Academic Press, 1989.

N. B. Abraham, F. T. Arecchi and L. A. Lugiato, ed., **Instabilities and Chaos in Quantum Optics II**, Plenum, 1988.

N. B. Abraham, T. Chyba, M. Coleman, R. S. Gioggia, N. Halas, L. M. Hoffer, S. N. Liu, M. Maeda and J. C. Wesson, in **Laser Physics**, ed. by J. D. Harvey and D. F. Walls, Springer-Verlag, 1983.

N. B. Abraham, D. Dangoisse, P. Glorieux, and P. Mandel, Opt. Soc. Am. B, **2** (1985) 23.

N. B. Abraham, L. A. Lugiato, P. Mandel, L. M. Narducci and D. K. Bandy, Opt. Soc. Am. B, **2** (1985) 35.

N. B. Abraham, L. A. Lugiato and L. M. Narducci, Opt. Soc. Am. B, **2** (1985) 7.

R. Abraham and J. Marsden, **Foundations of Mechanics**, Second Ed., Benjamin, 1978.

R. Abraham, J. Marsden and T. Ratiu, **Manifolds, Tensor Analysis, and Applications**, Addison-Wesley, 1983.

R. Abraham and S. Smale, in **Global Analysis**, ed. by S. S. Chern and S. Smale, 1970.

J. F. Adams, Ann. Math. **75** (1962) 603.

V. S. Afraimovich and Ya. B. Pesin, in **Mathematical Physics Reviews, Vol.6**, ed. by Ya. G. Sinai, Harwood Academic Pub., 1987.

V. S. Afraimovich and L. P. Sil'nikov, Sov. Math. Dokl., **15** (1974) 206.

A. K. Agarwal, K. Banerjee and J. K. Bhattacharjee, Phys. Lett. A, **119** (1986) 280.

J. K. Aggarwal and M. Vidyasagar, ed., **Nonlinear Systems: Stability Analysis**, Academic Press, 1977.

G. P. Agrawal, Appl. Phys. Lett., **49** (1986) 1013.

G. P. Agrawal, Phys. Rev. Lett., **59** (1987) 880.

A. M. Albano, J. Abounadi, T. H. Chyba, C. E. Searle, and S. Yong, Opt. Soc. Am. B, **2** (1985) 47.

Am. Math. Soc. Translation Ser. 1, Vol. 9 (1962).

L. W. Anacker and R. Kopelman, Phys. Rev. Lett., **58** (1987) 289.

U. an der Heiden, in **Differential-Difference Equations**, ed. by Collatz et al, Birkhauser Verlag, 1983.

U. an der Heiden and H.-O. Walther, J. Diff. Eqs., **47** (1983) 273.

R. D. Anderson, ed., **Symposium on Infinite Dimensional Topology**, Princeton, 1972.

R. F. S. Andrade, J. Math. Phys., **23** (1982) 2271.

A. A. Andronov and L. S. Pontryagin, Dokl. Akad. Nauk. SSSR, **14** (1937) 247.

A. A. Andronov, E. A. Vitt and S. E. Khaiken, **Theory of Oscillators**, Pergamon, 1966.

D. Anosov, Sov. Math. Dokl., **3** (1962) 1068.

H. A. Antosiewicz, in S. Lefschetz, 1958.

F. T. Arecchi, W. Gadomski, and R. Meucci, Phys. Rev. A, **34** (1986) 1617.

F. T. Arecchi, W. Gadomski, R. Meucci and J. A. Roversi, Opt. Communi., **70** (1989) 155.

F. T. Arecchi, R. Meucci, and W. Gadomski, Phys. Rev. Lett., **58** (1987) 2205.

F. Argoul, A. Arneodo and P. Richetti, Phys. Lett. A, **120** (1987) 269.

D. Armbruster, Z. Phys. B, **53** (1983) 157.

R. A. Armstrong and R. McGehee, J. Theor. Biol., **56** (1976) 499.

A. Arneodo, P. Coullet and C. Tresser, Phys. Lett. A, **79** (1980) 259.

L. Arnold, W. Horsthemake and R. Lefever, Z. Physik B, **29** (1978) 367.

V. I. Arnold, Russ. Math. Surv., **18** (1963) 9; 85.

V. I. Arnold, **Mathematical Methods of Classical Mechanics**, Springer-Verlag, 1978.

V. Arzt et al, Z. Phys., **197** (1966) 207.

U. Ascher, P. A. Markowich, C. Schmeiser, H. Steinruck and
    R. Weiss, SIAM J. Appl. Math., **49** (1989) 165.

M. F. Atiyah, Sem Bourbaki, Mai 1963.

M. F. Atiyah and I. M. Singer, Bull. Am. Math. Soc. **69**
    (1963) 422.

M. F. Atiyah and I. M. Singer, Ann. Math. **87** (1968) 484;
    546.

D. Auerbach, P. Cvitanovic, J.-P. Eckmann, G. Gunaratne and
    I. Procaccia, Phys. Rev. Lett., **58** (1987) 2387.

J. Auslander and P. Seibert, in **Nonlinear Differential
    Equations and Nonlinear Mechanics**, ed. by J. P. LaSalle
    and S. Lefschetz, Academic Press, 1963.

L. Auslander, L. Green and F. Hahn, ed., **Flows on
    Homogeneous Spaces**, Princeton, 1963.

L. Auslander and R. E. MacKenzie, **Introduction to
    Differential Manifolds**, McGraw-Hill, 1963.

A. Avez, A. Blaquiere and A. Marzollo, **Dynamical Systems and
    Microphysics**, Academic Press, 1982.

S. M. Baer and T. Erneux, SIAM J. Appl. Math., **46** (1986)
    721.

S. M. Baer, T. Erneux and J. Rinzel, SIAM J. Appl. Math., **49**
    (1989) 55.

P. Bak, Phys. Today, Dec. 1986, 38.

P. Bak, C. Tang and K. Wiesenfeld, Phys. Rev. Lett., **59**
    (1987) 381.

H. Baltes, **Inverse Scattering in Optics**, Springer-Verlag,
    1980.

D. K. Bandy, L. M. Narducci, and L. A. Lugiato, J. Opt. Soc.
    Am. B **2**, (1985) 148.

D. K. Bandy, L. M. Narducci, C. A. Pennise, and L. A.
    Lugiato, in **Coherence and Quantum Optics V**, ed. by L.
    Mandel and E. Wolf, Plenum, 1984.

C. Bardos and D. Bessis, ed., **Bifurcation Phenomena in
    Mathematical Physics and Related Topics**, D. Reidel,
    1980.

C. Bardos, J. M. Lasry, and M. Schatzman, ed., **Bifurcation
    and Nonlinear Eigenvalue Problems**, Springer-Verlag, 1980.

V. Bargmann, **Group Representations in Mathematical Physics**, Springer-Verlag, 1970.

J. D. Barrow, Phys. Rev. Lett., **46** (1981) 963.

J. D. Barrow, Gen. Rel. & Gravit., **14** (1982) 523.

J. D. Barrow, Phys. Rep., **85** (1982) 1.

K. Baumgartel, U. Motschmann, and K. Sauer, in **Nonlinear and Tubulent Processes in Physics**, ed. by R. Z. Sagdeev, Hardwood Academic Pub., 1984.

D. Baums, W. Elsasser, and E. O. Gobel, Phys. Rev. Lett., **63** (1989) 155.

A. D. Bazykin, Int. Inst. Appl. Syst. Analysis, Laxenburg, 1976.

T. Bedford and J. Swift, ed., **New Directions in Dynamical Systems**, Cambridge U. Press, 1988.

A. R. Bednarek and L. Cesari, Ed., **Dynamical Systems II**, Academic Press, 1982.

G. I. Bell and E. C. Anderson, Biophys. J., **7** (1967) 329.

E. Beltrami, **Mathematics for Dynamic Modeling**, Academic Press, 1987.

J. Bentley and N. B. Abraham, Opt. Commun. **41** (1982) 52.

E. Beretta and F. Solimano, SIAM J. Appl. Math., **48** (1988) 607.

P. Berge, Y. Pomeau and C. Vidal, **Order within Chaos**, Wiley, 1987.

M. S. Berger, **Nonlinearity and Functional Analysis**, Academic, 1977.

M. S. Berger and P. C. Fife, Comm. Pure & Appl. Math., **21** (1968) 227.

M. V. Berry, J. Phys. A, **8** (1975) 566.

M. V. Berry, J. Phys. A, **10** (1977) 2061.

M. V. Berry, I. C. Percival and N. O. Weiss, ed., **Dynamical Chaos**, Princeton Univ. Press, 1987.

M. V. Berry and C. Upstill, in **Progress in Optics, V.18**, ed. by E. Wolf, North-Holland, 1980.

V. I. Bespalov and V. I. Talanov, Sov. Phys. JETP Lett., **3** (1966) 307.

N. P. Bhatia and V. Lakshmikantham, Mich. Math. J., **12**

(1965) 183.

N. P. Bhatia and G. P. Szego, **Dynamical Systems: Stability Theory and Applications**, Springer-Verlag, 1967.

N. P. Bhatia and G. P. Szego, **Stability Theory of Dynamical Systems**, Springer-Verlag, 1970.

C. K. Biebricher, in **Evolutionary Biology, 16**, ed. by M. K. Hecht et al, Plenum Press, 1983.

O. Biham and W. Wenzel, Phys. Rev. Lett., **63** (1989) 819.

G. D. Birkhoff, **Dynamical Systems**, Am. Math. Soc., 1927.

J. S. Birman and R. F. Williams, Top., **22** (1983) 47.

J. S. Birman and R. F. Williams, Contemp. Math., **20** (1983) 1.

B. Birnir and P. J. Morrison, Univ. of Texas, Inst. for Fusion Studies, #243 (1986).

A. R. Bishop, L. J. Campbell and P. J. Channell, ed., **Fronts, Interfaces and Patterns**, North-Holland, 1984.

A. R. Bishop, D. Campbell, and B. Nicolaenko, ed., **Nonlinear Problems: Present and Future**, North-Holland, 1982.

A. R. Bishop, K. Fesser, P. S. Lomdahl, W. C. Kerr, M. B. Williams, and S. E. Trullinger, Phys. Rev. Lett., **50** (1983) 1095.

A. R. Bishop, D. W. McLaughlin, M. G. Forest and E. A. Overman II, Phys. Lett. A, **127** (1988) 335.

R. L. Bishop and R. J. Crittenden, **Geometry of Manifolds**, Academic Press, 1964.

D. J. Biswas and R. G. Harrison, Appl. Phys. Lett., **47** (1985) 198.

S. Bleher, E. Ott and C. Grebogi, Phys. Rev. Lett., **63** (1989) 919.

K. Bleuler et al (ed), **Differential Geometric Methods in Mathematical Physics**, Springer-Verlag, Vol.1 1977; Vol.2 1979.

K. J. Blow, N. J. Doran, and D. Wood, Opt. Lett., **12** (1987) 202.

E. J. Bochove, Opt. Lett., **11** (1986) 727.

Yu. L. Bolotin, V. Yu. Gonchar, V. N. Tarasov and N. A. Chekanov, Phys. Lett. A, **135** (1989) 29.

R. Bonifacio and L. A. Lugiato, Opt. Commun. **19** (1976) 172.

R. Bonifacio and L. A. Lugiato, Phys. Rev. A, **18** (1978) 1129.

R. Bonifacio and L. A. Lugiato, Lett. N. Cimento, **21** (1978) 505.

R. Bonifacio, C. Pellegrini, and L. M. Narducci, Opt. Communi., **50** (1984) 373.

R. Bott, **Lectures on Morse Theory**, Universität Bonn, 1960.

R. Bott and J. Milnor, Bull. Amer. Math. Soc. **64** (1958) 87.

J. L. Boulnois, A. Van Lergerghe, P. Cottin, F. T. Arecchi, and G. P. Puccioni, Opt. Communi., **58** (1986) 124.

D. G. Bourgin, **Modern Algebraic Topology**, MacMillan, 1963.

C. M. Bowden, M. Ciftan and H. R. Robl, ed. **Optical Bistability**, Plenum Press, 1981.

C. M. Bowden, H. M. Gibbs and S. L. McCall, ed. **Optical Bistability 2**, Plenum Press, 1984.

R. W. Boyd, M. G. Raymer and L. M. Narducci, ed. **Optical Instabilities**, Cambridge U. Press, 1986.

A. Brandstater, J. Swift, H. L. Swinney, A. Wolf, J. D. Farmer, E. Jen, and P. J. Crutchfield, Phys. Rev. Lett., **51** (1983) 1442.

T. Braun, J. A. Lisboa and R. E. Francke, Phys. Rev. Lett., **59** (1987) 613.

L. Brenig, Phys. Lett. A, **133** (1988) 378.

K. Briggs, Am. J. Phys., **55** (1987) 1083.

Th. Brocker, **Differentiable Germs and Catastrophes**, Cambridge U. Press, 1975.

D. C. Brown, **High-Peak-Power Nd:Glass Laser Systems**, Springer-Verlag, 1981.

M. Brown, Ann. Math. **75** (1962) 331.

R. Brown, E. Ott and C. Grebogi, Phys. Rev. Lett., **59** (1987) 1173.

W. Brunner and H. Paul, Opt. and Quant. Elect., **15** (1983) 87.

V. Brunsden and P. Holmes, Phys. Rev. Lett., **58** (1987) 1699.

P. Bryant and C. Jeffries, Phys. Rev. Lett., **53** (1984) 250.

J. R. Buchler and H. Eichhorn, ed., **Chaotic Phenomena in**

404

Astrophysics, N. Y. Academy of Sci., 1987.
J. Buchner and L. M. Zeleny, Phys. Lett. A, **118** (1986) 395.
D. Burghelea and N. Kuiper, Ann. Math. **90** (1969) 379.
D. Burghelea and R. Lashof, Trans. Am. Math. Soc. **196** (1974) 1.
D. Burghelea and R. Lashof, Trans. Am. Math. Soc. **196** (1974) 37.
W. L. Burke, **Applied Differential Geometry**, Cambridge Univ. Press, 1985.
A. V. Butenko, V. M. Shalaev, and M. I. Stockman, Z. Phys. D, **10** (1988) 81.
J. G. Byatt-Smith, SIAM J. Appl. Math., **47** (1987) 60.

M. Campanino, H. Epstein and D. Ruelle, Top., **21** (1982) 125.
H. J. Carmichael, Phil. Trans. R. Soc. Lond. A, **313** (1984) 433.
H. J. Carmichael, Phys. Rev. Lett., **52** (1984) 1292.
H. J. Carmichael, Opt. Communi., **53** (1985) 122.
J. Carr, **Applications of Centre Manifold Theory**, Springer-Verlag, 1981.
M. L. Cartwright, J. Inst. Elect. Eng., **95** (1948) 88.
L. W. Casperson, IEEE J. Quant. Elect., **QE-14** (1978) 756.
L. W. Casperson, Phys. Rev. A, **21** (1980) 911.
L. W. Casperson, Phys. Rev. A, **23** (1981) 248.
L. W. Casperson, J. Opt. Soc. Am. B, **2** (1985) 62.
L. W. Casperson, J. Opt. Soc. Am. B, **2** (1985) 73.
L. W. Casperson, J. Opt. Soc. Am. B, **2** (1985) 993.
L. W. Casperson, in **Optical Instabilities**, ed. by R. W. Boyd, M. G. Raymer, and L. M. Narducci, Cambridge U. Press, 1986.
L. W. Casperson, Opt. and Quant. Elect., **19** (1987) 29.
A. M. Castelfranco and H. W. Stech, SIAM J. Appl. Math., **47** (1987) 573.
L. Cesari, J. K. Hale, and J. P. LaSalle, ed., **Dynamical Systems. An International Symposium**, 2 Vols., Academic Press, 1976.
W.-C. C. Chan, SIAM J. Appl. Math., **47** (1987) 244.

S. Chandrasekar, **Hydrodynamic and Hydromagnetic Stability**,
Oxford Univ. Press, 1961. A Dover reprint edition is
also available.

H. Chate and P. Manneville, Phys. Rev. Lett., **58** (1987) 112.

L.-X. Chen, C.-F. Li, Q.-S. Hu, and J.-F. Li, J. Opt. Soc.
Am. B, **5** (1988) 1160.

Y. C. Chen, Appl. Phys. Lett., **44** (1984) 10.

Y. C. Chen, H. G. Winful, and J. M. Liu, Appl. Phys. Lett.,
47 (1985) 208.

Y.-J. Chen and A. W. Snyder, Opt. Lett., **14** (1989) 1237.

J.-L. Chern and J.-T. Shy, Opt. Communi., **69** (1988) 108.

S. S. Chern, **Topics in Differential Geometry**, Institute of
Advanced Study Lecture Notes, 1951.

S. S. Chern and R. Osserman, ed., **Differential Geometry**,
Sym. in Pure Math., Vol.27, 2 Parts, Am. Math. Soc.,
1975.

S. S. Chern and S. Smale, ed., **Global Analysis**, Sym. in Pure
Math., Vol.14-16, Am. Math. Soc., 1970.

A. A. Chernikov, R. Z. Sagdeev, D. A. Usikov, M. Yu Zakharov
and G. M. Zaslavsky, Nature, **326** (1987) 559.

A. A. Chernikov, R. Z. Sagdeev, and G. M. Zaslavsky, Phys.
Today, (Nov. 1988) 27.

P. R. Chernoff and J. Marsden, **Properties of Infinite
Dimensional Hamiltonian Systems**, Springer-Verlag, 1974.

N. G. Chetaev, Dokl. Akad. Nauk. SSSR, **1** (1934) 529.

P. Y. Cheung, S. Donovan and A. Y. Wong, Phys. Rev. Lett.,
61 (1988) 1360.

C. Chevalley, **Theory of Lie Groups**, Princeton Univ. Press,
1946.

C. Chevalley, **Theorie des Groupes de Lie, II**, Hermann and
Cie, 1951.

Y. Choquet-Bruhat, C. DeWitt and M. Dillard-Bleick,
Analysis, Manifolds, and Physics, North-Holland, 1977.

S. N. Chow and J. K. Hale, **Methods of Bifurcation Theory**,
Springer-Verlag, 1982.

J. Chrostowski and N. B. Abraham, ed., **Opticl Chaos**, Proc.
SPIE, **667** (1986).

L. O. Chua and R. N. Madan, IEEE Circuits & Dev., **4** (1988) 3.

S. Ciliberto and J. P. Gollub, J. Fluid Mech., **158** (1985) 381.

S. Ciliberta and M. A. Rubio, Phys. Rev. Lett., **58** (1987) 2652.

P. Collet and J.-P. Eckmann, **Iterated Maps on the Interval as Dynamical Systems**, Birkhauser, 1980.

P. Collet and J.-P. Eckmann, **Instabilities and Fronts in Extended Systems**, Princeton Univ. Press, 1990.

P. Collet, J.-P. Eckmann and L. Thomas, Commun. Math. Phys., **81** (1981) 261.

P. Collet, C. Tresser and A. Arneodo, Phys. Lett. A, **77** (1980) 327.

F. Conrad and N. Yebari, SIAM J. Appl. Math., **49** (1989) 134.

G. Contopoulos, in **Nonlinear Phenomena in Physics**, ed. by F. Claro, Springer-Verlag, 1985.

S. N. Coppersmith, Phys. Lett. A, **125** (1987) 473.

C. Corduneanu, Iasi Sect. I a Mat., **6** (1960) 47.

P. Coullet and C. Elphick, Phys. Lett. A, **121** (1987) 233.

P. Coullet, C. Elphick, L. Gil and J. Lega, Phys. Rev. Lett., **59** (1987) 884.

P. Coullet, C. Elphick and D. Repaux, Phys. Rev. Lett., **58** (1987) 431.

P. Coullet, C. Tresser and A. Arneodo, Phys. Lett. A, **72** (1979) 268.

P. Coullet, C. Tresser and A. Arneodo, Phys. Lett. A, **77** (1980) 327.

J. P. Crutchfield, Physica D, **10** (1984) 229.

J. P. Crutchfield, J. D. Farmer, N. H. Packard and R. S. Shaw, Sci. Am., Dec. (1986) 46.

J. P. Crutchfield and K. Kaneko, Phys. Rev. Lett., **60** (1988) 2715.

A. Cumming and P. S. Linsay, Phys. Rev. Lett., **59** (1987) 1633.

P. Cvitanovic, in **Instabilities and Dynamics of Lasers and Nonlinear Optical Systems**, ed. by R. W. Boyd, L. M.

Narducci and M. G. Reymer, Cambridge U. Press, 1985.
P. Cvitanovic, Phys. Rev. Lett., **61** (1988) 2729.

G. Daccord, Phys. Rev. Lett., **58** (1987) 479.
F. X. D'Amato, E. T. Siebert, and C. Roychoudhuri, in <u>Digest of Conference on Lasers and Electro-optics</u>, Optical Soc. of America, 1989, paper FL3.
M. V. Danileiko, A. L. Kravchuk, V. N. Nechiporenko, A. M. Tselinko, and L. P. Yatsenko, Sov. J. Quant. Elect., **16** (1986) 1420.
A. Dankner, Asterisque, **49** (1977) 19.
G. Dangelmayr and D. Armbruster, Proc. Lond. Math. Soc., **46** (1983) 517.
G. Dangelmayr and W. Guttinger, Geophys. J. Roy. Astro. Soc. **71** (1982) 79.
N. N. Degtyarenko, V. F. Elesin and V. A. Furmenov, Sov. Phys. Semicond., **7** (1974) 1147.
R. J. Deissler and K. Kaneko, Phys. Lett. A, **119** (1987) 397.
H. Dekker, Phys. Lett. A, **119** (1986) 10.
F. Delyon, Y.-E. Levy and B. Souillard, Phys. Rev. Lett., **57** (1986) 2010.
P. de Mottoni and L. Salvadori, ed., **Nonlinear Differential Equations**, Academic Press, 1981.
G. C. Dente, P. S. Durkin, K. A. Wilson, and C. E. Moeller, "Chaos in the coherence collapse of semiconductor lasers", preprint, 1989.
M. de Oliviera, Bull. Am. Math. Soc., **82** (1976) 786.
M. W. Derstine, H. M. Gibbs, F. A. Hopf, and J. F. Valley, in **Optical Bistability III**, ed. by H. M. Gibbs, P. Mandel, N. Peyghambarian and S. D. Smith, Springer-Verlag, 1986.
R. L. Devaney, **An Introduction to Chaotic Dynamical Systems**, Benjamin, 1986.
R. L. Devaney and L. Keen, ed., **Chaos and Fractals**, Am. Math. Soc., 1989.
R. H. Dicke, Science **129** (1959) 3349.
R. H. Dicke, **The Theoretical Significance of Experimental**

Relativity, Gordon and Breach, 1962.

C. R. Doering, J. D. Gibbon, D. D. Holm and B. Nicolaenko, Phys. Rev. Lett., **59** (1987) 2911.

M. Dorfle and R. Graham, Phys. Rev. A, **27** (1983) 1096.

S. R. Dunbar, SIAM J. Appl. Math., **46** (1986) 1057.

J.-P. Eckmann, Rev. Mod. Phys., **53** (1981) 643.

J.-P. Eckmann and D. Ruelle, Rev. Mod. Phys., **57** (1985) 617.

J. Eells, in Symposium Internacional de Topologia Algebraica, Mexico, 1958.

J. Eells, Bull. Am. Math. Soc. **72** (1966) 751.

J. Eells, in **Nonlinear Functional Analysis**, Am. Math. Soc., 1970.

J. Eells and K. D. Elworthy, in **Global Analysis**, ed. by S. S. Chern and S. Smale, Am. Math. Soc., 1970.

J. Eells and K. D. Elworthy, Actes. Congres intern. Math., 1970, Tome 2, 215.

M. Eigen, P. Schuster, K. Sigmund and R. Wolff, Biosystems, **13** (1980) 1.

S. Eilenberg and N. Steenrod, **Foundations of Algebraic Topology**, Princeton Univ. Press, 1952.

R. A. Elliott, R. K. DeFreez, T. L. Paoli, R. D. Burnham and W. Streifer, IEEE J-QE, **21** (1985) 598.

R. Ellis, **Lectures on Topological Dynamics**, Benjamin, 1969.

S. Ellner, Phys. Lett. A, **133** (1988) 128.

K. D. Elworthy, Bull. Am. Math. Soc., **74** (1968) 582.

K. D. Elworthy and A. J. Tromba, in **Nonlinear Functional Analysis**, Am. Math. Soc., 1970.

J. C. Englund, Phys. Rev. A, **33** (1986) 3606.

J. C. Englund, R. R. Snapp, and W. C. Schieve, in **Progress in Optics Vol. 21**, ed. by E. Wolf, Elsevier Sci. Pub., 1984.

I. R. Epstein, J. Theor. Biol., **78** (1979) 271.

T. Erneux, E. L. Reiss, J. F. Magnan and P. K. Jayakumar, SIAM J. Appl. Math., **47** (1987) 1163.

F. Estabrook and H. Wahlquist, J. Math. Phys. **16** (1975) 1.

B. Etkin, **Dynamics of Flight: Stability and Control**, Wiley,

1959.

J. D. Farmer, Physica D, **4** (1982) 366.

J. D. Farmer, A. Lapedes, N. Packard and B. Wendroff, Ed.,
   **Evolution, Games and Learning Models for Adaptation in
   Mechines and Nature**, Physica D, **22** (1986).

J. D. Farmer, E. Ott and J. A. Yorke, Physica D, **7** (1983)
   153.

J. D. Farmer and I. I. Satija, Phys. Rev. A, **31** (1985) 3520.

J. D. Farmer and J. J. Sidorowich, Phys. Rev. Lett., **59**
   (1987) 845.

F. Favre, D. le Guen and J. C. Simon, IEEE J. QE, **18** (1982)
   1712.

M. J. Feigenbaum, J. Stat. Phys., **19** (1978) 25.

M. J. Feigenbaum, Phys. Lett. A, **74** (1979) 375.

M. J. Feigenbaum, Los Alamos Science, **1** (1980) 1.

F. S. Felber and J. H. Marburger, Appl. Phys. Lett., **28**
   (1976) 731.

K. Fesser, A. R. Bishop and P. Kumar, Appl. Phys. Lett., **43**
   (1983) 123.

A. F. Filippov, Trans. Am. Math. Soc., Ser. 2, **42** (1964)
   199.

W. J. Firth, Phys. Rev. Lett., **61** (1988) 329.

W. J. Firth and S. W. Sinclair, J. Mod. Opt., **35** (1988) 431.

R. FitzHugh, Biophys. J., **1** (1961) 445.

J. A. Fleck, Jr., J. Appl. Phys., **37** (1966) 88.

J. A. Fleck, Jr., J. R. Morris, and M. D. Feit, Appl. Phys.,
   10 (1976) 129.

M. Fleming and A. Mooradian, IEEE J. QE, **17** (1981) 44.

A. C. Fowler, Stud. Appl. Math., **70** (1984) 215.

A. C. Fowler and M. J. McGuinness, Physica D, **5** (1982) 149.

R. F. Fox, J. Math. Phys., **13** (1972) 1196.

R. F. Fox and G. E. Uhlenbeck, Phys. Fluids, **13** (1970) 1893,
   2881.

V. Franceschini, Phys. Fluids, **26** (1983) 433.

V. Franceschini, Physica D, **6** (1983) 285.

J. M. Franks, Proc. Amer. Math. Soc., **37** (1973) 293.

J. M. Franks, Invent. Math., **24** (1974) 163.

J. M. Franks, **Homology and Dynamical Systems**, Am. Math.
   Soc., 1982.

J. M. Franks and R. F. Williams, Trans. Am. Math. Soc., **291**
   (1985) 241.

H. I. Freedman and P. Waltman, Math. Biosci., **33** (1977) 257.

E. Frehland, ed., **Symergetics - From Microscopic to
   Macroscopic Order**, Springer-Verlag, 1984.

K. O. Friedrichs and J. Stoker, Am. J. Math., **63** (1941) 839.

M. Funakoshi and S. Inoue, Phys. Lett. A, **121** (1987) 229.

A. L. Gaeta, R. W. Boyd, J. R. Ackerhalt, and P. W. Milonni,
   Phys. Rev. Lett., **58** (1987) 2432.

I. Galbraith and H. Haug, J. Opt. Soc. Am. B, **4** (1987) 1116.

J. Y. Gao and L. M. Narducci, Opt. communi., **58** (1986) 360.

J. Y. Gao, L. M. Narducci, H. Sadiky, M. Squicciarini, and
   J. M. Yuan, Phys. Rev. A, **30** (1984) 901.

J. Y. Gao, L. M. Narducci, L. S. Schulman, M. Squicciarini,
   and J. M. Yuan, Phys. Rev. A, **28** (1983) 2910.

J. Y. Gao, J. M. Yuan, and L. M. Narducci, Opt. Communi., **44**
   (1983) 201.

T. Gard and T. Hallman, Bull. Math. Biol., **41** (1979) 877.

C. W. Gardiner, **Handbook of Stochastic Methods**, 2nd ed.,
   Springer-Verlag, 1985.

L. Garrido, ed., **Dynamical Systems and Chaos**,
   Springer-Verlag, 1983.

D. J. Gauthier, M. S. Malcuit, and R. W. Boyd, Phys. Rev.
   Lett., **61** (1988) 1827.

D. J. Gauthier, P. Narum, and R. W. Boyd, Phys. Rev. Lett.,
   58 (1987) 1640.

T. Geisel and J. Niewetberg, Phys. Rev. Lett., **47** (1981)
   975.

T. Geisel and J. Niewetberg, Phys. Rev. Lett., **48** (1982) 7.

T. Geisel, A. Zacherl and G. Radons, Phys. Rev. Lett., **59**
   (1987) 2503.

D. P. George, Phys. Lett. A, **118** (1986) 17.

K. Germey, F.-J. Schutte and R. Tiebel, Opt. Communi., **69**

(1989) 438.

G. E. O. Giaeaglia, **Perturbation Methods in Nonlinear Systems**, Springer-Verlag, 1972.

H. M. Gibbs, **Optical Bistability: Controlling Light with Light**, Academic Press, 1985.

H. M. Gibbs, F. A. Hopf, D. L. Kaplan, and R. L. Shoemaker, Phys. Rev. Lett., **46** (1981) 474.

H. M. Gibbs, P. Mandel, N. Peyghambarian and S. D. Smith, ed. **Optical Bistability 3**, Springer-Verlag, 1986.

H. M. Gibbs, S. L. McCall and T. N. C. Venkatesan, Phys. Rev. Lett., **36** (1976) 1135.

C. G. Gibson et al, ed., **Topological Stability of Smooth Mappings**, Springer-Verlag, 1976.

M. Giglio, S. Musazzi, and U. Perini, Phys. Rev. Lett., **47** (1981) 243.

R. Gilmore, **Catastrophe Theory for Scientistis and Engineers**, Wiley, 1981.

R. S. Gioggia and N. B. Abraham, Phys. Rev. Lett., **51** (1983) 650.

R. S. Gioggia and N. B. Abraham, Opt. Commun., **47** (1983) 278.

R. S. Gioggia and N. B. Abraham, Phys. Rev. A, **29** (1984) 1304.

P. Glas, R. Muller, and A. Klehr, Opt. Communi., **47** (1983) 297.

P. Glas, R. Muller, and G. Wallis, Opt. Communi., **68** (1988) 133.

J. Gleick, **Chaos**, Viking, 1987.

L. Goldberg, H. F. Taylor, A. Dandridge, J. F. Weller and R. O. Miles, IEEE J. QE, **18** (1982) 555.

H. Goldstein, **Classical Mechanics**, Addison-Wesley, 1950.

J. A. Goldstone, in **Laser Handbook**, ed. by M. L. Stitch and M. Bass, North-Holland, 1985.

J. A. Goldstone and E. M. Garmire, IEEE J. Quant. Elect., **19** (1983) 208.

J. P. Gollub and S. V. Benson, J. Fluid Mech., **100** (1980) 449.

J. P. Gollub and A. R. McCarriar, Phys. Rev. A, **26** (1982) 3470.

J. P. Gollub, A. R. McCarriar and J. F. Steinman, J. Fluid Mech., **125** (1982) 259.

J. P. Gollub and J. F. Steinman, Phys. Rev. Lett., **45** (1980) 551.

M. Golubitsky and J. M. Guckenheimer, ed., **Multiparameter Bifurcation Theory**, Am. Math. Soc., 1986.

M. Golubitsky and V. Guillemin, **Stable Mappings and Their Singularities**, Springer-Verlag, 1973.

M. Golubitsky and W. F. Langford, J. Diff. Eqs., **41** (1981) 375.

M. Golubitsky and D. Schaeffer, Commun. Pure Appl. Math., **32** (1979) 21.

D. L. Gonzalez and O. Piro, Phys. Rev. Lett., **50** (1983) 870.

R. Graham, in **Synergetics: From Microscopic to Macroscopic Order**, ed. by E. Frehland, Springer-Verlag, 1984.

R. Graham, Phys. Lett. A, **99** (1983) 131.

R. Graham, Phys. Rev. Lett., **53** (1984) 2020.

R. Graham and H. Haken, Z. Phys., **213** (1968) 420.

R. Graham and H. Haken, Z. Phys., **237** (1970) 31.

R. Graham and A. Wunderlin, ed., **Lasers and Synergetics**, Springer-Verlag, 1987.

P. Grassberger and I. Procaccia, Physica D, **9** (1983) 189.

B. F. Gray, Proc. Roy. Soc. London A, **415** (1988) 1.

B. F. Gray and M. J. Roberts, Proc. Roy. Soc. London A, **416** (1988) 361.

B. F. Gray and M. J. Roberts, Proc. Roy. Soc. London A, **416** (1988) 391.

B. F. Gray and M. J. Roberts, Proc. Roy. Soc. London A, **416** (1988) 403.

B. F. Gray and M. J. Roberts, Proc. Roy. Soc. London A, **416** (1988) 425.

A. Z. Grazyuk and A. N. Oraevskii, in **Quantum Electronics and Coherent Light**, ed. by P. A. Miles, Academic Press, 1964.

A. Z. Grazyuk and A. N. Oraevskii, Radio Eng. Elect. Phys.,

9 (1964) 424.

C. Grebogi, E. Kostelich, E. Ott and J. A. Yorke, Phys.
Lett. A, **118** (1986) 448.

C. Grebogi, S. W. McDonald, E. Ott and J. A. Yorke, Phys.
Lett. A, **99** (1983) 415.

C. Grebogi, E. Ott and J. A. Yorke, Phys. Rev. Lett., **48**
(1982) 1507.

C. Grebogi, E. Ott and J. A. Yorke, Phys. Rev. Lett., **51**
(1983) 339.

C. Grebogi, E. Ott and J. A. Yorke, Physica D, **7** (1983) 181.

C. Grebogi, E. Ott and J. A. Yorke, Phys. Rev. Lett., **57**
(1986) 1284.

M. Greenberg, **Lectures on Algebraic Topology**, Benjamin,
1967.

J. M. Greene, J. Math. Phys., **20** (1979) 1183.

J. M. Greene, R. S. MacKay, F. Vivaldi and M. J. Feigenbaum,
Physica D, **3** (1981) 468.

H.-R. Gregorius, Internat. J. Systems Sci., **8** (1979) 863.

W. Greub, S. Halperin and R. Vanstone, **Connections,
Curvature and Cohomology**, 3 vols, Academic, 1972.

M. Grmela and J. Marsden, ed., **Global Analysis**,
Springer-Verlag, 1979.

D. M. Grobman, Dokl. Akad. Nauk SSSR, **128** (1959) 880.

D. M. Grobman, Mat. Sb. (NS), **56** (1962) 77.

G. Grynberg, E. Le Bihan, P. Verkerk, P. Simoneau, J. R. R.
Leite, D. Bloch, S. Le Boiteux, and M. Ducloy, Opt.
Communi., **67** (1988) 363.

Y. Gu, D. K. Bandy, J. M. Yuan, and L. M. Narducci, Phys.
Rev. A, **31** (1985) 354.

Y. Gu, M.-W. Tung, J. M. Yuan, D. H. Feng, and L. M.
Narducci, Phys. Rev. Lett., **52** (1984) 701.

J. Guckenheimer, SIAM J. Math. Anal., **15** (1984) 1.

J. Guckenheimer and P. Holmes, **Nonlinear Oscillations,
Dynamical Systems, and Bifurcations of Vector Fields**,
Springer-Verlag, 1983.

J. Guckenheimer, J. Moser, and S. E. Newhouse, **Dynamical
Systems**, Birkhauser, 1980.

414

J. Guckenheimer and R. F. Williams, Pub. Math. IHES, **50** (1980) 73.

M. R. Guevara, L. Glass and A. Shrier, Science, **214** (1981) 1350.

G. H. Gunaratne, P. S. Linsay and M. J. Vinson, Phys. Rev. Lett., **63** (1989) 1.

G. H. Gunaratne and I. Procaccia, Phys. Rev. Lett., **59** (1987) 1377.

M. E. Gurtin and R. C. MacCamy, Arch. Rat. Mech. Anal., **54** (1974) 281.

A. G. Gurtovnick, Izv. Vyss. Uchebn. Zaved. Radiofiz., **1** (1958) 83.

W. Guttinger and H. Eikemeier, ed., **Structural Stability in Physics**, Springer-Verlag, 1979.

G. Gyorgyi and N. Tishby, Phys. Rev. Lett., **62** (1989) 353.

P. Hadley and M. R. Beasley, Appl. Phys. Lett., **50** (1987) 621.

W. Hahn, **Stability of Motion**, Springer-Verlag, 1967.

J. Hainzl, Aiam J. Appl. Math., **48** (1988) 170.

H. Haken, Z. Phys., **181** (1964) 96.

H. Haken, **Laser Theory**, Springer-Verlag, 1970.

H. Haken, Phys. Lett. A, **53** (1975) 77.

H. Haken, Rev. Mod. Phys., **47** (1975) 67.

H. Haken, **Synergetics**, Third Ed., Springer-Verlag, 1983.

H. Haken, **Advanced Symergetics**, Springer-Verlag, 1983.

H. Haken, in **Optical Instabilities**, ed. by R. W. Boyd, M. G. Raymer and L. M. Narducci, Cambridge U. Press, 1986.

H. Haken and H. Sanermann, Z. Phys., **176** (1963) 47.

J. K. Hale, **Theory of Functional Differential Equations**, Springer-Verlag, 1977.

J. K. Hale, **Topics in Dynamical Bifurcation Theory**, Am. Math. Soc., 1981.

J. K. Hale, **Asymptotic Behavior of Dissipative Systems**, Am. Math. Soc., 1988.

J. K. Hale and J. P. LaSalle, ed., **Differential Equations and Dynamical Systems**, Academic Press, 1967.

J. K. Hale, L. T. Magalhaes, W. M. Oliva, and K. P.
Rybakowski, **An Introduction to Infinite Dimensional
Dynamical Systems-Geometric Theory**, Springer-Verlag,
1984.

J. K. Hale and J. Scheurle, J. Diff. Eqs., **56** (1985) 142.

C. C. Hamakiotes and S. A. Berger, Phys. Rev. Lett., **62**
(1989) 1270.

M. Hamermesh, **Group Theory and Its Application to Physical
Problems**, Addison-Wesley, 1962.

B.-L. Hao, ed., **Chaos**, World Scientific Pub., 1984.

B.-L. Hao, ed., **Directions in Chaos**, 2 Vols., World
Scientific Pub., 1988.

D. P. Hardin, P. Takac, and G. F. Webb, SIAM J. Appl. Math.,
48 (1988) 1396.

K. P. Harikrishnan and V. M. Nandakumaran, Phys. Lett. A,
133 (1988) 305.

R. G. Harrison and I. A. Al-Saidi, Opt. Communi., **54** (1985)
107.

R. G. Harrison and D. J. Biswas, Phys. Rev. Lett., **55** (1985)
63.

R. G. Harrison, W. J. Firth, and I. A. Al-Saidi, Phys. Rev.
Lett., **53** (1984) 258.

R. G. Harrison, W. J. Firth, C. A. Emshary, and I. A.
Al-Saidi, Phys. Rev. Lett., **51** (1983) 562.

R. G. Harrison, W.-P. Lu, and P. K. Gupta, Phys. Rev. Lett.,
63 (1989) 1372.

P. Hartman, **Ordinary Differential Equations**, Wiley, 1964.

S. Hastings and J. Murray, SIAM J. Appl. Math., **28** (1975)
678.

R. Hauck, F. Hollinger and H. Weber, Opt. Communi., **47**
(1983) 141.

J. W. Havstad and C. L. Ehlers, Phys. Rev. A, **39** (1989) 845.

K. Hayata and M. Koshiba, Opt. Lett., **13** (1988) 1041.

R. D. Hazeltine, D. D. Holm, J. E. Marsden, and P. J.
Morrison, Inst. for Fusion Studies Report 139 (1984).

S. Helgason, **Differential Geometry and Symmetric Spaces**,
Academic Press, 1962.

D. W. Henderson, Top. **9** (1969) 25.

S. T. Hendow and M. Sargent, III, Opt. Commun., **43** (1982) 59.

S. T. Hendow and M. Sargent, III, J. Opt. Soc. Am. B, **2** (1985) 84.

M. Henon, Commun. Math. Phys., **50** (1976) 69.

C. H. Henry, IEEE J. Lightwave Tech., **4** (1986) 288.

C. H. Henry, IEEE J. Lightwave Tech., **4** (1986) 298.

C. H. Henry and R. F. Kazarinov, IEEE J. QE, **22** (1986) 294.

J. Herath and K. Fesser, Phys. Lett. A, **120** (1987) 265.

B. M. Herbst and M. J. Ablowitz, Phys. Rev. Lett., **62** (1989) 2065.

R. Hermann, **Lie Groups for Physicists**, Benjamin, 1966.

R. Hermann, **The Geometry of Nonlinear Differentail Equations, Backlund Transformations, and Solitons, Part A**, Math. Sci. Press, 1976.

R. Hermann, **The Geometry of Nonlinear Differential Equations, Backlund Transformations, and Solitons, Part B**, Math. Sci. Press, 1977.

H. Hermes, in **Differential Equations and Dynamical Systems**, ed. by J. K. Hale and J. P. LaSalle, Wiley, 1967.

N. J. Hicks, **Notes on Differential Geometry**, Van Nostrand, 1971.

R. C. Hilborn, Phys. Rev. A, **31** (1985) 378.

L. W. Hillman, R. W. Boyd, J. Krasinski, and C. R. Stroud, Jr., in **Optical Bistability II**, ed. by C. M. Bowden, H. M. Gibbs and S. L. McCall, Plenum, 1984.

L. W. Hillman, J. Krasinski, R. W. Boyd and C. R. Stroud, Jr., Phys. Rev. Lett., **52** (1984) 1605.

L. W. Hillman, J. Krasinski, K. Koch and C. R. Stroud, Jr., Opt. Soc. Am. B, **2** (1985) 211.

F. J. Hilterman, Geophys. **40** (1975) 745.

P. Hilton, ed., **Structural Stability, the Theory of Catastrophes, and Applications in the Sciences**, Springer-Verlag, 1976.

M. W. Hirsch, Tran. Am. Math. Soc. **93** (1959) 242.

M. W. Hirsch, **Differential Topology**, Springer-Verlag, 1976.

M. W. Hirsch, SIAM J. Math. Anal., **13** (1982) 167.
M. W. Hirsch, SIAM J. Math. Anal., **13** (1982) 233.
M. W. Hirsch, SIAM J. Math. Anal., **16** (1985) 423.
M. W. Hirsch, Nonlinearity, **1** (1988) 51.
M. W. Hirsch, Systems of Differential Equations that are
      Competitive or Cooperative. V: Convergence in
      3-Dimensional Systems,   (1989)   .
M. W. Hirsch, Systems of Differential Equations that are
      Competitive or Cooperative. IV: Structural Stability in
      3-Dimensional Systems,   (1990)   .
M. W. Hirsch, Convergent Activation Dynamics in Continuous
      Time Networks,   (1990)   .
M. W. Hirsch, J. Palis, C. Pugh and M. Shub, Inventiones
      Math., **9** (1970) 121.
M. W. Hirsch and C. Pugh, in **Global Analysis**, ed. by S.
      Smale and S. S. Chern, Am. Math. Soc., 1970.
M. W. Hirsch and S. Smale, **Differential Equations, Dynamical
      Systems, and Linear Algebra**, Academic Press, 1974.
G. Hochschild, **The Structure of Lie Groups**, Holden-Day,
      1965.
J. Hofbauer, Monatsh. Math., **91** (1981) 233.
J. Hofbauer, SIAM J. Appl. Math., **44** (1984) 762.
J. Hofbauer and P. Schuster, in **Stochastic Phenomena and
      Chaotic Behavior in Complex Systems**, ed. by P.
      Schuster, Springer-Verlag, 1984.
J. Hofbauer, P. Schuster and K. Sigmund, J. Theor. Biol., **81**
      (1979) 609.
J. Hofbauer, P. Schuster and K. Sigmund, J. Math. Biol., **11**
      (1981) 155.
L. M. Hoffer, G. L. Lippi, N. B. Abraham and P. Mandel, Opt.
      Communi., **66** (1988) 219.
J. Hoffnagle, R. G. DeVoe, L. Reyna and R. G. Brewer, Phys.
      Rev. Lett., **61** (1988) 255.
A. V. Holden, ed., **Chaos**, Princeton U. Press, 1986.
P. J. Holmes, Phil. Trans. Roy. Soc. London A, **292** (1979)
      419.
P. J. Holmes, J. Fluid Mech., **162** (1986) 365.

P. J. Holmes, in **New Directions in Dynamical Systems**, ed. by T. Bedford and J. Swift, Cambridge U. Press, 1988.

P. J. Holmes and F. C. Moon, J. Appl. Mech., **105** (1983) 1021.

P. J. Holmes and D. A. Rand, Quart. J. Appl. Math., **35** (1978) 495.

P. J. Holmes and R. F. Williams, Arch. Rat. Mech. Anal., **90** (1985) 115.

E. Hopf, Ber. Verh. Sachs. Akad. Wiss. Leipzig, Math. Phys., **94** (1942) 1.

E. Hopf, Ber. Verh. Sachs. Akad. Wiss. Leipzig, Math. Phys., **95** (1943) 3.

F. A. Hopf, D. L. Kaplan, H. M. Gibbs, and R. L. Shoemaker, Phys. Rev. A, **25** (1982) 2172.

F. A. Hopf, D. L. Kaplan, M. H. Rose, L. D. Sanders and M. W. Derstine, Phys. Rev. Lett., **57** (1986) 1394.

L. Hormander, Ark. Math. **5** (1964) 425.

L. Hormander, ed., **Seminar on Singularities of Solutions of Linear Partial Differential Equations**, Princeton, 1979.

W. Horsthemke and D. K. Kondepudi, ed., **Fluctuations and Sensitivity in Nonequilibrium Systems**, Springer-Verlag, 1984.

W. Horsthemke and R. Lefever, Phys. Lett. A, **64** (1977) 19.

W. Horsthemke and M. Malek-Mansour, Z. Physik B, **24** (1976) 307.

G.-H. Hsu, E. Ott and C. Grebogi, Phys. Lett. A, **127** (1988) 199.

B. A. Huberman and J. P. Crutchfield, Phys. Rev. Lett., **43** (1979) 1743.

B. A. Huberman, J. P. Crutchfield and N. H. Packard, Appl. Phys. Lett., **37** (1980) 750.

J. L. Hudson, M. Hart, and D. Marinko, J. Chem. Phys., **71** (1979) 1601.

J. L. Hudson, J. C. Mankin and O. E. Rössler, in P. Schuster, [1984].

D. Husemoller, **Fibre Bundles**, 2nd Ed., Springer-Verlag, 1975.

V. Hutson and W. Moran, J. Math. Biol., **15** (1982) 203.

V. Hutson and G. T. Vickers, Math. Biosci., **63** (1983) 253.

T. Hwa and M. Kardar, Phys. Rev. Lett., **62** (1989) 1813.

J. L. Ibanez and M. G. Velarde, J. Physique Lett., **38** (1979) L465.

K. Ikeda, Opt. Commun., **30** (1979) 257.

K. Ikeda, H. Daido and O. Akimoto, Phys. Rev. Lett., **45** (1980) 709.

K. Ikeda, K. Kondo and O. Akimoto, Phys. Rev. Lett., **49** (1982) 1467.

K. Ikeda and K. Matsumoto, Phys. Rev. Lett., **62** (1989) 2265.

K. Ikeda and M. Mizuno, Phys. Rev. Lett., **53** (1984) 1340.

E. F. Infante, in **Dynamical Systems II**, ed. by A. R. Bednarek and L. Cesari, Academic Press, 1982.

G. Iooss and D. D. Joseph, **Elementary Stability and Bifurcation Theory**, Springer-Verlag, 1980.

M. C. Irwing, **Smooth Dynamical Systems**, Academic Press, 1980.

K. Ito, Proc. Imp. Acad. Tokyo, **20** (1944) 519.

K. Ito, Mem. Amer. Math. Soc., **4** (1951).

K. Ito, Nagoya Math. J., **3** (1951) 55.

A. F. Ize, ed., **Functional Differential Equations and Bifurcation**, Springer-Verlag, 1980.

J. Janich, Math. Ann. **161** (1965) 129.

M. Jansen, J. J. Yang, S. S. Ou, J. Wilcox, D. Botez, L. Mawst and W. W. Simmons, in Digest of Conference on Lasers and Electro-optics, Optical Soc. of America, 1989, paper FL2.

C. Jeffries and K. Wiesenfeld, Phys. Rev. A, **31** (1985) 1077.

N. H. Jensen, P. L. Christiansen and O. Skovgaard, IEE Proc., **135** (1988) 285.

D. W. Kahn, **Introduction to Global Analysis**, Academic Press, 1980.

R. E. Kalman, Proc. Nat. Acad. Sci., **49** (1963) 201.

K. Kaneko, Phys. Rev. Lett., **63** (1989) 219.

H. Kantz and P. Grassberger, Physica D, **17** (1985) 75.

J. Kaplan and J. Yorke, in **Functional Differential Equations and Approximation of Fixed Points**, ed. by H. O. Peitgen and H. O. Walther, Springer-Verlag, 1979.

R. Kapral and P. Mandel, Phys. Rev. A, **32** (1985) 1076.

W. L. Kath, Stud. Appl. Math., **65** (1981) 95.

R. L. Kautz and J. C. Macfarlane, Phys. Rev. A, **33** (1986) 498.

H. Kawaguchi, Appl. Phys. Lett., **45** (1984) 1264.

H. Kawaguchi and K. Otsuka, Appl. Phys. Lett., **45** (1984) 934.

K. Kawasaki and T. Ohta, Physica A, **116** (1983) 573.

S. R. Kay and S. K. Scott, Proc. Roy. Soc. London A, **418** (1988) 345.

J. D. Keeler and J. D. Farmer, Robust Space-Time Intermittency, (1987).

J. B. Keller and S. Autman, ed., **Bifurcation Theory and Nonlinear Eigenvalue Problems**, Benjamin, 1969.

J. B. Keller and I. Papadakis, **Wave Propagation and Underwater Acoustics**, Springer-Verlag, 1977.

A. Kelley, J. Diff. Eqns., **3** (1967) 546.

J. L. Kelley, **General Topology**, Van Nostrand, 1955.

M. A. Kervaire, Comment. Math. Helv. **35** (1961) 1.

D. A. Kessler and H. Levine, Phys. Rev. Lett., **57** (1986) 3069.

P. A. Khandokhin, Ya. I. Khanin and I. V. Koryukin, Opt. Communi., **65** (1988) 367.

G. Khitrova, J. F. Valley and H. M. Gibbs, Phys. Rev. Lett., **60** (1988) 1126.

J. I. Kifer, Math. USSR Izvestija, **8** (1974) 1083.

C. Kikuchi, J. Lambe, G. Makhov and R. W. Terhune, J. Appl. Phys., **30** 1959) 1061.

R. H. Kingston, **Detection of Optical and Infrared Radiation**, Springer-Verlag, 1978.

K. Kirchgassner and H. Kielhofer, Rocky Mount. Math. J., **3** (1973) 275.

W. Klingenstein and W. Schmid, Phys. Rev. B, **20** (1979) 3285.

W. Klische, H. R. Telle and C. O. Weiss, Opt. Lett., **9** (1984) 561.

W. Klische, C. O. Weiss and B. Wellegehausen, Phys. Rev. A, **39** (1989) 919.

M. M. Klosek-Dygas, B. J. Matkowsky and Z. Schuss, SIAM J. Appl. Math., **48** (1988) 1115.

E. Knobloch and N. O. Weiss, Phys. Lett., **85A** (1981) 127.

E. Knobloch and M. R. E. Proctor, Proc. Roy. Soc. A, **415** (1988) 61.

S. Kobayashi and K. Nomizu, **Foundations of Differential Geometry**, Interscience, Vol.1 1963; Vol.2 1969.

A. L. Koch, J. Theor. Biol., **44** (1974) 387.

A. N. Kolmogorov, Dokl. Akad. Nauk. SSSR, **98** (1954) 527.

N. Kopell and D. Ruelle, SIAM J. Appl. Math., **46** (1986) 68.

A. Korpel, Opt. Communi., **61** (1987) 66.

N. C. Kothari, Opt. Communi., **62** (1987) 247.

J. Kotus, M. Krych and Z. Nitecki, **Global Structural Stability of Flows on Open Surfaces**, Am. Math. Soc., 1982.

M. A. Krasnosel'skii, V. Sh. Burd, and Yu. S. Kolesov, **Nonlinear Almost Periodic Oscillations**, Wiley, 1973.

M. Kubicek and M. Marek, **Computational Methods in Bifurcation Theory and Dissipative Structures**, Springer-Verlag, 1983.

N. H. Kuiper, Top. 3 (1965) 19.

N. H. Kuiper, in **Manifolds-Tokyo 1973**, Univ. of Tokyo Press, 1975.

I. Kupka, Contributions to Diff. Eqs., **2** (1963) 457.

I. Kupka, Contributions to Diff. Eqs., **3** (1964) 411.

Y. Kuramoto, ed., **Chaos and Statistical Methods**, Springer-Verlag, 1984.

Y. Kuramoto and T. Yamada, Prog. Theor. Phys., **56** (1976) 679.

J. Kurzweil, Czech. Math. J., **5** (1955) 382.

H. J. Kushner, SIAM J. Appl. Math., **47** (1987) 169.

A. Lahiri and T. Nag, Phys. Rev. Lett., **62** (1989) 1933.

W. E. Lamb, Jr., Phys. Rev., **134A** (1964) 1429.

L. D. Landau and E. M. Lifshitz, **Fluid Mechanics**, Pergamon, 1959.

P. T. Landsberg and A. Pimpale, J. Phys. C, **9** (1976) 1243.

S. Lang, **Introduction to Differentiable Manifolds**, Interscience, 1962.

P. Langevin, Compt. Rend., **146** (1908) 530.

J. P. LaSalle and S. Lefschetz, **Stability by Liapunovs' Direct Method with Applications**, Academic, 1961.

J. P. LaSalle and S. Lefschetz, ed., **Nonlinear Differential Equations and Nonlinear Mechanics**, Academic Press, 1963.

R. Lashof, Ann. Math. **81** (1965) 565.

C. T. Law and A. E. Kaplan, Opt. Lett., **14** (1989) 734.

N. M. Lawandy and K. Lee, Opt. Communi., **61** (1987) 137.

N. M. Lawandy and D. V. Plant, Opt. Communi., **59** (1986) 55.

N. M. Lawandy and J. C. Ryan, Opt. Communi., **63** (1987) 53.

N. M. Lawandy, M. D. Selker and K. Lee, Opt. Communi., **61** (1987) 134.

M. LeBerre, E. Ressayre, A. Tallet, K. ai, H. M. Gibbs, M. C. Rushford, N. Peyghambarian, J. Opt. Soc. Am. B, **1** (1984) 591.

C.-H. Lee, T.-H. Yoon and S.-Y. Shin, Appl. Phys. Lett., **46** (1985) 95.

K. K. Lee, OMEGA Technical Note No. 74, Lab. for Laser Energetics, Univ. of Rochester, 1977.

K. K. Lee, to be submitted for publication (1990).

S. Lefschetz, **Introduction to Topology**, Princeton Univ. Press, 1949.

S. Lefschetz, ed., **Contributions to the Theory of Nonlinear Oscillations I**, Princeton U. Press, 1950.

S. Lefschetz, ed., **Contributions to the Theory of Nonlinear Oscillations II**, Princeton U. Press, 1950.

S. Lefschetz, ed., **Contributions to the Theory of Nonlinear Oscillations III**, Princeton U. Press, 1956.

S. Lefschetz, ed., **Contributions to the Theory of Nonlinear**

Oscillations IV, Princeton U. Press, 1958.

S. Lefschetz, Bol. Soc. Mat. Mex., 3 (1958) 25.

S. Lefschetz, ed., Contributions to the Theory of Nonlinear Oscillations V, Princeton U. Press, 1960.

S. Lefschetz, Differential Equations: Geometric Theory, Second Ed., Wiley, 1962.

S. Lefschetz, Stability of Nonlinear Control Systems, Academic Press, 1965.

S. H. Lehnigk, Stability Theorems for Linear Motions, Prentice Hall, 1966.

D. Lenstra, B. H. Verbeek and A. J. den Boef, IEEE J. QE, 21 (1985) 674.

J. A. Leslie, Top. 6 (1967) 263.

S. A. Levin, Am. Nat., 104 (1970) 413.

H. Li and N. B. Abraham, IEEE J QE, 25 (1989) 1782.

T.-Y. Li and J. A. Yorke, Am. Math. Monthly, 82 (1975) 985.

A. M. Liapunov, Probleme General de la Stabilite du Mouvement, 1907 French translation of a Russian Memoire dated 1892, reprinted by Princeton U. Press, 1947.

A. J. Lichtenberg and M. A. Lieberman, Regular and Stochastic Motion, Springer-Verlag, 1983.

J. Lighthill, Proc. Roy. Sco. A, 407 (1986) 35.

M. Lindberg, S. W. Koch and H. Haug, J. Opt. Soc. Am. B, 3 (1986) 751.

F. H. Ling and G. W. Bao, Phys. Lett. A, 122 (1987) 413.

P. S. Linsay, Phys. Rev. Lett., 47 (1981) 1349.

G. L. Lippi, N. B. Abraham, J. R. Tredicce, L. M. Narducci, G. P. Puccioni and F. T. Arecchi, in Optical Chaos, ed. by J. Chrostowski and N. B. Abraham, SPIE, 667 (1986) 41.

G. L. Lippi, J. R. Tredicce, N. B. Abraham and F. T. Arecchi, Opt. Communi., 53 (1985) 129.

B. X. Liu, L. J. Huang, K. Tao, C. H. Shang and H.-D. Li, Phys. Rev. Lett., 59 (1987) 745.

K. L. Liu and K. Young, J. Math. Phys., 27 (1986) 502.

R.-W. Liu and R. J. Leake, in Differential Equations and Dynamical Systems, ed. by J. K. Hale and J. P. LaSalle,

Academic Press, 1967.

N. G. Lloyd and S. Lynch, Proc. Roy. Soc. A, **418** (1988) 199.

E. N. Lorenz, J. Atmos. Sci., **20** (1963) 130.

W. Lu and W. Tan, Opt. Communi., **61** (1987) 271.

Y. C. Lu, **Singularity Theory**, Springer-Verlag, 1976.

D. G. Luenberger, **Introduction to Dynamical Systems**, Wiley, 1979.

L. A. Lugiato, in **Progress in Optics, 21**, ed. by E. Wolf, North-Holland, 1984.

L. A. Lugiato, M. L. Asquini and L. M. Narducci, SPIE, **667** (1986) 132.

L. A. Lugiato, R. J. Horowicz, G. Strini and L. M. Narducci, Phil. Trans. R. Soc. Lond. A, **313** 291.

L. A. Lugiato and L. M. Narducci, in **Coherence and Quantum Optics V**, ed. by L. Mandel and E. Wolf, Plenum, 1984.

L. A. Lugiato, L. M. Narducci, D. K. Bandy and C. A. Pennise, Opt. Commun., **43** (1982) 281.

L. A. Lugiato, L. M. Narducci, E. V. Eschenazi, D. K. Bandy and N. B. Abraham, Phys. Rev. A, **32** (1985) 1563.

L. A. Lugiato, L. M. Narducci and M. F. Squicciarini, Phys. Rev. A, **34** (1986) 3101.

L. A. Lugiato, C. Oldano, C. Fabre, E. Giacobino and R. J. Horowicz, N. Cimento D, **10** (1988) 959.

L. A. Lugiato, G.-L. Oppo, M. A. Pernigo, J. R. Tredicce, L. M. Narducci, and D. K. Bandy, Opt. Communi., **68** (1988) 63.

M. Machacek, Phys. Rev. Lett., **57** (1986) 2014.

M. C. Mackey and U. an der Heiden, J. Math. Biol., **19** (1984) 211.

R. S. MacKay and J. B. J. van Zeijts, Nonlinearity, **1** (1988) 253.

G. Makhov, L. G. Cross, R. W. Terhune and J. Lambe, J. Appl. Phys., **31** (1960) 936.

G. Makhov, C. Kikuchi, J. Lambe and R. W. Terhune, Phys. Rev., **109** (1958) 1399.

B. Malgrange, **The Preparation Theorem for Differentiable**

Functions, Differential Analysis, Oxford U. Press, 1964.

I. G. Malkin, Prikl. Mat. Meh., **8** (1944) 241.

I. G. Malkin, **Theorie der Stabilitat einer Bewegung**, translated from 1952 Russian edition, R. Oldenburg, 1959.

L. Mandel and E. Wolf, Phys. Rev., **124** (1961) 1696.

P. Mandel, Phys. Rev. A, **21** (1980) 2020.

P. Mandel, Opt. Communi., **44** (1983) 400.

P. Mandel, Opt. Communi., **45** (1983) 269.

P. Mandel, J. Opt. Soc. Am. B, **2** (1985) 112.

P. Mandel, Opt. Communi., **53** (1985) 249.

P. Mandel, Dynamics versus Static Stability, preprint, 1986.

P. Mandel and N. B. Abraham, Opt. Communi., **51** (1984) 87.

P. Mandel and G. P. Agrawal, Opt. Communi., **42** (1982) 269.

P. Mandel and T. Erneux, Phys. Rev. Lett., **53** (1984) 1818.

P. Mandel and R. Kapral, Subharmonic and chaotic Bifurcation Structure in Optical Bistability, preprint, 1984.

P. Mandel and X.-G. Wu, J. Opt. Soc. Am. B, **3** (1986) 940.

P. Mandel and H. Zeghlache, Opt. Communi., **47** (1983) 146.

M. Marachkov, Bull. Soc. Phys.-Mat., Kazan, **12** (1940) 171.

S. B. Margolis and B. J. Matkowsky, SIAM J. Appl. Math., **48** (1988) 828.

L. Markus, **Lectures in Differentiable Dynamics**, AMS, 1971.

J. Marsden, **Applications of Global Analysis in Mathematical Physics**, Publish or Perish, 1974.

J. Marsden, Ed., **Fluids and Plasmas: Geometry and Dynamics**, Am. Math. Soc., 1984.

J. Marsden, D. Ebin, and A. Fischer, in **Proc. 13th Biannual Seminar of the Canadian Math. Congress, Vol. 1**, ed. by J. R. Vanstone, Can. Math. Cong., 1972.

J. Marsden and M. F. McCracken, **The Hopf Bifurcation and its Applications**, Springer-Verlag, 1976.

D. Martland, in **Neural Computing Architectures**, ed. by I. Aleksander, MIT Press, 1989.

J. L. Massera, Ann. Math. **50** (1949) 705.

J. L. Massera, Ann. Math. **64** (1956) 182; **68** (1958) 202.

B. J. Matkowsky and G. I. Sivashinsky, SIAM J. Appl. Math., **35** (1978) 465.

R. M. May, Nature, **261** (1976) 459.

R. M. May and W. J. Leonard, SIAM J. Appl. Math., **29** (1975) 243.

G. Mayer-Kress, ed., **Dimensions and Entropies in Chaotic Systems**, Springer-Verlag, 1986.

N. H. McClamroch, **State Models of Dynamical Systems**, Springer-Verlag, 1980.

S. McDonald, C. Grebogi, E. Ott and J. A. Yorke, Physica D, **17** (1985) 125.

A. G. McKendrick, Proc. Edin. Math. Soc., **44** (1926) 98.

D. W. McLaughlin, J. V. Moloney and A. C. Newell, Phys. Rev. Lett., **54** (1985) 681.

J. B. McLaughlin and P. C. Martin, Phys. Rev. A, **12** (1975) 186.

E. J. McShane, **Integration**, Princeton U. Press, 1944.

V. K. Mel'nikov, Trans. Moscow Math. Soc., **12** (1963) 1.

V. K. Mel'nikov, Phys. Lett. A, **118** (1986) 22.

R. Meucci, A. Poggi, F. T. Arecchi and J. R. Tredicce, Opt. Communi., **65** (1988) 151.

T. Midavaine, D. Dangoisse and P. Glorieux, Phys. Rev. Lett., **55** (1985) 1989.

F. A. Milner, Num. Meth. Part. Diff. Eqs., **4** (1988) 329.

J. W. Miles, J. Fluid Mech., **75** (1976) 419.

J. Miles, Phys. Lett. A, **133** (1988) 295.

J. Milnor, Ann. of Math. **64** (1956) 399.

J. Milnor, **Differential Topology**, Princeton Univ. mineo. notes, 1958.

J. Milnor, **Morse Theory**, Princeton University Press, 1963.

J. Milnor, **Topology From the Differentiable Viewpoint**, Univ. of Virginia Press, 1965.

J. Milnor and J. Stasheff, **Characteristic Classes**, Princeton Univ. Press, 1974.

P. W. Milonni, M. L. Shih and J. R. Ackerhalt, **Chaos in Laser-Matter Interactions**, World Sci. Pub., Inc., 1987.

P. W. Milonni and J. H. Eberly, **Lasers**, Wiley, 1988.

N. Minorsky, **Nonlinear Oscillations**, Van Nostrand, 1962.

J. V. Moloney, Phys. Rev. Lett., **53** (1984) 556.

J. V. Moloney, M. R. Belic and H. M. Gibbs, Opt. Commun., **41** (1982) 379.

J. V. Moloney, W. Forysiak, J. S. Uppal and R. G. Harrison, Phys. Rev. A, **39** (1989) 1277.

J. V. Moloney, J. S. Uppal and R. G. Harrison, Phys. Rev. Lett., **59** (1987) 2868.

F. C. Moon, **Chaotic Vibrations**, Wiley, 1987.

F. C. Moon, Phys. Lett. A, **132** (1988) 249.

F. C. Moon, M. M. Yanoviak and R. Ware, Appl. Phys. Lett., **52** (1988) 1534.

Y. Morimoto, Phys. Lett. A, **134** (1988) 179.

B. Morris and F. Moss, Phys. Lett. A, **118** (1986) 117.

P. J. Morrison and S. Eliezer, Spontaneous symmetry breaking and neutral stability in the noncanonical Hamiltonian formalism, Institute for Fusion Studies Rep. #212 (1986).

M. Morse, **The Calculus of Variations in the Large**, Am. Math. Soc. 1934.

J. Moser, Nachr. Akad. Wiss. Gottingen Math. Phys. Kl., **2** (1962) 1.

J. Moser, J. Diff. Equa., **5** (1969) 411.

J. Moser, **Stable and Random Motions in Dynamical Systems**, Princeton, 1973.

J. Moser, ed., **Dynamical Systems, Theory and Applications**, Springer-Verlag, 1975.

J. Moser, SIAM J. Appl. Math., **28** (1986) 459.

A. P. Mullhaupt, Phys. Lett. A, **122** (1987) 403.

J. R. Munkres, **Elementary Differential Topology**, Princeton Univ. Press, 1966.

K. Nakamura, Phys. Lett. A, **134** (1988) 173.

M. Nakazawa, K. Suzuki, and H. A. Haus, IEEE J-QE, **25** (1989) 2036.

M. Nakazawa, K. Suzuki, H. Kubota and H. A. Haus, IEEE J-QE, **25** (1989) 2045.

L. M. Narducci and N. B. Abraham, **Laser Physics and Laser Instabilities**, World Scientific Pub., 1988.

L. M. Narducci, D. K. Bandy, L. A. Lugiato and N. B. Abraham, in **Coherence and Quantum Optics V**, ed. by L. Mandel and E. Wolf, Plenum, 1984.

L. M. Narducci, H. Sadiky, L. A. Lugiato and N. B. Abraham, Opt. Communi., **55** (1985) 370.

L. M. Narducci, J. R. Tredicce, L. A. Lugiato, N. B. Abraham and D. K. Bandy, Phys. Rev. A, **33** (1986) 1842.

P. Narum, D. J. Gauthier and R. W. Boyd, in **Optical Bistability 3**, ed. by H. M. Gibbs, P. Mandel, N. Peyghambarian and S. D. Smith, Springer-Verlag, 1986.

C. Nash and S. Sen, **Topology and Geometry for Physicists**, Academic Press, 1983.

P. Naslin, **The Dynamics of Linear and Nonlinear Systems**, Gordon & Breach, 1965.

A. H. Nayfeh and D. T. Mook, **Nonlinear Oscillations**, Wiley, 1979.

Ju. I. Neimark and N. A. Fufaev, **Dynamics of Nonholonomic Systems**, Am. Math. Soc., 1972.

E. Nelson, **Topics in Dynamics I: Flows**, Princeton U. Press, 1969.

V. V. Nemytskii and V. V. Stepanov, **Qualitative Theory of Differential Equations**, Princeton, 1960.

N. Neskovic and B. Perovic, Phys. Rev. Lett., **59** (1987) 308.

A. C. Newell, ed., **Nonlinear Wave Motion**, Am. Math. Soc., 1974.

A. C. Newell, D. A. Rand and D. Russell, Phys. Lett. A, **132** (1988) 112.

S. E. Newhouse, in **Global Analysis**, ed. by S. S. Chern and S. Smale, 1970.

S. E. Newhouse, Ann. N. Y. Acad. Sci., **316** (1979) 121.

S. E. Newhouse and J. Palis, in **Global Analysis**, ed. by S. S. Chern and S. Smale, 1970.

S. E. Newhouse, D. Ruelle and F. Takens, Commun. Math. Phys., **64** (1978) 35.

P. K. Newton, J. Math. Phys., **29** (1988) 2245.

C. Nicolis and G. Nicolis, Phys. Rev. A, **34** (1986) 2384.

Z. Nitecki, **Differentiable Dynamics**, MIT Press, 1971.

Z. Nitecki, J. Math. Anal. Appl., **72** (1979) 446.

J. Nittmann, H. E. Stanley, E. Touboul and G. Daccord, Phys. Rev. Lett., **58** (1987) 619.

S. Novak and R. G. Frehlich, Phys. Rev. A, **26** (1982) 3660.

R. M. Noyes, J. Chem. Phys., **69** (1978) 2514.

K. Nozaki and N. Bekki, Phys. Rev. Lett., **50** (1983) 1226.

H. E. Nusse, SIAM J. Appl. Math., **47** (1987) 498.

H. E. Nusse and J. A. Yorke, Phys. Lett. A, **127** (1988) 328.

J. F. Nye, Proc. Roy. Soc. London, **A361** (1978) 21.

P. O'Connor, J. Gehlen and E. J. Heller, Phys. Rev. Lett., **58** (1987) 1296.

K. Ohe and H. Tanaka, J. Phys. D, **21** (1999) 1391.

T. Ohtsuki and T. Keyes, Phys. Rev. A, **36** (1987) 4434.

A. Yu. Okulov and A. N. Oraevsky, J. Opt. Soc. Am. B, **3** (1986) 741.

H. Olesen, J. H. Osmundsen and B. Tromborg, IEEE J. QE, **22** (1986) 762.

O. H. Olsen and M. r. Samuelsen, Phys. Lett. A, **119** (1987) 391.

L. F. Olson and H. Degn, Nature, **271** (1977) 177.

H. Omori, Trans. Am. Math. Soc., **179** (1973) 85.

L. Onsager and S. Machlup, Phys. Rev., **91** (1953) 1512.

G.-L. Oppo, J. R. Tredicce and L. M. Narducci, Opt. Communi., **69** (1989) 393.

A. N. Oraevskii, Sov. J. Quant. Elect., **14** (1984) 1182.

A. N. Orayvskiy, in **Research on Laser Theory**, ed. by A. N. Orayevskiy, Nova Science Pub., 1988.

L. A. Orozco, H. J. Kimble, A. T. Rosenberger, L. A. Lugiato, M. L. Asquini, M. Brambilla and L. M. Narducci, Phys. Rev. A, **39** (1989) 1235.

K. Otsuka and K. Ikeda, Phys. Rev. Lett., **59** (1987) 194.

K. Otsuka and H. Iwamura, Phys. Rev. A, **28** (1983) 3153.

K. Otsuka and H. Kawaguchi, Phys. Rev. A, **29** (1984) 2953.

K. Otsuka and H. Kawaguchi, Phys. Rev. A, **30** (1984) 1575.

E. Ott, Rev. Mod. Phys., **53** (1981) 655.

A. Pacault and C. Vidal, ed., **Synergetics: Far from Equilibrium**, Springer-Verlag, 1979.

G. Paladin and A. Vulpiani, Phys. Lett. A, **118** (1986) 14.

R. S. Palais, **Lectures on the Differential Topology of Infinite Dimensional Manifolds**, Brandeis Univ., 1964.

R. S. Palais, ed., **Seminar on the Atiyah-Singer Index Theorem**, Princeton, 1965.

R. S. Palais, Top., **3** (1965) 271.

R. S. Palais, Top. **5** (1966) 1.

R. S. Palais, Top. **5** (1966) 115.

R. S. Palais, **Foundations of Global Non-linear Analysis**, Benjamin, 1968.

R. S. Palais, in **Nonlinear Functional Aanlysis**, Am. Math. Soc., 1970.

R. S. Palais, Act. Cong. Int. Math. **2** (1971) 243.

J. Palis and W. de Melo, **Geometric Thoery of Dynamical Systems. An Introduction**, Springer-Verlag, 1982.

J. Palis and S. Smale, in **Global Analysis**, ed. by S. S. Chern and S. Smale, 1970.

J. I. Palmore and J. L. McCauley, Phys. Lett. A, **122** (1987) 399.

V. Pankovic, Phys. Lett. A, **133** (1988) 267.

E. Papantonopoulos, T. Uematsu and T. Yanagida, Phys. Lett. B, **183** (1987) 282.

F. Papoff, D. Dangoisse, E. Poite-Hanoteau and P. Glorieux, Opt. Communi., **67** (1988) 358.

M. Paramio and J. Sesma, Phys. Lett. A, **132** (1988) 98.

E. Pardoux and V. Wihstutz, SIAM J. Appl. Math., **48** (1988) 442.

C. Pare, M. Piche and P.-A. Belanger, in **Optical Instibitities**, ed. by R. W. Boyd, M. G. Raymer and L. M. Narducci, Cambridge U. Press, 1986.

U. Parlitz and W. Lauterborn, Phys. Lett. A, **107** (1985) 351.

I. E. Parra and J. M. Vega, SIAM J. Appl. Math., **48** (1988) 854.

E. Patzak, H. Olseen, A. Sugimura, S. Saita and T. Mukai, Electron. Lett., **19** (1983) 938.

E. Patzak, A. Sugimura, A. Saita, T. Mukai and H. Olesen, Electron. Lett., **19** (1983) 1026.

J. Peetre, Math. Scand. **8** (1960) 116.

M. M. Peixoto, Top., **1** (1962) 101.

M. M. Peixoto, Ed., **Dynamical Systems**, Academic Press, 1973.

M. M. Peixoto and C. C. Pugh, Ann. Math. **87** (1968) 423.

R. Perez and L. Glass, Phys. Lett. A, **90** (1982) 441.

K. P. Persidski, Mat. Sb. **40** (1933) 284.

F. Peter and H. Weyl, Math. Ann. **97** (1927) 737.

P. Phelan, J. O'Gorman, J. McInerney and D. Heffernan, Appl. Phys. Lett., **49** (1986) 1502.

M. W. Phillips, H. Gong, A. I. Ferguson and D. C. Hanna, Opt. Communi., **61** (1987) 215.

P. E. Phillipson, Phys. Lett. A, **133** (1988) 383.

G. Picard and T. W. Johnston, Phys. Rev. Lett., **48** (1982) 1610.

A. Pimpale, P. T. Landsberg, L. L. Bonilla and M. G. Velarde, J. Phys. Chem. Solids, **42** (1981) 873.

F. A. E. Pirani, D. C. Robinson, and W. F. Shadwick, **Local Jet-Bundle Formulation of Backlund Transformations**, Reidel, 1979.

O. Piro and M. Feingold, Phys. Rev. Lett., **61** (1988) 1799.

L. M. Pismen, Phys. Rev. Lett., **59** (1987) 2740.

R. E. Plant, SIAM J. Appl. Math., **40** (1981) 150.

B. S. Poh and T. E. Rozzi, IEEE J-QE, **17** (1981) 723.

H. Poincaré, J. de Math. **7** (1881) 375.

H. Poincaré, ibid, **8** (1882) 251.

H. Poincaré, J. de Math. Pure et Appl. **1** (1885) 167.

Y. Pomeau and P. Manneville, Commun. Math. Phys., **74** (1980) 77, 189.

L. S. Pontryagin, **Ordinary Differential Equations**, Addison-Wesley, 1962.

L. S. Pontryagin, **Topological Groups**, 2nd Ed., Gordon and Breach, 1966.

V. M. Popov, Aut. Rem. Control, **22** (1962) 857.

V. M. Popov, **Hyperstability of Control Systems**, Springer-Verlag, 1973.

M. B. Priestley, **Nonlinear and Nonstationary Time Series Analysis**, Academic Press, 1988.

I. Prigogine and I. Stengers, **Order out of Chaos**, Bantam, 1984.

I. Procaccia and R. Zeitak, Phys. Rev. Lett., **60** (1988) 2511.

G. P. Puccioni, A. Poggi, W. Gadomski, J. R. Tredicce and F. T. Arecchi, Phys. Rev. Lett., **55** (1985) 339.

G. P. Puccioni, M. V. Tratnik, J. E. Sipe and G. L. Oppo, Opt. Lett., **12** (1987) 242.

G. R. W. Quispel, Phys. Lett. A, **118** (1986) 457.

J. I. Ramos, Numeri. Meth. Partial Diff. Eqs., **3** (1987) 241.

D. A. Rand, Nonlinearity, **1** (1988) 181.

D. A. Rand and L. S. Young, Ed., **Dynamical Systems and Turbulence, Warwick 1980**, Springer-Verlag, 1981.

L. E. Reichl and W. M. Zheng, Phys. Rev. A, **29** (1984) 2186.

L. E. Reichl and W. M. Zheng, Phys. Rev. A, **30** (1984) 1068.

H. Risken, Z. Phys., **186** (1965) 85.

H. Risken, **The Fokker-Planck Equation**, Springer-Verlag, 1984.

H. Risken, in **Optical Instabilities**, ed. by R. W. Boyd, M. G. Raymer and L. M. Narducci, Cambridge U. Press, 1986.

H. Risken, in **Quantum Optics IV**, ed. by J. D. Harvey and D. F. Walls, Springer-Verlag, 1986.

H. Risken, in **Optical Bistability, Instability and Optical Computing**, ed. by H. Y. Zhang and K. K. Lee, World Sci. Pub., Inc., 1988.

J. W. Robbin, Ann. Math., **94** (1971) 447.

P. H. Robinowitz, ed., **Applications of Bifurcation Theory**, Academic Press, 1977.

C. Robinson, in Springer Math. Lecture Notes, Vol.468 (1975).

C. Robinson, Bol. Soc. Bras. Mat., **6** (1975) 129.

C. Robinson, J. Diff. Eq., **22** (1976) 28.

C. Robinson, Rocky Mount. J. Math., **7** (1977) 425.

C. Robinson, J. Diff. Eq., **37** (1980) 1.

C. Rogers and W. F. Shadwick, **Backlund Transformations and Their Applications**, Academic Press, 1982.

R. W. Rollins and E. R. Hunt, Phys. Rev. Lett., **49** (1982) 1295.

R. W. Rollins and E. R. Hunt, Phys. Rev. A, **29** (1984) 3327.

S. Rosenblat and D. S. Cohen, Stud. Appl. Math., **63** (1980) 1.

S. Rosenblat and D. S. Cohen, Stud. Appl. Math., **64** (1981) 143.

H. H. Rosenbrock, **State-Space and Multivariable Theory**, Wiley, 1970.

O. E. Rössler, Phys. Lett. A, **57** (1976) 397.

O. E. Rössler, Z. Naturforsch., **31A** (1976) 259.

O. E. Rössler and J. L. Hudson, in **Symposium on Chemical Applications of Topology and Graph Theory**, Elsevier, 1983.

N. Rouche, P. Habets, and M. Laloy, **Stability Theory by Liapunov's Direct Method**, Springer-Verlag, 1977.

H. L. Royden, **Real Analysis**, Macmillan, 1963.

V. Roytburd, J. Diff. Eqs., **56** (1985) 40.

T. E. Rozzi and K. A. Shore, J. Opt. Soc. Am. B, **2** (1985) 237.

D. Ruelle, Physica Scripta, **T9** (1985) 147.

D. Ruelle, Phys. Rev. Lett., **56** (1986) 405.

D. Ruelle, J. Stat. Phys., **44** (1986) 281.

D. Ruelle, Proc. R. Soc. A, **413** (1987) 5.

D. Ruelle, **Elements of Differentiable Dynamics and Bifurcation Theory**, Academic Press, 1989.

D. Ruelle, **Chaotic Evolution and Strange Attractors**, Cambridge U. Press, 1989.

D. Ruelle and F. Takens, Commun. Math. Phys., **20** (1971) 167.

H. Rüssman, Nachr. Akad. Wiss. Gottingen Math. Phys. Kl., ** (1970) 67.

J. Sacher, W. Elsasser and E. O. Gobel, Phys. Rev. Lett., **63** (1989) 2224.

R. Z. Sagdeev, Ed., **Nonlinear and Turbulent Processes in Physics**, 3 Vols., Harwood Academic, 1984.

F. M. A. Salam, SIAM J. Appl. Math., **47** (1987) 232.

L. Salvadori, Rend. dell' Accad. Naz. Lincei, **53** (1972) 35.

L. M. Sander, Sci. Am., **256** (1987) 94.

J. M. Sanz-Serna and F. Vadillo, SIAM J. Appl. Math., **47** (1987) 92.

R. Saravanan, O. Narayan, K. Banerjee and J. K. Bhattacharjee, Phys. Rev. A, **31** (1985) 520.

M. Sargent, IV, M. O. Scully, and W. Lamb, Jr., **Laser Physics**, Addison-Wesley, 1974.

A. N. Sarkovskii, Ukr. Math. J., **16** (1964) 61.

I. I. Satija, Phys. Rev. Lett., **58** (1987) 623.

D. H. Sattinger, **Branching in the Presence of Symmetry**, Soc. Ind. Appl. Math., 1983.

P. T. Saunders, **An Introduction to Catastrophe Theory**, Cambridge U. Press, 1980.

W. Scharpf, M. Squicciarini, D. Bromley, C. Green, J. R. Tredicce and L. M. Narducci, Opt. Communi., **63** (1987) 344.

A. Schenzle and H. Brand, Phys. Rev. A, **20** (1979) 1628.

R. A. Schimitz, K. R. Graziani and J. L. Hudson, J. Chem. Phys., **67** (1977) 3040.

W. A. Schlup, J. Phys. A, **11** (1978) 1871.

H. J. Scholz, T. Yamada, H. Brand and R. Graham, Phys. Lett. A, **82** (1981) 321.

A. Schremer, T. Fujita, C. F. Lin and C. L. Tang, Appl. Phys. Lett., **52** (1988) 263.

P. Schuster, ed., **Stochastic Phenomena and Chaotic Behaviour in Complex Systems**, Springer-Verlag, 1984.

P. Schuster, K. Sigmund and R. Wolff, J. Diff. Eqs., **32** (1979) 357.

P. Schuster, K. Sigmund and R. Wolff, J. Math. Analy. Appl., **78** (1980) 88.

I. B. Schwartz, Phys. Rev. Lett., **60** (1988) 1359.

435

J. T. Schwartz, **Differential Geometry and Topology**, Gordon & Breach, 1968.

K. Seeger, **Semiconductor Physics**, Springer-Verlag, 1982.

G. Seeley, T. Keyes and T. Ohtsuki, Phys. Rev. Lett., **60** (1988) 290.

B. Segard, B. Macke, L. A. Lugiato, F. Prati and M. Brambilla, Phys. Rev. A, **39** (1989) 703.

P. Seibert, in **Nonlinear Differential Equations and Nonlinear Mechanics**, ed. by J. P. LaSalle and S. Lefschetz, Academic Press, 1963.

J.-P. Serre, **Lie Algebras and Lie Groups**, Benjamin, 1965.

V. M. Shalaev and M. I. Stockman, Z. Phys. D, **10** (1988) 71.

F. R. Sharpe and A. J. Lotka, Phil. Mag., **21** (1911) 435.

T. M. Shay, J. D. Dobbins and Y. C. Chung, Top. Meeting on Laser and Optical Remote Sensing, Sept. 1987.

M.-L. Shih and P. W. Milonni, Opt. Communi., **49** (1984) 155.

I. Shimada and T. Nagashima, Prog. Theor. Phys., **59** (1978) 1033.

T. Shimizu and N. Morioka, Phys. Lett. A, **66** (1978) 182.

K. A. Shore, IEEE J.-QE, **21** (1985) 1249.

K. A. Shore, SPIE, **667** (1986) 109.

K. A. Shore, IEE Proc., **134** (1987) 51.

K. A. Shore and M. W. McCall, IEE Proc., **136** (1989) 14.

K. A. Shore and T. E. Rozzi, IEEE J.-QE, **20** (1984) 246.

M. Shub, Bull. Am. Math. Soc., **78** (1972) 817.

A. Siegman, Presentation at the 19th Winter Colloquium on Quantum Electronics, Snowbird, Utah, Jan. 1989.

K. Sigmund and P. Schuster, in **Stochastic Phenomena and Chaotic Behaviour in Complex Systems**, ed. by P. Schuster, Springer-Verlag, 1984.

L. P. Sil'nikov, Sov. Math. Dokl., **6** (1965) 163.

L. P. Sil'nikov, Math. Sb. USSR, **10** (1970) 91.

I. M. Singer and J. A. Thorpe, **Lecture Notes on Elementary Topology and Geometry**, Scott and Foresman, 1967.

D. Singer, SIAM J. Appl. Math, **35** (1978) 260.

L. Sirovich and J. D. Rodriguez, Phys. Lett. A, **120** (1987) 211.

F. Skiff, F. Anderegg, T. N. Good, P. J. Paris, M. Q. Tran, N. Rynn and R. A. Stern, Phys. Rev. Lett., **61** (1988) 2034.

S. Smale, Ann. Math., **74** (1961) 199.

S. Smale, Ann. Scuola Normale Superiore Pisa, **18** (1963) 97.

S. Smale, Bull. Amer. Math. Soc., **73** (1967) 747.

S. Smale, in **Dynamical Systems**, ed. by M. M. Peixoto, Academic Press, 1973.

S. Smale, J. Math. Biol., **3** (1976) 5.

L. A. Smith, Phys. Lett. A, **133** (1988) 283.

J. Smoller, **Shock Waves and Reaction-Diffusion Equations**, Springer-Verlag, 1983.

L. Sneddon and K. A. Cox, Phys. Rev. Lett., **58** (1987) 1903.

J. So, J. Theor. Biol., **80** (1979) 185.

H. Sompolinsky and A. Crisanti, Phys. Rev. Lett., **61** (1988) 259.

E. H. Spanier, **Algebraic Topology**, McGraw-Hill, 1966.

C. Sparrow, **The Lorenz Equations: Bifurcations, Chaos, and Strange Attractors**, Springer-Verlag, 1982.

C. Sparrow, in **Opticl Instabilities**, ed. by R. W. Boyd, M. G. Raymer and L. M. Narducci, Cambridge U. Press, 1986.

D. C. Spencer and S. Iyanaga, ed., **Global Analysis**, Princeton, 1969.

M. Spivak, **Differential Geometry**, 5 vols, Publish or Perish, 2nd ed., 1979.

N. Sri Namachchivaya and S. T. Ariaratnam, SIAM J. Appl. Math., **47** (1987) 15.

W.-H. Steeb, J. Phys. A, **10** (1977) L221.

N. Steenrod, **The Topology of Fibre Bundles**, Princeton Univ. Press, 1951.

H. Steinruck, SIAM J. Appl. Math., **47** (1987) 1131.

J. Stensby, SIAM J. Appl. Math., **47** (1987) 1177.

S. Sternberg, Am. J. Math., **79** (1957) 809.

S. Sternberg, Am. J. Math., **80** (1958) 623.

I. Stewart, in **New Directions in Dynamical Systems**, ed. by T. Bedford and J. Swift, Cambridge U. Press, 1988.

E. Stiefel, Comment. Math. Helv. **17** (1944/5) 165.

D. A. Stone, **The Exponential Map at an Isolated Singular Point**, Am. Math. Soc., 1982.

B. Straughan, R. E. Ewing, P. G. Jacobs and M. J. Djomehri, Numeri. Meth. Partial Diff. Eqs., **3** (1987) 51.

W. J. Stronge and D. Shu, Proc. Roy. Soc. A, **418** (1988) 155.

C. R. Stroud, Jr., K. Koch and S. Chakmakjian, in **Optical Instabilities**, ed. by R. W. Boyd, M. G. Raymer and L. M. Narducci, Cambridge U. Press, 1986.

J. W. Swift and K. Wiesenfeld, Phys. Rev. Lett., **52** (1984) 705.

P. Szepfalusy and T. Tel, Phys. Rev. A, **34** (1986) 2520.

M. Tachikawa, F.-L. Hong, K. Tanii and T. Shimizu, Phys. Rev. Lett., **60** (1988) 2266.

F. Takens, J. Diff. Eqs., **14** (1973) 476.

A. Tallet, M. Le Berre and E. Ressayre, in **Optical Bistability, Instability, and Optical Computing**, ed. by H. Y. Zhang and K. K. Lee, World Sci. Pub., 1988.

W. Y. Tam, J. A. Vastano, H. L. Swinney and W. Horsthemke, Phys. Rev. Lett., **61** (1988) 2163.

C. Tang and P. Bak, Phys. Rev. Lett., **60** (1988) 2347.

C. Tang, K. Wiesenfeld, P. Bak, S. Coppersmith and P. Littlewood, Phys. Rev. Lett., **58** (1987) 1161.

M. Tang and S. Wang, Appl. Phys. Lett., **48** (1986) 900.

R. Temam, **Navier-Stokes Equations and Nonlinear Functional Analysis**, SIAM, 1983.

J. Testa, J. Perez and C. Jeffries, Phys. Rev. Lett., **48** (1982) 714.

B. Thedrez and J. G. Provost, Opt. Lett., **14** (1989) 958.

R. Thom, **Stabilite Structurelle et Morphogenese**, Addison-Wesley, 1973.

J. M. T. Thompson, Phil. Tran. Roy. Soc. London A, **292** (1979) 1.

J. M. T. Thompson and R. Ghaffari, Phys. Lett. A, **91** (1982) 5.

J. M. T. Thompson and H. B. Stewart, **Nonlinear Dynamics and Chaos**, Wiley, 1986.

A. Thorpe, **Elementary Topics in Differential Geometry**,

Springer-Verlag, 1979.

A. Thyagaraja, Phys. Fluids, **22** (1979) 2093.

R. W. Tkach and A. R. Chraplyvy, Electron. Lett., **21** (1985) 1081.

K. Tomita, Phys. Reports, **86** (1982) 113.

K. Tomita and I. Tsuda, Phys. Lett., **71A** (1979) 489.

J. R. Tredicce, N. B. Abraham, G. P. Puccioni and F. T. Arecchi, Opt. Communi., **55** (1985) 131.

J. R. Tredicce, G. L. Lippi, F. T. Arecchi and N. B. Abraham, Phil. Trans. R. Soc. Lond. A, **313** (1984) 411.

J. R. Tredicce, L. M. Narducci, D. K. Bandy. L. A. Lugiato and N. B. Abraham, Opt. Commun., **56** (1986) 435.

J. R. Tredicce, E. J. Quel, A. M. Ghazzawi, C. Green, M. A. Pernigo, L. M. Narducci and L. A. Lugiato, Phys. Rev. Lett., **62** (1989) 1274.

J. Treiber and R. I. Kitney, Phys. Lett. A, **134** (1988) 108.

S. Trillo, S. Wabnitz, R. H. Stolen, G. Assanto, C. T. Seaton and G. I. Stegeman, Appl. Phys. Lett., **49** (1986) 1224.

K. Y. Tsang, Phys. Rev. Lett., **57** (1986) 1390.

S. L. Tucker and S. O. Zimmerman, SIAM J. Appl. Math., **48** (1988) 549.

J. S. Turner, J. C. Roux, W. D. McCormick and H. L. Swinney, Phys. Lett., **85A** (1981) 9.

H. Turschner, Phys. Lett. A, **90** (1982) 385.

D. K. Umberger, J. D. Farmer and I. I. Satija, Phys. Lett. A, **114** (1986) 341.

I. Vaisman, **Cohomology and Differential Forms**, Marcel Dekker, 1973.

J. F. Valley, H. M. Gibbs, M. W. Derstine, R. Pon, K. Tay, M. LeBerre, E. Ressayre, and A. Tallet, in **Optical Bistability III**, ed. by H. M. Gibbs, P. Mandel, N. Peyghambarian and S. D. Smith, Springer-Verlag, 1986.

R. Van Buskirk and C. Jeffries, Phys. Rev. A, **31** (1985) 3332.

M. Varghese and J. S. Thorp, Phys. Rev. Lett., **60** (1988) 665.

C. Vidal and A. Pacault, ed., **Non-Equilibrium Dynamics in Chemical Systems**, Springer-Verlag, 1984.

S. Wabnitz, Phys. Rev. Lett., **58** (1987) 1415.

S. Wabnitz and G. Gregori, Opt. Communi., **59** (1986) 72.

R. W. Walden and G. Ahlers, Phys. Rev. Lett., **44** (1980) 445.

C. T. C. Wall, **Surgery on Compact Manifolds**, Academic Press, 1970.

A. H. Wallace, **An Introduction to Algebraic Topology**, Pergamon Press, 1957.

A. H. Wallace, **Differential Topology. First Step**, Benjamin, 1968.

P.-Y. Wang, J.-H. Dai and H.-J. Zhang, Acta Phys. Sinica, **34** (1985) 581.

P.-Y. Wang, H.-J. Zhang and J.-H. Dai, Acta Phys. Sinica, **34** (1985) 1233.

F. W. Warner, **Foundations of Differentiable Manifolds and Lie Groups**, Scott Foresman, 1971.

G. Wasserman, **Stability of Unfoldings**, Springer-Verlag, 1974.

T. Wazewski, Ann. Soc. Polonaise Math., **23** (1950) 112.

E. Weinstock and C. Rorres, SIAM J. Appl. Math., **47** (1987) 589.

C. O. Weiss, N. B. Abraham and U. Hubner, Phys. Rev. Lett., **61** (1988) 1587.

C. O. Weiss and J. Brock, Phys. Rev. Lett., **57** (1986) 2804.

C. O. Weiss and W. Klische, Opt. Communi., **51** (1984) 47.

C. O. Weiss, W. Klische, P. S. Ering and M. Cooper, Opt. Communi., **52** (1985) 405.

S. Weiss and B. Fischer, Opt. Lett., **14** (1989) 1213.

J. E. Wesfreid and S. Zaleski, **Cellular Structures in Instabilities**, Springer-Verlag, 1984.

J. H. Westcott, Proc. Roy. Soc. London A, **407** (1986) 89.

D. Whitley, Bull. Lond. Math. Soc., **15** (1983) 177.

H. Whitney, Duke Math. J. **1** (1935) 514.

H. Whitney, Ann. of Math. **37** (1936) 645.

H. Whitney, Ann. of Math. **45** (1944) 220; 247.

K. Wiesenfeld, J. Stat. Phys., **38** (1985) 1071.

K. Wiesenfeld and P. Hadley, Phys. Rev. Lett., **62** (1989) 1335.

K. Wiesenfeld, E. Knobloch, R. F. Miracky and J. Clarke, Phys. Rev. A, **29** (1984) 2102.

K. Wiesenfeld and I. Satija, Phys. Rev. B, **36** (1987) 2483.

S. Wiggins, SIAM J. Appl. Math., **48** (1988) 262.

J. L. Willems, **Stability Theory of Dynamical Systems**, Wiley, 1970.

R. F. Williams, in **Global Analysis**, ed. by S. S. Chern and S. Smale, 1970.

R. F. Williams, Expanding Attractors, (1971) 169.

R. F. Williams, Pub. Math. IHES, **50** (1980) 59.

R. F. Williams, Ergod. Th. Dynam. Sys., **4** (1983) 147.

C. R. Willis, in **Coherence and Quantum Optics IV**, ed. by L. Mandel and E. Wolf, Plenum, 1978.

T. J. Willmore, **An Introduction to Differential Geometry**, Oxford Univ. Press, 1959.

H. G. Winful, Opt. Lett., **11** (1986) 33.

D. Wintgen, Phys. Rev. Lett., **58** (1987) 1589.

D. Wolf, ed., **Noise in Physical Systems**, Springer-Verlag, 1978.

W. A. Wolovich, **Linear Multivariable Systems**, Springer-Verlag, 1974.

T. Q. Wu and C. O. Weiss, Opt. Communi., **61** (1987) 337.

X.-G. Wu and P. Mandel, J. Opt. Soc. Am. B, **3** (1986) 724.

E. Yablonovitch, Phys. Rev. Lett., **58** (1987) 2059.

V. A. Yacuborich, Dokl. Akad. Nauk. SSSR, **143** (1962) 1304.

S. Yamaguchi and M. Suzuki, IEEE J. QE, **19** (1983) 1514.

Y. Yamaguchi and K. Sakai, Phys. Lett. A, **131** (1988) 499.

K. Yano and S. Ishihara, **Tangent and Cotangent Bundles**, Marcel Dekker, 1973.

Y. Yao, Phys. Lett. A, **118** (1986) 59.

T. Yazaki, S. Takashima and F. Mizutani, Phys. Rev. Lett.,

    58 (1987) 1108.

W. J. Yeh and Y. H. Kao, Phys. Rev. Lett., **49** (1982) 1888.

T. H. Yoon, C.-H. Lee and S.-Y. Shin, IEEE J-QE, **25** (1989) 1993.

G. W. Young and S. H. Davis, SIAM J. Appl. Math., **49** (1989) 152.

T. Yoshizawa, **Stability Theory by Liapunov's Second Method**, Math. Soc. of Japan, 1966.

P. Yu and K. Huseyin, SIAM J. Appl. Math., **48** (1988) 229.

J.-M. Yuan, in **Directions in Chaos, Vol. 1**, ed. by B.-L. Hao, World Sci. Pub., 1987.

J.-M. Yuan, M. Tung, D. H. Feng and L. M. Narducci, Phys. Rev. A, **28** (1983) 1662.

J. Yumoto and K. Otsuka, Phys. Rev. Lett., **54** (1985) 1806.

A. Zardecki, Phys. Lett. A, **90** (1982) 274.

E. C. Zeeman, in P. Hilton 1976.

E. C. Zeeman, in **New directions in Applied Mathematics**, ed. by P. J. Hilton and G. S. Young, Springer-Verlag, 1981.

E. C. Zeeman, Nonlinearity, **1** (1988) 115.

H. Zeghlache and P. Mandel, J. Opt. Soc. Am. B, **2** (1985) 18.

H. Zeghlache and P. Mandel, Phys. Rev. A, **37** (1988) 470.

H. Zeghlache, P. Mandel, N. B. Abraham and C. O. Weiss, Phys. Rev. A, **38** (1988) 3128.

W.-Z. Zeng and B.-L. Hao, Chinese Phys. Lett., **3** (1986) 285.

L. A. Zenteno, K. E. Aldous and G. H. C. New, J. Mod. Opt., **35** (1988) 1281.

H.-Y. Zhang and K. K. Lee, **Optical Bistability, Instability, and Optical Computing**, World Scientific Pub., 1988.

J.-Y. Zhang, H. Haken and H. Ohno, J. Opt. Soc. Am. B, **2** (1985) 141.

Y.-C. Zhang, Phys. Rev. Lett., **59** (1987) 2125.

B. Zhou, T. J. Kane, G. J. Dixon, and R. L. Byer, Opt. Lett., **10** (1985) 62.

Index

444

446

Group of diffeomorphisms, 49

Hausdorff space, 31
Hessian, 60, 191
Heteroclinic orbits, 162
Homeomorphic, 35
Homeomorphism, 35
Homoclinic cycles, 162
Homoclinic orbits, 162
Homogeneous, 75
Homologous, 41, 42
Homology group, 42
Homomorphism, 25
Homotopically trivial, 43
Hopf bifurcation, 273
Hyperbolic, 179, 188, 193
Hyperbolic attractor, 250
Hyperbolic linear automorphisms, 182, 184
Hyperbolic structure with respect to f, 250
Hyperbolic subset of X, 251
Hyperbolic toral automorphism, 151

Image, 25
Imbedding, 51, 125
Immersion, 51, 125
Indecomposable, 261
Induced bundle, 105
Inner product, 95
Integrable, 56, 168
Integral curve, 54, 168
Integral flow, 168
Integral manifold, 56
Interior of A, 28
Interior point, 28
Invariant k-form, 98
Invariant set, 160
Involutive, 56

447

Ultimately bounded,2834
Unfoldings, 277
Uniform attractor, 202, 233, 236
Uniform attractor relative to U, 236
Uniform stability with respect to x, 209
Uniformly approximate, 239
Uniformly asymptotically stable, 203, 219, 233
Uniformly attractive, 201
Uniformly globally asymptotically stable, 203
Uniformly globally attractive, 203
Uniformly stable, 201
Uninvadable, 369
Universal, 278
Universal bundle over G, 106
Universal covering space of X, 44
Unstable, 199, 201
Unstable manifolds, 246
Unstable set, 246, 250

van der Pol's equation, 171
Vector bundle of a manifold M, 84
Vector bundles, 89
Vector field, 50
Versal, 278
Votka-Volterra equations, 359

Weak attractor, 202, 233, 236
Weak attractor relative to U, 236
Weakly continuous, 130
Weakly attracted, 202
Whitney C^∞ topology, 118
Whitney (or direct) sum, 89, 104